D0076189

HUMANS AND ANIMALS

HUMANS AND ANIMALS

A Geography of Coexistence

Julie Urbanik and Connie L. Johnston, Editors

An Imprint of ABC-CLIO, LLC
Santa Barbara, California • Denver, Colorado

Library of Congress Cataloging-in-Publication Data

Names: Urbanik, Julie, 1971– editor. | Johnston, Connie L., editor.
Title: Humans and animals : a geography of coexistence / Julie Urbanik and Connie L. Johnston, editors.
Other titles: Humans and animals (Santa Barbara, Calif.)
Description: Santa Barbara, California : ABC-CLIO, an Imprint of ABC-CLIO, LLC, [2017] | Includes bibliographical references and index.
Identifiers: LCCN 2016016455 | ISBN 9781440838347 (hard copy : alk.paper) | ISBN 9781440838354 (EISBN)
Subjects: LCSH: Human-animal relationships—Encyclopedias. | Human geography—Encyclopedias. | Zoogeography—Encyclopedias.
Classification: LCC QL85 .H86 2017 | DDC 590—dc23
LC record available at https://lccn.loc.gov/2016016455

ISBN: 978-1-4408-3834-7
EISBN: 978-1-4408-3835-4

21 20 19 18 17 1 2 3 4 5

This book is also available as an eBook.

ABC-CLIO
An Imprint of ABC-CLIO, LLC

ABC-CLIO, LLC
130 Cremona Drive, P.O. Box 1911
Santa Barbara, California 93116-1911
www.abc-clio.com

This book is printed on acid-free paper ∞

Manufactured in the United States of America

CONTENTS

Primary Documents

PREFACE

As co-editors of the first one-volume encyclopedia to address the geography of human-animal coexistence for a general audience, we are excited to have a role in sharing knowledge of a field about which we are passionate. As geographers, we are used to multifaceted complexity, but at times we were surprised by the challenge of bringing together for this volume an area of study that ranges across disciplines as diverse as anthropology, art, biology, cultural studies, economics, ethology, geography, history, law, literature, politics, science studies, and veterinary sciences. We feel that we successfully met this challenge and hope that our readers will as well.

As with any work of this kind, tough decisions must be made on relevant topics to include. We have done our utmost to exercise good judgment in this regard, and the resulting scope of this work is a cohesive presentation of the spectrum of human-animal relations, organized alphabetically around 150 topics. These include specific species, biological concepts, philosophical concepts, social movements, specialized fields, and different categories of relations. Every effort has been made to include as global a perspective as possible, in recognition of cultural/spatial variety even within broad categories like pets or religion.

The contributors were tasked to write succinct yet comprehensive entries for a worldwide, English-speaking audience. Entries provide accessible, jargon-free overviews of topics so that readers may gain an understanding of key terms, relevant histories, geographic locations or variations, and explanations of any controversies. The selected images allow readers to see human-animal relations in the world visually, while they are learning to "see" them textually.

We have also provided one set of supplemental materials. The set includes a selection of excerpts from 20 key documents ranging from foundational single-author books, to court decisions, government legislation, and international treaties. These primary sources exemplify the different ways in which human-animal relations are articulated by different social bodies.

We hope this volume will not only serve as a reference but also as a starting point for deeper engagement. To that end, we have provided further reading suggestions with each entry, a master bibliography, a glossary of terms for quick clarifications, and a full index for ease of locating specific topics.

This volume would not have been possible without the vision, guidance, and support of our editor, Julie Dunbar, at ABC-CLIO. We are grateful to her for providing us with this opportunity, and we appreciate everything we have learned from her. We also extend our deepest thanks to our incredible group of contributors. Their expertise and enthusiasm (along with patience and sense of humor) have ensured that this encyclopedia will become an essential reference for anyone with a budding interest in the coexistence of humans and animals. This work would not have been possible without them.

INTRODUCTION

It is practically impossible to move through your day without encountering animals in one form or another. They might be on your plate, snuggled next to you in bed, talking to you in advertisements, television, or film, or you might hear them flying overhead on your way to work or drive by their remains on the side of the road. You might hunt them, photograph them, draw them, get a tattoo of them, see them as having souls or not, avoid them at all costs, or take a family picnic to the zoo to surround yourself with them. The fact is, animals are everywhere humans are—from dust mites enjoying a snack of your dead skin, to the companion animals and wildlife in your neighborhood, to the livestock and wildlife that live farther away from you. Animals also live in more places than do humans—able to survive in the deep ocean trenches or in the harsh cold of the Arctic and Antarctica. Indeed, humans are but one of millions of species of animal and, as animals ourselves, our very survival as a species is intimately connected to these others. How does a person begin to make sense of the many ways we have relationships with all the different nonhuman species when it involves considerations of ethics, biology, economics, cultural difference, and the larger planetary environment?

This encyclopedia provides a first step and ongoing guide for those looking to explore these intricate relationships. Recent decades have seen a dramatic rise in scholarly interest in human-animal relationships, and this volume is a concentration and synthesis of that work. Scholars are interested in these relationships for many of the same reasons anyone else might be—for example, they care about a particular species and want to learn more about it, they want to understand why some people eat pigs and others don't, or they may be curious about how the new neighborhood being built down the street will impact wildlife. Some scholars may be animal advocates themselves—meaning they act politically (e.g., writing letters, protesting, making animal-friendly consumer choices, participating in policy-making or legislation) to support what they believe are ethical ways of interacting with other species, while others are focused more on solving scientific questions about behavior, habitat, or conservation, or studying how and why relations between humans and animals are the way they are. The umbrella term for those scholars studying these topics is Human-Animal Studies (HAS), and this encyclopedia is a contribution to this field. HAS encompasses work from fields as diverse as anthropology, biology, geography, history, literature, medicine, philosophy, and veterinary medicine, to name but a few. There are now undergraduate and graduate programs in various fields of HAS, along with dozens of research journals and specialty groups developing around the world.

Indeed, the rise of public and scholarly interest in human-animal relations has come at a time when our one, human, species is having tremendous impact on all other species. This time period is now being referred to as the Anthropocene—or the Age of Humans—because humans have become the primary driver of actions shaping life on Earth. At the same time that scientists are learning more about our impact on animal species in the world, they are also learning more about the amazing experiences and capabilities of animals themselves. While many people around the world have grown increasingly fond of animals with the growth of visual media, they have also grown increasingly concerned about their treatment from exposés by animal advocacy organizations and dire warnings from conservation scientists. Until now, there has not been a resource for those looking to sort out their own views on human-animal relations without struggling with jargon in scholarly writing or feeling hesitant to engage with the activist stance of animal advocacy groups. As educators, we know that individuals first engaging with the topic of human-animal relations can face an overwhelming amount of material, so we believe that this one-volume work on the topic is an excellent way to open the door to many different people.

As geographers, it was essential for us to ground the volume in the perspective of the subfield of animal geography. Geography as a whole is interested in where, how, and why earthly phenomena happen, as well as the connections between phenomena. Geography is about the relations in and between places and across space. It is simultaneously about the specific and the general patterns of human and natural life. For animal geographers, the focus is on where, how, and why we have the relations that we do with other species, both historically and in the present day.

There are two main ways all human-animal interaction is geographical. The first way is linked to the specific, as it is clear that every individual human-animal interaction happens in a particular location—in a place. Therefore, we need to understand those specific locations and their relationships to more general spatial patterns to get a deeper understanding of why a relationship is happening as it is. The second way human-animal relations are fundamentally geographical is because each relation sits at the center of a constellation of conceptual linkages. This means that we not only enact human-animal relations in a physical way, but that we also create human-animal relations based on how we conceptually "place" animals in human social structures. It is for both reasons that mapping human-dog relations, for example, can be both quite complex and extraordinarily fascinating. In some places dogs are food, in others they are pets, yet in still other places they are workers or entertainers. We can also note that even among those who consider dogs to be pets, or companions, some may still conceptualize the proper place of the dog as being in the yard, not in a human's bed. While many people regard kicking, starving, or fighting dogs to clearly be cruel, these same people may have no problem with getting their dogs' tails, ears, or vocal cords removed to satisfy human taste or convenience preferences. These acts can also arguably be seen as cruel because they also cause pain and, additionally, remove key ways that dogs express themselves and experience their world. Mapping where and how conceptions of cruelty are linked to treatment of dogs helps us understand the contexts in which dogs themselves are being recognized as beings who subjectively experience their world and have dog-specific needs and desires. Animal geographers, then, understand the human-dog relationship, like any human-animal one, by bringing together and analyzing a wide range of varied material.

The word "coexistence" was essential in our book's title for us to convey a second framework for the volume. Nonhuman species that live on, with, and around us have been both visible and invisible to us since our species began. As parts of human societies, they have been and continue to be both visible and invisible, intentionally excluded and included. Understanding human-animal relations is essential in today's world. Many scholars who are researching and writing about human-animal relations, both inside and outside of geography, are not only concerned with the quality of their research for advancing the field but also with an unwavering belief that humans owe it to our fellow animals to learn how to better live alongside them and to show them the respect they deserve as, like us, inhabitants of planet Earth with their own value. There is a belief in something not only profoundly powerful about sharing our lives with so many wondrous other animals—each with their own ways of being in the world—but also our being inextricably linked with them in terms of our well-being and survival. For example, without bees to pollinate them, a multitude of crops would not produce food. Although many people believe it is wrong and choose not to eat animals, for some they are an important protein source. As social beings ourselves, for many who would otherwise be alone, animals provide vital companionship and love. Although thought to be unethical by many, a number of humans have been helped by animals used in medical research and science. And finally, what would the world be like without animals to stop us in our tracks with their fierceness, mystery, beauty, or silliness? We are better equipped than at any time in history to reflect on our capacities as human animals to fundamentally alter the planet, nonhuman animal lives, and, by default, our own. Asking questions about our relations to other species enables us to make choices about how to productively evolve in these relations. And getting to know animals—as themselves and not just as not-human—connects us to them in ways that can fundamentally deepen our appreciation for our shared lives and places in the world.

We approached the opportunity to edit this encyclopedia with excitement and ambition, but also a sense of limitation. Our conversations about which entries to include, important documents to excerpt, best ways to organize, and so on quickly highlighted the tiny slice of human and animal coexistence that we would be able to represent. Our aim, therefore, has been to capture a broad representation of important aspects of this coexistence, not only in the present day and in the English-speaking world but also historically and globally. That being said, we (and many of our contributors) recognize that there will be differences of opinion on the entries and documents that we have selected.

Any limitations aside, we believe this encyclopedia achieves the aim of providing a foundational collection of entries, primary documents, and a key readings bibliography for new scholars and/or the general public seeking an engaging and accessible reference for their questions about human-animal relations. It is our hope that those who pick up this encyclopedia with a curiosity about only one topic will find themselves following references to other pages, or reading other entries that caught their eye. We especially hope that this book will excite the interest of readers to learn more about the amazing geography of our human coexistence with other animals, and will therefore be a springboard to further exploration.

A

Advertising, Animals in

When an animal such as a dog or a cat is included in advertisements for products that they need or use, such as cat litter or dog treats, it makes sense to have them there, scurrying around or eagerly eating. When what is being sold is for humans, however, their inclusion is less easily explained. Certain species of animals are frequently used in advertising, and there are several common ways of using them to symbolize aspects of human life.

The first animal symbols made by humans likely date to art in ancient Paleolithic caves and on rock outcroppings. While not advertisements per se, they may have been used to indicate, for example, important spiritual sites or plentiful hunting or fishing areas. Animals are employed in mythological tales and stories, such as Aesop's fables, in order to address the big questions of life (where did we come from? what is love? what is death? what is proper behavior toward others?).

In addition to their functions providing food, clothing, tools, chemicals, and companions, animals—and their likenesses—are used to convey cultural meanings. Considering the thousands of years during which humans and animals have been interacting, it is no surprise that animals are also used in advertising. Advertising, defined as time and space paid for to sell something, is designed to tell us something about a product or service, but more so, what that product or service can do for us.

Animals in advertisements, whether real or fictional, animate or inanimate, draw on culturally specific, shared understandings of what is believed about a particular species and how those attributes can be used to say something about a product or service. For example, while we know raccoons don't really sit in chairs (e.g., La-Z-Boy® recliners), dogs don't drive cars (e.g., Subaru®), and polar bears don't consume soft drinks (e.g., Coca-Cola®), portrayals of them in these ads have come to seem so normal and natural we often fail to see the constructed nature of them. Like other communication tools, animals are used to link a particular set of qualities with a brand's image. Animal characters are the most commonly used trade characters because they are efficient communicators of human qualities, characteristics, and values. For example, at least in the United States, beavers are considered industrious, turtles steadfast, and monkeys funny; a dove suggests peace; and an elephant signifies memory. Thus, the shared meaning a species has in a culture translates to its use in the culture's advertising.

Gidget, the Taco Bell dog, during a photo session in October 1998 in Los Angeles, California. Dogs are the most popular animal used in advertising in the United States. Advertisers choose animals they believe will both appeal to their target audiences and become associated with a particular brand. (Vern Evans/Getty Images)

The most popular animal used in advertising in the United States is the dog, with the likes of Budweiser® (beer), Trivago® (a travel website), Taco Bell® (fast food), Geico® (insurance), and Bush's Baked Beans® all including canines in ads. No doubt this is because dogs and humans have an ancient relationship, and dogs figure prominently in many people's everyday lives. Other globally popular advertising animals are birds, horses, cows, bulls, fish, cats, insects, elephants, mice, and rabbits. In print and television advertising, research shows that the choice to use anthropomorphized (i.e., human characteristics ascribed to other species) animals is related to the type of product (Spears, Mowen, John, and Chakraborty 1996).

Furthermore, advertising that uses animals portrays them in at least six (not mutually exclusive) ways: as tools (transporting humans, working for them, as food), loved ones (active members of family life), symbols (as images that stand in for the brand), nuisances (problems to be solved), allegories (playing human-type roles), and in nature (flying, climbing, jumping, doing what comes natural to them) (Lerner and Kalof 1999). Animals can convey both positive and negative characteristics and can reinforce gender, class, and racial stereotypes. Advertisers choose which representations they believe will most resonate with the target audience, who will quickly associate something about the animal to the product.

A primary consideration in using animals in advertising is whether they should be seen as more or less like humans. Anthropomorphized portrayals are often used in advertising to minimize tension, discuss awkward topics, or provide emotional distance, and are designed to reach a general audience. Anthropomorphized animals speak, wear clothing, display human-attributed emotions, or do something only humans do, such as vote, drive a car, or use toilet paper. Nonanthropomorphized portrayals are used when the point of including an animal is to demonstrate the work they do for us (tools), how they are acceptable as food, or for recreation. This strategy is used to reinforce the species barrier between them and us. Also, the more a part of nature animals are meant to be, such as in travel advertising, the less likely they are to be anthropomorphized.

Other considerations factor in as well. For example, when dogs are used, we see friendly yellow Labrador retrievers, beagles, and Irish setters, not fearsome pit bulls, in ads for home products and comfy clothing. However, if an advertisement were for a product with an image of being tough, strong, and fearless, a pit bull might be the perfect choice. Similarly, a cheetah or rabbit would be used to signify speed, whereas a goat or turtle would not. Furthermore, animals in advertising serve two primary functions that appeal to different sexes. The first is using animals to represent a desired quality such as strength or loyalty. This function has been shown to appeal primarily to men. The second function is relational, showing animals and people interacting, which appeals mostly to women (Magdoff and Barnett 1989).

Debra Merskin

See also: Animals; Anthropomorphism; Dogs; Race and Animals

Further Reading

Baker, S. 2001. *Picturing the Beast: Animals, Identity, and Representation*. Chicago: University of Illinois Press.

Lerner, J. E., and Kalof, L. 1999. "The Animal Text: Message and Meaning in Television Advertisements." *Sociological Quarterly* 40(4): 565–586.

Magdoff, J., and Barnett, S. 1989. "Self-Imaging and Animals in Ads." In Hoage, R. J., ed., *Perceptions of Animals in American Culture*. Washington, DC: Smithsonian Institution Press.

Phillips, B. J. 1996. "Advertising and the Cultural Meaning of Animals." *Advances in Consumer Research* 23: 354–360.

Spears, N., Mowen, E., John, C., and Chakraborty, G. 1996. "Symbolic Role of Animals in Print Advertising: Content Analysis and Conceptual Development." *Journal of Business Research* 37: 87–95.

Advocacy

Although animals are often understood to be mere instruments for human use, there is a longstanding resistance to this worldview with the argument that nonhuman species deserve a certain level of respect as more than objects, and as beings related to us. The animal advocacy (or "animal protection") movements of today can generally be divided into two categories: animal welfare and animal rights/liberation, although there is overlap. Animal welfare groups tend to agree with mainstream opinions that animals may be used for human benefit in the form of food, experimentation, entertainment, and companions, but maintain that certain levels of treatment and care must be met. The animal rights/liberation position, however, proposes a more radical shift—demanding nonhuman species no longer be seen as property or tools for humans, with animal rights theory proposing more legal rights, and animal liberation promoting an increase in the status of nonhuman animals that is less centered on a discussion of rights, if at all. The animal rights position is chiefly derived from the work of philosopher Tom Regan and his 1983 book *The Case for Animal Rights*, but the utilitarian Peter Singer's 1975 work, *Animal Liberation*, is widely credited with creating the modern animal liberation movement, of which demands for legal rights are a component.

Early animal welfare movements emerged in the United Kingdom with the 1824 establishment of the Society for the Prevention of Cruelty to Animals (now the Royal Society for the Prevention of Cruelty to Animals, or RSPCA), which at the time was largely concerned with the treatment of working animals such as carriage horses. Its American equivalent, ASPCA, established in 1866, is today one of the largest animal advocacy groups in the world.

Animal welfare advocacy today often focuses on the treatment of pets, especially dogs and cats, and on improving conditions for animals used for food and various types of research. Animal welfare campaigns include advocating for legislation that bans extreme confinement for farmed animals, such as the 2008 California referendum Proposition 2 ("Prop 2"), which requires farmers to phase out farming practices and housing that prevent animals from turning around, fully extending their limbs, or lying down. Legislation is a popular tool with animal welfare advocates, who often push for harsher penalties and criminalization of animal mistreatment. They also protest using animals for entertainment such as rodeos and circuses, and also hunting. As part of these campaigns, activists use social pressure through media campaigns, civil disobedience, protest, and releasing undercover footage of animal mistreatment.

Although animal rights and liberation activists often engage in the above activities, they also work to reduce or even eliminate consumption of animal products such as meat, milk, eggs, fur, leather, and wool, as well as cosmetics, medical devices, and drugs that are tested on animals. Less frequently, activists vandalize buildings, steal devices such as knives used to kill and gut animals, and remove animals from captivity.

Although there are significant differences between animal welfare and animal rights/liberation positions, overlaps exist, with prominent animal rights groups like People for the Ethical Treatment of Animals (PETA) and Mercy For Animals (MFA) working alongside welfare-oriented organizations such as the Humane Society of the United States (HSUS) and ASPCA toward increased legal protection, harsher criminal penalties, and banning certain practices.

The Nonhuman Rights Project, spearheaded by animal law attorney Steven Wise (1952–), is pursuing the legal recognition of four captive chimpanzees as "persons" under common law, rather than as property—an argument that could be expanded to other great apes, elephants, and dolphins. Using *habeas* law—a legal action asserting unlawful imprisonment brought about by a third party on behalf of someone without legal standing, such as slaves—the project aims to establish that at least some nonhuman animals have rights to bodily liberty and bodily integrity. The animal species prioritized—dolphins, whales, great apes, and elephants—are chosen on the grounds that they are scientifically established as being self-aware.

Recent campaigns by both animal welfare and rights/liberation activists have aimed at banning bull hooks (pointed hooks used to direct circus elephants), increasing penalties for dogfighting and other sports in which animals are or may be killed, and developing less painful methods for killing farmed animals.

Today, animal protection movements exist all over the world. People for Animals, the largest animal welfare organization in India, led by Maneka Sanjay Gandhi (1956–), operates shelters, rescues animals from neglect, and works toward legal changes such as successfully lobbying to have all packaged food in India labeled as vegetarian or nonvegetarian. Notably, India's constitution stipulates that all citizens have a duty of compassion toward animals. Palestine Animal League works for stray animals, hosts humane education events, and pursues greater legal protection for animals in Gaza and the West Bank.

China has very few animal protection laws, and does not have a longstanding animal welfare movement. Although many religious traditions in China, such as Taoism and Buddhism, include statements about compassion for animals, the ruling Communist Party has regarded compassion for animals as opposed to their political program. Since the 1990s, however, compassion for animals has grown. Both domestic and international animal welfare organizations like Animals Asia are working to combat practices such as dog and cat meat consumption and bear bile farming—a practice in which bears are kept in cages with open wounds to harvest their stomach acids because of their value to traditional medicine.

Drew Robert Winter

See also: Ethics; Humane Education; People for the Ethical Treatment of Animals (PETA); Rights; Welfare

Further Reading

Donaldson, S., and Kymlicka, W. 2011. *Zoopolis: A Political Theory of Animal Rights*. Oxford: Oxford University Press.

Donovan, J., and Adams, C. J. 2007. *The Feminist Care Tradition in Animal Ethics: A Reader*. New York: Columbia University Press.

Kelch, T. G. 2014. "Cultural Solipsism, Cultural Lenses, Universal Principles, and Animal Advocacy." *Pace Environmental Law Review* 31(2), 403–472.

Phelps, N. 2007. *The Longest Struggle: Animal Protection from Pythagoras to PETA.* New York: Lantern Books.

Regan, T. 1983. *The Case for Animal Rights.* Oakland, CA: The University of California Press.

Singer, P. 1975. *Animal Liberation: A New Ethics for Our Treatment of Animals.* New York: HarperCollins.

Agency

Agency is an individual or group's ability to exert power and choice in the world. Individual or collective agency can be limited or enabled by material (e.g., walls or money), institutional (rules and norms), or cultural (customs and traditions) modes of societal structure. *Animal agency* refers specifically to a nonhuman animal's ability to exercise individual or collective power and choice, usually in the context of human-dominated society. The question of whether animals have agency is significant in the context of animal ethics, law, advocacy, husbandry, and conservation.

The idea of human agency has roots in early Greek philosophy (ca. 400–320 BCE), which explored the role of human intention, obligation, and accountability in Western society, and 19th-century sociology (the study of human social behavior). Philosophers have used the idea of agency to address questions about action and causality (what action caused what result) and the nature of free will (individual capacity to choose). Philosopher René Descartes' (1596–1650) statement *cogito, ergo sum* ("I think, therefore I am") implied that if a person could think, a person had agency. Sociologist Émile Durkheim (1858–1917) challenged this simplistic idea of thought as agency, arguing that social structures influence thought and decisions. For example, social class and gender influence what choices are available to individuals and how individuals respond to those choices. Free will implies accountability for choices made, but because all people do not have the same power, opportunities, and choices, they may not have the same ability to act according to what is right or wrong (defined as *moral agency*).

Agency can be active or passive. *Active agency* refers to individuals doing, constructing, or controlling with conscious, goal-directed intention, often using rational thought (making conscious sense of the world using logic). Resistance or purposeful inaction is also a form of active agency. For example, when African American Rosa Parks (1913–2005) famously refused to give up her seat on the bus to a white passenger in 1955 in the U.S. state of Alabama, she exercised agency by resisting societal rules that existed at that time. Her resistance contributed to the collective action that developed into the U.S. Civil Rights Movement. *Passive agency* is demonstrated when an individual unintentionally exerts power or unknowingly participates in events that determine or affect power. Nonhuman individuals often demonstrate passive agency. For example, a single jaguar occupying Southwestern Arizona forced the protection of over 750,000 acres of critical habitat (legally protected habitat mandated by the U.S. Endangered Species Act) in 2014. This habitat protection limits the agency of ranchers and developers to make choices about how they use the land. In this example, the jaguar acted as an unintentional agent of legal and land management change. Another example of passive agency occurred in 2009, when a flock of Canada geese collided with a jet, forcing its emergency landing on the Hudson River. Birds cause airports an estimated $1.2 billion per year in damage and lost flight time, and airports spend millions more attempting to manage the behavior of birds.

Animals also exhibit active agency. Domestic animals such as sheep and horses will jump fences, a donkey may refuse to move under a heavy load, and working animals may not cooperate with their handlers. Wild horses strategically scatter mares in their bands to avoid capture during round-ups, and captured horses notoriously resist being tamed. Wild animals such as coyotes often maintain their traditional territories and travel routes despite human development.

These acts are often perceived by animal handlers and observers as intentional acts of resistance, but scientists often attribute this kind of active animal agency to biological instinct, such as what drives beavers to build dams or a termite colony to transform a landscape when they construct a mound. Despite growing evidence that animals can have complex thought and emotional processes, the extension of agency to the nonhuman is a challenge, as it remains unknown exactly how animals think. We do not know the depth of animal self-awareness, or know with certainty if they act with intention. We do not know to what extent animals make conscious choices, or what factors enable or restrict an individual animal to act on its will. Moreover, we cannot generalize about animal agency. We cannot assume that what is true of a cow is true of a wolf, and we cannot assume that what is true of one wolf is true of another.

Emerging studies on nonhuman agency in many areas of modern thought and academic disciplines, including geography, anthropology, and political theory, challenge the anthropocentric (human-centered) focus on agency and expand the concept of agency to nonhuman animals (even proposing the agency of inanimate objects and technological creations).

The broadening of agency to nonhuman categories challenges previously limiting assumptions, including the restriction of agency to humans, and the notion that rationality and intentionality are necessary for agency. The extension of agency to the nonhuman is particularly important in the inquiry of Human-Animal Studies, Animal Geography, Animal Ethics, Ethology, and other intellectual fields focusing on nonhuman animals in society.

Anita Hagy Ferguson

See also: Advocacy; Animal Geography; Animal Law; Cruelty; Domestic Violence and Animal Cruelty; Ethics; Ethology; Factory Farming; Geography; Human-Animal Studies; Human-Wildlife Conflict; Husbandry; Intelligence

Further Reading

Appleby, M., Mench, J., Olsson, I. A., and Hughes, B. 2011. *Animal Welfare*, 2nd ed. Cambridge, MA: CABI.

DeMello, M. 2012. *Animals and Society: An Introduction to Human-Animal Studies*. New York: Columbia University Press.

Ingold, T. 2011. *Being Alive: Essays on Movement, Knowledge and Description*. New York: Routledge.

Latour, B. 2005. *Reassembling the Social: An Introduction to Actor-Network-Theory*. New York: Oxford University Press.

Riley, A. T. 2015. *The Social Thought of Emile Durkheim*. Los Angeles: Sage.

Whatmore, S. 2002. *Hybrid Geographies: Natures Cultures Spaces*. London: Sage Publications.

Wolch, J., and Emel, J., eds. 1998. *Animal Geographies: Place, Politics, and Identity in the Nature-Culture Borderlands*. New York: Verso.

Animal Assisted Activities

Animal assisted activities (AAA) are casual interactions between animals, their human handlers, and clients. AAA is used to provide motivational, educational, and/or recreational opportunities to enhance quality of life. Historically, animals have held significant support roles in human society, improving health, social interactions, and overall well-being. AAA builds upon this relationship as a method to provide emotional, social, and cognitive outlets and support for humans.

The importance of animals as companions to humans is documented going back thousands of years. A 12,000-year-old human skeleton with the skeleton of a puppy in its hands was found in Israel (Davis and Valla 1978). Tombs excavated in ancient Egypt contain the mummies of important pets and animal totems of gods (Pitt Rivers Museum 2002). Ancient Greeks believed that dogs' saliva had healing properties (Becker 2013). In the present day, we are learning how animals have a calming effect on humans, reduce anxiety, and improve self-esteem (Help Guide 2015).

AAA encompasses many approaches to partnering with nonhuman animals to improve the lives of clients. Dogs are the typical partner in animal assisted activities, but horses, cats, and rabbits are becoming more common. AAA can take place in almost any environment, and involve one or more clients. The clients meet the animal with direct supervision of the animal handler. The handlers are trained in supporting the specific animal they are partnering with and in interacting with clients. Animals are selected for participation in AAA based on preferred behaviors and temperament.

Unlike animal assisted therapy (AAT), key factors of AAA sessions are not individualized to the specific needs of a client. They are, instead, composed of spontaneous content and do not have predetermined goals. Sessions can be any length of time, and are determined by handlers' assessment of the animal partner and by client preferences. Handlers do not need to be therapeutically licensed professionals and are not required to keep records about the visit.

There are numerous forms and functions of AAA. It is used in prisons, juvenile homes, hospices, retirement homes, treatment centers, homeless shelters, schools, and hospitals. Teachers use animals to teach and transfer skills through the care of small animals like hamsters, rabbits, and fish. AAA can be brought to the classroom with larger animals, like dogs, to facilitate ongoing learning with the animal as an educational tool. Additionally, human and dog teams visit university campuses during finals to provide support to stressed students through animal interaction. Patients living in long-term care benefit from changes to daily schedule and a friendly visitor. Furthermore, there is growing support for pet-friendly assisted living communities to improve the quality of life of residents.

AAA can be used to assist participants in the development of specific skills. Reading-aloud programs, during which participants read out loud to animals, benefit people who are learning to read, have a reading disorder, or are uncomfortable reading. It encourages participants to gain experience and confidence by reading to a nonjudgmental animal, usually a dog but also with horses. Prison-Based Animal Programs (PAP) partner inmates with rescue animals. Inmates care for and train dogs or wild horses to improve their likelihood of adoption. If support sessions for inmates are provided outside the scope of training the animals, then this component falls into animal assisted therapy.

There are many benefits of participating in AAA. It aids in development of motor skills like coordination, balance, and posture. It has positive, psychosocial effects such as decreased feelings of loneliness, despair, isolation, and fear. AAA can decrease incidents of aggression and problem behavior, while having positive impact on self-esteem. AAA can facilitate human social interaction that increases levels of communication and change the relationship dynamic of those who participate in them. The presence of an animal serves as a social lubricant and is a neutral topic to facilitate shared experiences of pet ownership.

Several studies aim to better understand the impact of AAA on the health of humans. One study documented that cardiac care patients who had pets lived longer than their non-pet-owning counterparts. Interactions with animals are shown to decrease stress, depression, and anxiety, while increasing feelings of comfort (Pichot 2006, 4). A study that looked at the impact of a dog's presence on children found that, even without touching or petting the dog, children's blood pressure decreased (Friedmann et al. 1983, 44).

Concerns about the possible negative impacts to humans require consideration. Some studies argue that AAA has no effect on the elderly and can possibly decrease morale and health in some populations. Health concerns include the transmission of disease and fleas, client allergies, fear/dislike of animals, negative emotional consequences of the death of an animal, or incorrect perception of client ownership of the animal. Specific concerns for the welfare of the animals include limited access to water, high temperatures at facilities, high expectations of the length of time for visits, overall stress of work, safety from aggressive client behavior, and the potential for accidental harm to the animal. Animal partners have the capacity to feel complex emotions; therefore, the welfare of animal partners is paramount to discussions about implications and overall benefits of AAA.

Andy VanderLinde

See also: Animal Assisted Therapy; Service Animals

Further Reading

Animal Assisted Intervention International. 2013. "Animal Assisted Intervention International." Accessed March 20, 2015. www.animalassistedintervention.org

Becker, M. 2013. "5 Surprising Things You Didn't Know about Dogs." Accessed April 14, 2015. www.vetstreet.com/dr-marty-becker/5-surprising-things-you-didnt-know-about-dogs

Davis, S., and Valla, F. 1978. "Evidence for Domestication of the Dog 12,000 Years Ago in the Natufian of Israel." *Nature* 276: 608–610.

Friedmann, E., Katcher, A., Thomas, S., Lynch, J., and Messent, P. 1983. "Social Interaction and Blood Pressure: The Influence of Animal Companions." *Journal of Nervous and Mental Disease* 171: 461–465.

Help Guide. 2015. "The Health Benefits of Dogs (and Cats)." Accessed April 14, 2015. www.helpguide.org/articles/emotional-health/the-health-benefits-of-pets.htm

Interactive Autism Network. 2011. "Dogs, Horses and ASD: What Are Animal-Assisted Therapies?" Accessed March 20, 2015. www.iancommunity.org/cs/articles/asds_and_animal_assisted_therapies

Pet Partners. 2012. "Animal-Assisted Activities Overview." Accessed March 20, 2015. www.petpartners.org/page.aspx?pid=319

Pichot, T. 2012. *Animal-Assisted Brief Therapy: A Solution-Focused Approach.* New York: Routledge.

Pitt Rivers Museum. 2002. "Animals and Belief: Ancient Egyptian Religion." Accessed April 14, 2015. www.prm.ox.ac.uk/AnimalMummification.html

Animal Assisted Therapy

Animal assisted therapy (AAT) is an intervention that partners animals with therapists as a treatment option to meet specific health goals for a client. Since the 1700s, AAT has been used as an alternative to traditional therapies, like in-office talk therapy. AAT is shown to improve therapeutic relationships and help clients

The use of animals in human health environments is becoming more mainstream. For many people, animals are easier than humans to interact with, and in the right environment animal assisted therapies can open the door to healing. (AP Photo/Efrem Lukatsky)

reach therapeutic goals more quickly than traditional therapies. Despite the apparent benefit to humans, the ethics of using animals in AAT continues to be debated.

For over three centuries, humans have partnered with animals to provide AAT. Horses were used in therapy for various illnesses as early as the 1700s. However, the first clearly documented case of the use of animals in a therapeutic setting to teach self-control appears in 1792, at the York Retreat in England. In 1860, Florence Nightingale (1820–1910) wrote about her observations of the beneficial relationship between companion animals and chronically ill patients. She is credited as the first clinician to document the positive role of animals on the health of patients. Sigmund Freud's (1856–1939) dogs attended therapy sessions because he claimed that they had a "special sense that enabled them to judge his patient's character" (Latham 2009). In 1969, Boris Levinson (1908–1984) began the work of popularizing and mainstreaming the idea of partnering with animals in therapeutic practices.

AAT is a treatment option for individuals with physical, social, emotional, or cognitive challenges and does not follow a single theoretical approach. AAT is the general category of interventions that use animals to reach specific therapeutic goals. Unlike animal assisted activities (AAA), AAT sessions have specific, stated goals; are of pre-determined lengths; and are scheduled. Each session is documented, along with progress toward meeting therapeutic goals.

There are several organizations that provide training and certification in AAT internationally. These organizations typically specialize in providing training and guidelines for the partnership with a particular species. Equine Assisted Growth and Learning Association (EAGALA) and Professional Association of Therapeutic Horsemanship International (PATH Intl) provide training for working in equine assisted interventions. Pet Partners offers training and certification for working with dogs in a therapeutic setting. Additionally, the growing support for AAT resulted in the integration of AAT studies at universities, including Prescott College in Arizona and the University of Denver in Colorado.

Around the world, AAT is used to improve therapeutic relationships and facilitate emotional growth. Mental health outcomes include increased verbal interactions, attention skills, self-esteem, and reduced anxiety. Physical health outcomes include improved fine motor skills, balance, flexibility, and muscle strength. AAT can broadly be divided into programs that use the movement of the animals and those that involve human/animal relationships. Dolphin Assisted Therapy (DAT) uses facilitated swimming and interaction with dolphins to reach patients' goals. Hippotherapy uses movement of horses as a physical, occupational, or speech therapy treatment strategy to address impairments, functional limitations, and disabilities, often through mounted work with the horse.

Animal Facilitated Counseling uses the presence of an animal to build rapport and trust. Animal partners enable the counselor to interact with clients who are withdrawn and do not respond to traditional therapies. Equine Facilitated Psychotherapy and Equine Assisted Therapy (EAT) are interactive processes in which a licensed mental health professional and an equine professional partner with horses to address psychotherapy goals. It usually does not include riding.

Programs for autism have documented the positive effects of AAT, specifically that it aids youth with autism in learning to bond and form social attachments. There is ongoing research to substantiate anecdotal support of AAT. One study documented drops in stress hormones and blood pressure and increases in healthy social hormones after time with a therapy dog. AAT is also shown to improve behavior, reduce depression, and assist in treating symptoms of Alzheimer's, Post Traumatic Stress Disorder (PTSD), and trauma (Altschiller 2011).

Critics of AAT cite the need for improved quality of research, including randomized trials and long-term follow-ups, especially regarding Equine Assisted Therapy (Anestis et al. 2014). Although there are numerous studies touting the success of AAT, they often involve small sample populations and short-term analysis of results. Critics seek to establish the limitations and reach of AAT's efficacy and effectiveness to ensure clients are fully informed of treatment expectations while participating in AAT and are receiving the best treatment possible for their needs.

Ethical questions arise with the use of animals in therapy, because they are voiceless, and it is difficult to guarantee their physical and emotional welfare. Generally, work with domestic animals such as horses and dogs is considered an ethical partnership, because these animals would not exist without a close relationship with humans. However, programs that involve the use of wildlife, such as elephants, monkeys, and dolphins, are often the focus of criticism.

Betsy Smith, one of the original proponents of Dolphin Assisted Therapy (DAT), and the Whale and Dolphin Conservation Society called for the end to the practice after reflecting on the negative ethical implications of the use of dolphins in therapy and the potential for harm to humans. The methods for obtaining dolphins for DAT can be cruel and even fatal to the dolphins. Being restricted from engaging in normal behavior can result in dolphins' illness and premature death (Whale and Dolphin Conservation Society 2014). Furthermore, DAT exposes human participants to serious risk of physical harm, including bites, bruises, scratches, abrasions, and broken bones, and potential risks to health, such as disease transmission. These concerns are shared and generalized to AAT practices that use wildlife as the animal partner.

Andy VanderLinde

See also: Animal Assisted Activities; Service Animals

Further Reading

Altschiller, D. 2011. *Animal-Assisted Therapy.* Santa Barbara, CA: ABC-CLIO.

American Humane Association. 2013. "Animal-Assisted Therapy." Accessed March 20, 2015. www.americanhumane.org/interaction/programs/animal-assisted-therapy

Anestis, M., Anestis, J., Zawilinski, L., Hopkins, T., and Lilienfeld, S. 2014. "Equine-Related Treatments for Mental Disorders Lack Empirical Support: A Systematic Review of Empirical Investigations." *Journal of Clinical Psychology* 70(12): 1115–1132.

ASPCA. 2015. "ASPCA Animal Assisted Therapy Program." Accessed March 20, 2015. https://www.aspca.org/nyc/aspca-animal-assisted-therapy-programs

EAGALA. 2010. "EAGALA." Accessed March 20, 2015. www.eagala.org

Latham, T. 2009. "Dogs, Man's Best Therapist: Dogs Can Be More Effective Than Either Therapy or Medication." Accessed April 9, 2015. www.psychologytoday.com/blog/therapy-matters/201104/dogs-man-s-best-therapist

PATH International. 2015. "PATH International." Accessed March 20, 2015. www.pathintl.org

Pet Partners. 2012. "Animal-Assisted Therapy Overview." Accessed March 20, 2015. www.petpartners.org/page.aspx?pid=320

Prescott College. 2013. "Equine Assisted Mental Health (EAMH)—Graduate and Post-MS Certificate." Accessed April 14, 2015. www.prescott.edu/academics/master-of-science-counseling/areas-of-concentration/masters-counseling-equine-assisted-mental-health.html#

University of Denver. 2015. "Animals & Human Health Certificate." Accessed April 14, 2015. www.du.edu/socialwork/licensureandcpd/animalsandhumanhealth.html

Whale and Dolphin Conservation Society. 2014. "The Case Against Dolphin-Assisted Therapy." Accessed March 20, 2015. www.us.whales.org/issues/case-against-dolphin-assisted-therapy

Animal Cultures

Scientific evidence is mounting that some animals use tools, live by moral codes, use complex communication systems, and have culture. These findings fit squarely within Charles Darwin's (1809–1882) theory of evolution, which predicts that differences between humans and other animals are in degree, not kind. Yet there is an ongoing debate about the nature and sufficiency of the evidence for culture among animals. Some scholars aren't convinced that ecological and genetic explanations for animal behavior have been ruled out in all cases, while others define culture in ways that exclude nonhuman animals. The unresolved debate makes this an active, exciting field of study, with new discoveries and important advances appearing regularly.

Culture has been defined in many different ways since the first anthropological definitions were published in the 1860s, but at the heart of the concept is the idea that culture is learned. People learn culture by observing and interacting with other people; in this process, they construct shared systems of meaning and shared norms of behavior. Debates about whether or not animals have culture hinge on how culture is defined. Some have argued that culture distinguishes humans from animals; they tend to focus on definitions of culture as complex systems of meaning. Others have argued that some animals do have culture; they tend to define culture as shared behaviors acquired through social learning. This broader definition of culture allows scholars to address interesting questions about the different kinds of culture across species, the role culture plays in helping different species adapt to the environment, and the evolution of culture across species in combination with genetic evolution.

Scholars from a variety of academic disciplines study animal cultures using two major methodologies. Primatologists (who study apes and monkeys) and cetologists (who study whales and dolphins) tend to take an ethnographic approach. Developed by anthropologists for studying human cultures, this approach involves observing animal behavior and interaction over time, usually in the wild. It produces both qualitative data (detailed descriptions of animal behavior) and quantitative data (systematic and comparative observations of animal behavior over time). Comparative psychologists (who study psychology across species) tend to take an experimental approach, which involves designing and carrying out experiments on animal behavior, often in captivity. Ethologists (biologists who study animal behavior) use both approaches: they conduct ethnographic work to document the range and kinds of animal behavior and then design experiments to more fully understand these behaviors.

Scholars taking the ethnographic approach to studying animal cultures look for evidence that animals use learned behaviors to engage with their environment and with each other. In a famous early study, Japanese primatologists inspired by Kinji Imanishi (1902–1992) identified Japanese macaques as individuals and studied their social interactions, enabling them to trace the spread of behaviors within troops. When one macaque took provisioned potatoes and washed them in seawater, for example, others in her troop soon learned to follow her example to get cleaner, tastier potatoes. They traced the spread of stone handling—which appears to be a form of play or relaxation—through another troop.

Ethnographic studies also provide evidence that shared behavior varies from one social group to another within the same species. Extensive research on chimpanzee behavior at multiple sites across Africa by William McGrew, Andrew Whiten, and others has revealed over three dozen behaviors, both ecological (like techniques used to fish for termites with sticks) and social (like grooming techniques), that vary between groups. Dolphin foraging behavior, as documented by Janet Mann and others, is also highly diverse across social groups and includes many different cooperative hunting strategies, a multistaged method for processing cuttlefish, and even tool use among two Australian groups whose members regularly put sponges on their noses to protect themselves while rooting around on the seafloor for fish.

Scholars taking the experimental approach aim to prove that animals use one or more social learning processes. These include stimulus or local enhancement (drawing another's attention to a particular object or place), imitation (observing and imitating another's behavior), and active teaching. Experiments can also rule out genetic or ecological factors, leaving culture as the explanation for behavior. Experiments on wild, coral-reef fish, for example, have shown that French grunts removed from one group and introduced into another learned the schooling sites and migration routes of their new host group. Likewise, when one group was entirely removed by researchers and another group was introduced to its vacated habitat, the newcomers did not follow the patterns of the original inhabitants but developed their own, showing that their movements were not entirely determined by ecological factors.

Scholars have used both approaches to study communication systems among animals as forms of culture. Experimental studies of birdsong provided the earliest scientific evidence for animal culture, as ethnologists demonstrated that some bird species inherit their songs genetically, while others learn their songs from more experienced members of their species. Ethnographic studies of humpback whales by Roger Payne and Katharine Payne, among others, provide an especially powerful example of animal culture. All males within a breeding population share the same complex song, but individuals introduce changes that others in their group then adopt, so that the song slowly changes over the course of months and years. In the South Pacific, entire song types migrate from east to west as humpbacks adopt songs from their eastern neighbors, then pass them on to their western neighbors.

Wendi A. Haugh

See also: Chimpanzees; Communication and Language; Dolphins; Ethology; Great Apes; Multispecies Ethnography; Personhood; Research and Experimentation

Further Reading

De Waal, F. 2001. *The Ape and the Sushi Master*. New York: Basic Books.

Laland, K., and Galef, B. 2009. *The Question of Animal Culture*. Cambridge, MA: Harvard University Press.

Whitehead, H., and Rendell, L. 2015. *The Cultural Lives of Whales and Dolphins*. Chicago: University of Chicago Press.

Animal Geography

Animal Geography is defined as "the study of where, when, why and how nonhuman animals intersect with human societies" (Urbanik 2012, 38). It is a subfield of the academic discipline of Geography and of the multidisciplinary research field of Human-Animal Studies. Its areas of focus have expanded in three major waves since the start of the 20th century: from an initial focus on mapping wild species to today's focus on cultural practices (e.g., fighting, animals in media) and animal subjectivities (how do animals experience their worlds?). For animal geographers, understanding where a human-animal interaction takes place—and why—is fundamental to 1) gaining a deeper understanding of how and why humans have the relations that they do with other species and 2) revealing the inseparability of other species from human societies.

Geographers have always been interested in animals because the goal of geography (the name of which comes from the Greek for "earth description" or "earth writing") has been to discover, describe, and interpret all phenomena on the planet. Historically, however, geographic work on animals consisted only of simple descriptions of the types of animals being encountered in different parts of the world. As geography developed into a formal, scholarly discipline in the late 19th century in Europe and the United States, geographers began to focus in a more systematic way on animals. This first wave of animal geography, called zoogeography, focused on mapping the ranges and types of wild species on the planet. These historical maps remain useful today, as geographers, biologists, and conservation scientists use them to assist in understanding how animal ranges have increased or decreased over the years.

A second wave of animal geography was developing by the mid-20th century as part of the growing geographical focus on cultural ecology (how humans adapt to their environments). This wave added to the first by studying domesticated animals and also by including the human relationship. For example, geographer Carl Sauer (1889–1975) researched the ways in which nonnomadic pastoral groups (sedentary animal herders) in Mexico reshaped the landscape through grazing, building pens, and protecting their animals.

Toward the end of the 20th century, several processes converged, shaping the third wave or the "new" animal geography of today: the exponential increase in scientific understanding of the impact humans are having on the planet, new understandings of animals and animal behaviors, the increase in political movements around animal advocacy, and the increasing love of animals around the world as evidenced by the growing number of pets and visibility of animals in popular culture. Animal geographers began to realize that the spaces and places of human-animal encounters, relations, and practices went far beyond mapping wild species or the human-domesticated animal pastoralist relationship.

Animal geographers today focus on practices connecting humans and animals and on specific animals or species alone. Their methods include quantitative (calculations using large-scale data sets), field-based (direct observation, basic counting and mapping), and qualitative (in-depth, small sample sets) such as interviews and reviews of archival material. There are four main conceptual approaches. The first is historical and explores where and how animals intersected with human societies in the past and how changes in relations have occurred over time. The second explores how economic practices impact animals both wild and domesticated. The third examines ethical and/or political conflicts around animals to see where, how, and why animals are included or excluded from human societies. The final area of focus is the cultural landscape, which refers to how human-animal relations are (in)visible in our daily lives. In any of these approaches, animal geographers may also focus on the experiential lives of individual animals/species.

To illustrate the wide spectrum of animal geography today, we can use the cultural practice of meat eating and the domesticated dog as two examples. With an historical approach, animal geographers might focus on where geographically people have eaten meat, or where specific dog breeds originated and why (e.g., for herding or companionship?). Economic analysis might include studying the scale of meat production (how many animals being produced, under what conditions), or the growth of the modern pet economy through the rise of dog daycares, dog sitters, dog groomers, and dog supplies. Ethical/political approaches might include studies about the relationship between meat eating and climate change, or cultural attitudes toward eating and farming dogs. In terms of the cultural landscape, geographers might study the places where meat eating occurs—such as religious festivals or restaurants—or explore how changing attitudes toward pets are visible in our everyday lives through the rise of modern dog parks and dog-friendly public spaces.

Animal geographers are interested in animals' perceptions of their own experiences. Making visible an animal's experience of its life is an essential part of better understanding the animal side of human-animal relations. For example, with respect to farmed animals, geographers are exploring how farming practices such as electronic milking equipment or access to the outdoors help or hinder what we understand about a cow's experiences and needs as a cow. With regard to dogs, geographers have studied the ways in which breeding for specific traits (e.g., shortened nose) or removing specific parts (e.g., docking tails) impact their health and communication needs. Geographers have also examined the different ways in which humans and dogs have learned to communicate with each other in the home, revealing how much of the human-animal relationship is produced by *both* humans and animals.

Julie Urbanik

See also: Biogeography; Ethics; Geography; Human-Animal Studies; Multispecies Ethnography; Pastoralism; Social Construction; Zoogeomorphology

Further Reading

Buller, H. 2014. "Animal Geographies I." *Progress in Human Geography* 38: 308–318.

Buller, H., and Morris, C. 2003. "Farm Animal Welfare: A New Repertoire of Nature-Society Relations or Modernism Re-embedded?" *Sociologica Ruralis* 43(3): 216–236.

Philo, C., and Wilbert, C., eds. 2000. *Animal Spaces, Beastly Places: New Geographies of Human-Animal Relations.* New York: Routledge.

Sauer, C. 1952. *Seeds, Spades, Hearths and Herds.* New York: American Geographical Society.

Tuan, Y. 1984. *Dominance and Affection: The Making of Pets.* New Haven, CT: Yale University Press.

Urbanik, J. 2012. *Placing Animals: An Introduction to the Geography of Human-Animal Relations.* Lanham, MD: Rowman & Littlefield Publishers.

Wolch, J., and Emel, J., eds. 1998. *Animal Geographies: Place, Politics, and Identity in the Nature-Culture Borderlands.* New York: Verso.

Animal Law

Animal Law is a broad term which at present encompasses the body of law, or jurisprudence, concerning various rights which may be asserted by humans over or on behalf of animals. The main focus of Animal Law concerns animal welfare standards and restrictions upon human actions that may affect animals. The classification of animals as property, or "things," means that the law excludes animals from being subjects in society, or citizens who are entitled to legal rights and protections, and therefore from the class of "legal persons," or those who have legal standing to assert even limited rights. The battle for legal personhood (or to view animals as subjects rather than objects) for animals, and the resulting legal standing of animals to assert those rights and protections, is at the forefront of Animal Law today.

For thousands of years, various cultures have enacted laws regulating human conduct toward animals. Examples of early animal laws include hunting bans and meat-eating restrictions by the Indian emperor Ashoka (304–232 BCE); Hebrew and Islamic ritual slaughter practice laws (still adhered to today) designed to limit animal suffering by specifying the method of killing; and the *Edicts on Compassion for Living Things* enacted by the Japanese ruler Tokugawa Tsunayoshi (1646–1709), also known as "the Dog Shogun," in 1690, for the benefit of dogs. Other cultures held animals accountable for the consequences of their actions. For example, medieval European animal trials expressly recognized personhood in naming animals as criminal defendants capable of intentional action, and also held them directly liable for crimes such as murder, bestiality, theft, and killing other animals. In fact, animal defendants sometimes appeared in court and were on several occasions represented by counsel.

Western cultures, influenced by Greek and Roman philosophers, historically have considered animals as chattel (personal property) and rejected the idea of legal and moral obligations to, or rights which may be asserted by, animals. For example, the Prussian philosopher Immanuel Kant (1724–1804) based such views upon the belief that animals are not rational and do not possess free will, and therefore had no intrinsic value (value in themselves) and so should not be afforded legal rights and protections. The idea that animals have natural rights that should be respected by humans came into being in 19th-century Britain and America, resulting in the establishment of various humane organizations (e.g., the Royal Society for the Prevention of Cruelty to Animals), which emphasized anti-cruelty statutes and criminal enforcement. The first Western animal protection law was England's *An Act to prevent the cruel and improper treatment of Cattle* (1822), which set punishments for those who beat certain types of farm animals.

Animal Law did not exist as a separate specialization until recently. In 1973, Henry Mark Holzer (1933–), an America lawyer, filed the landmark lawsuit *Jones v. Butz* (374 F.Supp. 1284 (D.C.N.Y. 1974)), which advocated for the animals' interest in challenging an exception under the federal *Humane Methods of Livestock Slaughter Act of 1958*. Holzer ultimately lost, but only after the case went all the way to the U.S. Supreme Court. His lawsuit is credited with creating the emerging field of Animal Law, which he actively promoted through outreach and by establishing a professional journal, the *Animal Rights Law Reporter*. Prior to the *Jones* case, cases involving animals were usually viewed as subsets of property, contract, trusts and estates—and, occasionally, criminal—law and were not directly concerned with the interests of the animals. Consequently, no specific body of law that considered the extent of human obligations to animals existed until recently.

Cases subsequent to *Jones* include challenges to zoo conditions (*Jones v. Beame* (380 N.E.2d 277 (N.Y. 1978))) and aerial shooting of goats on federal land (*Animal Lovers Volunteer Association, Inc. v. Weinberger* (765 F.2d 937 (C.A.9 (Cal.), 1985))). Additionally, *TVA v. Hill* (437 U.S. 153 (1978)), an environmental case, considered the rights of the snail darter (a species of fish) to occupy critical habitat threatened by the construction of a dam. Throughout these early cases, there was an attempt to balance the previously unrestrained rights of humans with the newly arising legal protections afforded to animals and animal interests, environmental concerns, and the law's concern for species preservation as a whole. However, there was little to no recognition of animals' rights or allowing the assertion of protections for individual animals.

There are two major currents in Animal Law today. The first is the struggle on behalf of certain animals for legal personhood and legal standing, which has taken the form of legislative efforts, *habeas corpus* (literally, "present the body") petitions, and administrative law changes with which to assert protections and remedies through animal guardians. For example, *habeas corpus* petitions have been filed in Austria, Argentina, Brazil, and the United States, seeking to assert legal standing by chimpanzees as persons to gain release from confinement, primarily from animal research facilities, zoos, commercial exhibition, and private captivity, to more suitable conditions such as wildlife preserves. These legal challenges, asserted by self-appointed guardians who claim to represent the animals' interests, have been mainly based upon the "capabilities approach," which asserts that the mental capabilities of these nonhumans are such that they could, like a human, exercise free will and choose a better environment. The other main area of Animal Law is the struggle to protect animals from the effects of human-induced global warming and habitat destruction. This takes the form of legislation aimed at preserving wildlife by limiting human activities, such as polluting, construction and habitat encroachment, hunting, and interference with migration routes, that affect animals and their habitat.

John T. Maher

See also: Advocacy; Agency; Animal Liberation Front (ALF); Chimpanzees; Climate Change; Cruelty; Dolphins; Endangered Species; Ethics; Great Apes; Habitat Loss; Institutional Animal Care and Use Committees (IACUCs); Personhood; Rights

Further Reading

Berman, P. S. 1994. "Rats, Pigs, and Statues on Trial: The Creation of Cultural Narratives in the Prosecution of Animals and Inanimate Objects." *New York University Law Review* 69: 288.

Bodart-Bailey, B. M. 2006. *The Dog Shogun: The Personality and Policies of Tokugawa Tsunayoshi*. Honolulu: University of Hawaii Press.

Evans, E. P. 1987. *The Criminal Prosecution and Capital Punishment of Animals*. London: Faber & Faber Ltd.

Tischler, J. 2008. "The History of Animal Law, Part I (1972–1987)." *Stanford Journal of Animal Law and Policy*: 1–49.

Wise, S. 2014. *Rattling the Cage: Toward Legal Rights for Animals*. Boston: Da Capo Press.

Animal Liberation Front (ALF)

The Animal Liberation Front (ALF) is a leaderless, decentralized animal rights organization that engages in direct action (the use of public forms of protest instead of negotiations), often through illegal means. ALF members believe that no animal should be exploited for food, entertainment, or science and that nonhuman lives should not be seen as private property, meaning that animals should not be owned. The ALF's mission is to "abolish institutionalized animal exploitation because it assumes that animals are property" (ALF n.d.). The ALF operates through small groups of individuals, called "cells," that operate independently of one another, without a hierarchical chain of command. This allows ALF members to avoid identifying other members if they are questioned. Due to this structure and anonymity, the ALF is able to operate underground. Active in dozens of countries, the ALF works to remove animals from factory farms, laboratories and testing facilities, and zoos, while inflicting economic damage on the institutions and organizations that promote animal exploitation for profit or entertainment. Animals liberated by ALF members are placed in sanctuaries or homes where they can live out their natural lives. Although it has been classified as a domestic terrorist organization in the United States, one of the ALF's tenets is to take all precautions against harming any human or nonhuman animal. The Animal Liberation Front is an important topic area for animal studies, especially in characterizing the different levels of opposition to the use of animals' lives for financial or personal gain.

The Animal Liberation Front emerged in 1974, but its roots extend back to the 1960s. Founded in 1963, the Hunt Saboteurs Association physically interfered with animal hunts in the United Kingdom throughout the 1960s. Inspired by the Hunt Saboteurs, in 1971 British activist Ronnie Lee started a group called the Band of Mercy, which focused on not only sabotaging hunters' vehicles, but also on protesting animal testing (vivisection) in laboratories. These groups primarily used direct action. Lee began to support the use of arson and other forms of property destruction as the main tactics used in the group's mission.

In 1974, Lee, along with fellow-activist Cliff Goodman, created the Animal Liberation Front. The new group was incredibly successful, and more than £250,000 ($397,000) worth of damage was attributed to the ALF in its first year of operation. It was not until the 1980s that the Animal Liberation Front moved to North America, and it did not gain much traction there until the 1990s. To enact meaningful change, the ALF focuses on direct economic threat placed upon companies that use animals in research or entertainment. Together with the Environmental Liberation Front (the ALF's sister organization), the ALF is estimated to be responsible for around $43 million in damages between 1996 and 2002.

One of the first high-profile rescues that the ALF committed in the United States was in September 1985, when they raided a laboratory at the University of California at Riverside. Members of the ALF removed a stump-tailed macaque monkey named Britches who had been separated from his mother at birth and had his eyes sewn shut for a study that tested the effects of sensory deprivation on young monkeys. During the raid, activists rescued Britches along with 467 mice, cats, opossums, pigeons, rabbits, and rats, and also committed $700,000 worth of damage to equipment. Britches was taken to a sanctuary where he spent the remainder of his life.

As a result of the raid, the University of California stopped several research programs and no longer allows monkeys' eyes to be sewn shut. The ALF still recognizes the rescue of Britches as one of their most successful.

According to philosopher Steven Best, there are several ways to understand the ALF in the United States. First, as an organization, the ALF operates as part of a new social movement that places attention on the historically ignored issue of animal rights and welfare. Second, this animal liberation movement is focused on stopping all nonhuman animals from being categorized as legal property. The movement in general, and the ALF in particular, argue that we should end all exploitation of animals and should place greater importance on life rather than on any financial gains. Finally, and most importantly according to the ALF, the animal liberation movement can be compared to the U.S. antislavery and abolitionist movement of the 19th century. Although this is a controversial and often highly contested perspective, the ALF hopes that society will accept the immorality of using animals for economic, scientific, or personal gain, just as we now see slavery as a historical and moral injustice.

Today, the ALF is classified as a domestic terrorism organization in both the United States and the United Kingdom. The British government established a task force, the National Extremism Tactical Coordination Unit, in 2004 to investigate and monitor the activities of the ALF and other domestic terrorism organizations. In 2005, the ALF was listed by the U.S. Department of Homeland Security as a domestic terrorism organization and as a significant national security threat. However, the ALF maintains that its tactics are nonviolent, citing the fact that no ALF actions have resulted in any deaths. In 2006, the U.S. Congress passed the Animal Enterprise Terrorism Act (AETA), which prohibits a person from engaging in any conduct with the intent of damaging or interfering with the operations of an animal enterprise or business. This act gives the Department of Justice greater authority to target animal rights protesters more generally, including those that are not violent in nature, suggesting that the concern over the ALF is not going to diminish anytime soon.

Stefanie Georgakis Abbott

See also: Advocacy; People for the Ethical Treatment of Animals (PETA); Research and Experimentation; Rights; Vivisection

Further Reading

Animal Liberation Front (ALF). n.d. "Mission Statement." Accessed June 21, 2015. http://www.animalliberationfront.com/ALFront/mission_statement.htm

Best, S., and Nocella, A. J., eds. 2004. *Terrorists or Freedom Fighters?* New York: Lantern Books.

Guillermo, K. S. 1993. *Monkey Business.* Washington, DC: National Press Books.

Molland, N. *Thirty Years of Direct Action.* Accessed December 5, 2011. http://www.animalliberationfront.com/ALFront/Premise_History/30yearsofDirectAction.htm

Newkirk, I. 2000. *Free the Animals: The Story of the Animal Liberation Front.* New York: Lantern Books.

Animals

Including single-celled organisms, there are millions of nonhuman species on Earth, ranging from the tiniest bacteria to enormous mammals such as the blue whale. Animal species have extraordinarily diverse physical capacities—from being able to live miles deep in oceans (giant squid) to building skyscrapers (humans). While it was long thought that humans were the only animals with emotional and intelligence capacities, scientific research is rapidly expanding our understanding of how these capacities manifest in other animal species. As human animals we must recognize that our own survival is tied to that of other animals and that we have ethical responsibility to them.

Scientists estimate that there are approximately 1.3 million animal species currently on Earth (Mora et al. 2011). Taxonomists are biologists who classify species, and while they debate many details about the classification system, they do recognize a general system that goes from the general to the specific. The most general are three domains of life. Two, Archaea and Bacteria, include single-celled organisms without nuclei, while the third, Eukaryote, contains all multicellular organisms with nuclei, including animals and plants. Within the Eukaryote domain are kingdoms, of which Animalia is one. Defining characteristics of animals include being heterotrophic (eating and digesting other organisms for food), being able to move voluntarily, having a life cycle that includes developing from an embryo through sexual reproduction, and having sensory organs with which to experience their environment. Within the Animalia kingdom are five main phyla. This category differentiates between Cnidaria (invertebrates—without backbone), Chordata (vertebrates—with backbone), Arthropods (insects, spiders), Molluscs (octopus, snails), and Echinoderms (starfish, sea urchins). Within the phylum of Chordata we find the classes of mammals, birds, amphibians, reptiles, and bony and cartilaginous fish. Classifications continue into the smaller groups of order, family, genus, and finally species. For example, a human would be classified this way: domain—*Eukaryote;* kingdom—*Animalia;* phylum—*Chordata;* class—*Mammalia;* order—*Primate;* family—*Hominidae*; genus—*Homo;* species—*Sapiens.*

Each species (not just animals) on Earth occupies an ecological niche. An ecological niche relates to the role that each species plays in its environment, including how a particular species population affects its environment through both consuming resources and being resources for predators or parasites. It also includes how a species population responds to available resources and threats, by, for example, growing in number when resources are abundant and predators, parasites, and diseases are few. There is great flexibility in these systems of give-and-take, resulting in adaptation and evolution; but there is also fragility, resulting in extinction.

Biodiversity in the animal kingdom is an integral part of what are called "ecosystem services" that make life on Earth possible. For example, one important ecosystem service that animals provide, and that humans cannot live without, is pollination. Many animals, but especially insects, are responsible for pollinating 80 percent of all flowering plant species, which contributes significantly to the world's crop production, our food supply, and plant-derived medicines (FAO 2015).

Not only are we learning more about nonhuman animals' crucial contributions to the existing order and health of our planet, but also about many species' mental capacities. For years, many scientists held that other animals lacked sentience (possessing the capability to be aware of sensations and emotions such as pain, suffering, happiness, joy, empathy, and grief) and that instead they only reacted instinctively to stimuli. Although this opinion is still held by some, scientists now have substantial evidence that many animals—for example, dogs, elephants, and dolphins, to name just a few—*are* sentient. Such evidence is in line with Charles Darwin's (1809–1882) demonstration of the evolutionary continuity of species, which indicates gradual variation, rather than sharp distinctions, between them. For example, in 1960, when anthropologist Jane Goodall (1934–) began her studies of chimpanzees, humans were thought to be the only animals with the intelligence to make and use tools. However, Goodall proved this false while observing chimpanzees fashion sticks into tools to extract termites from their mounds. Today, we know that many other animals, such as capuchin monkeys, raccoons, and octopuses, also use tools. Additionally, many species, such as crows and orangutans, display intelligence more generally, for example, through conceptualizing similarities and differences, memory, self-awareness, deception, complex communication, and learning from others.

Indeed, research is showing that all animals have some type of intelligence and sentient capacity—even if we do not yet have the ability to precisely measure these. In other words, species have the intelligence and sentience they need to carry out their own lives, which may or may not manifest in the same way human intelligence and sentience do. Recognition, and equally important, *acceptance*, of animal sentience is significant because it raises important ethical questions about how we treat other animals.

The major concern for the well-being and survival of nonhuman animal species is the impact on them from the current lifestyle of the human animal. Many now call our geologic time period the Anthropocene (Age of Humans), which recognizes the human animal as the primary driver of environmental change. For example, in the last 40 years, vertebrate species populations have decreased by 52 percent due mainly to habitat loss (WWF 2014). Experts warn we may be entering the "sixth mass extinction" of species, with unknown consequences, because of human population growth and unsustainable rates of consumption. The previous five extinction events occurred before the arrival of humans, but their detrimental impact on lifeforms is documented, almost as a warning, in the fossil record.

Sarah M. Bexell, Stephanie Johnson, and Courtney Brown

See also: Anthropomorphism; Biodiversity; Biogeography; Communication and Language; Emotions, Animal; Evolution; Humans; Intelligence; Sentience; Species; Taxonomy; Zoogeomorphology; Zoology

Further Reading

Bekoff, M., and Goodall, J. 2008. *The Emotional Lives of Animals: A Leading Scientist Explores Animal Joy, Sorrow, and Empathy—and Why They Matter.* Novato, CA: New World Library.

Ceballos, G., Ehrlich, P. R., Barnosky, A. D., Garcia, A., Pringle, R. M., and Palmers, T. M. 2015. "Accelerated Modern Human-Induced Species Losses: Entering the Sixth Mass Extinction." *Science Adv.* 1, e1400253.

Cox, C. B., and Moore, P. D. 2010. *Biogeography: An Ecological and Evolutionary Approach.* Hoboken, NJ: Wiley.

Food and Agriculture Organization (FAO) of the United Nations. 2015. "Biodiversity for a World without Hunger: Pollinators." Accessed November 19, 2015. http://www.fao.org/biodiversity/components/pollinators/en

Jane Goodall Institute. 2015. "Dr. Goodall Long Bio." Accessed October 30, 2015. http://www.janegoodall.org/wp-content/uploads/the-Jane-Goodall-Institute_JaneGoodall_LongBio.pdf

Mora, C., Tittensor, D. P., Adl, S., Simpson, A. G. B., and Worm, B. 2011. "How Many Species Are There on Earth and in the Ocean?" *PLoS Biology* 9(8) e1001127.

World Wildlife Fund. 2014. "Living Planet Report." Accessed January 15, 2015. http://wwf.panda.org/about_our_earth/all_publications/living_planet_report/living_planet_index2/

Animal Welfare Act. *See* Institutional Animal Care and Use Committees (IACUCs); Research and Experimentation; Welfare; Primary Documents

Anthropogenic. *See* Climate Change

Anthropomorphism

Anthropomorphism is defined as the attribution of human-like traits or emotions to nonhuman objects, deities, or animals. From the Greek terms *anthropōs* (human) and *mōrphe* (shape), the concept originally appeared in the writings of Xenophanes of Colophon, a sixth-century-BCE Greek philosopher, who criticized the use of human characteristics to describe deities by stating: "But if horses or oxen or lions had hands or could draw with their hands and accomplish such works as men, horses would draw the figures of gods as similar to horses, and the oxen as similar to oxen, and they would make the bodies of the sort which each of them had" (Lesher 1992, 25). Anthropomorphism is used in everyday conversation and marketing (e.g., the Geico® gecko) and, according to some psychologists, occurs automatically in human judgment to better understand and control one's environment. However, its use in scientific inquiry to explain mental states of nonhuman animals has long been a topic of debate. This is due to humans' inability to verbally communicate directly with animals and a lack of tools to measure subjective mental and emotional states. In contrast, animal advocates have pointed out a contradiction between denying the similarities between humans and other animals but then using animals as models for humans in the biological sciences and psychology.

The motivation behind the use of anthropomorphic descriptions of nonhuman animals, deities, or objects is not completely known. However, psychologist Nicholas Epley and colleagues have proposed a number of possible explanations. One explanation, known as "elicited agent knowledge," cites the inability of humans to examine nonhuman experience without drawing from their own experience as the first step to investigation. The second explanation, "effectance motivation," attempts to increase control over one's environment. In this case, anthropomorphism is used as a strategy to increase understanding and control of an environment by attributing human-like emotions to nonhuman objects or animals with which an observer is unfamiliar. Finally, "sociality motivation" may include anthropomorphism to increase the sense of inclusion, given the natural avoidance of solitude that seems to be present in human behavior.

Many people regularly anthropomorphize nonhuman animals and objects in their daily lives. In the biological sciences, however, there are longstanding debates over the usefulness or harm associated with incorporating anthropomorphism into scientific inquiry, most prominently within the fields of psychology and ethology (the study of animal behavior). Charles Darwin (1809–1882) embraced the concept in *The Descent of Man, and Selection in Relation to Sex* (1871, 77) by saying, "The fact that lower animals are excited by the same emotions as ourselves is so well established that it will not be necessary to weary the reader with many details." He argued for evolutionary continuity between different species' mental experiences in his 1872 book, *The Expression of the Emotions in Man and Animals.* Darwin believed the extension of human-like mental and emotional states to nonhuman animals was appropriate, given behavioral and anatomical similarities passed on through evolution.

In contrast, early critics of anthropomorphism, like psychologist John B. Watson (1878–1958), believed there was no place for the study of mental states in humans or animals because they could not be measured directly. Watson is known as the "father of behaviorism," a branch of psychology that became standard for studying human behavior in the early to mid-1900s by focusing on "the prediction and control of human action and not with an analysis of [mental states]" (Watson 1919, ix). Behaviorism's widespread influence extended to the study of animal behavior, as seen in the writings of Niko Tinbergen (1907–1988), a Nobel Prize winner and pioneer of modern ethology. In his book, *The Study of Instinct*, Tinbergen wrote,

"[T]he ethologist does not want to deny the possible existence of [mental and emotional states] in animals, he claims it is futile to present them as causes, since they cannot be observed by scientific methods" (Tinbergen 1951, 5). Despite its popularity in the early and mid-20th century, however, strictly behavioristic approaches to studying animal behavior seemed to dwindle following the release of zoologist Donald Griffin's (1915–2003) book, *The Question of Animal Awareness: Evolutionary Continuity of Mental Experience* (1976), which mirrored Darwin's position on evolutionary continuity. More recently, Frans de Waal (1948–), a primatologist and ethologist, has suggested that "anthropodenial," or the intentional *exclusion* of biological and behavioral similarities between humans and animals from scientific hypotheses and research findings, may be more harmful to scientific discovery than their inclusion. However, psychologist Clive D. L. Wynne's (1961–) recent essay on anthropomorphism as an inappropriate, nonobjective study method suggests that the debate is far from over.

While there is still no consensus on how anthropomorphism should be used in scientific inquiry, if at all, some scholars have argued for a more "critical" anthropomorphism wherein the existence of nonhuman mental experience is acknowledged as a starting point for forming a hypothesis, but objectively studied using variables such as species-specific behavior, natural history, or physiology to draw conclusions. For example, an adult dog that consistently chases its tail may be interpreted by a nonscientist as "happy and playful." Critical anthropomorphism might hypothesize that the dog is exhibiting positive, playful behavior through tail chasing (as is often seen in young wolf pups in the wild). However, behavioral surveillance and physiological measures associated with stress may actually reveal that the constant tail chasing is, instead, an abnormal attempt to cope with a stressful environment or experience.

Christopher J. Byrd

See also: Emotions, Animal; Empathy; Evolution; Intelligence; Sentience

Further Reading

Darwin, C. 1871. *The Descent of Man, and Selection in Relation to Sex.* New York: A. L. Burt Company.

Darwin, C. 1872. *The Expression of the Emotions in Man and Animals.* London: John Murray.

De Waal, F. 1997. "Are We in Anthropodenial?" *Discover* 18: 50–53.

Epley, N., Waytz, A., and Cacioppo, J. T. 2007. "On Seeing Human: A Three-Factor Theory of Anthropomorphism." *Psychological Review* 114(4): 864–886.

Griffin, D. R. 1981. *The Question of Animal Awareness: Evolutionary Continuity of Mental Experience.* New York: The Rockefeller University Press.

Tinbergen, N. 1951. *The Study of Instinct.* Oxford: Clarendon Press.

Watson, J. B. 1919. *Psychology from the Standpoint of a Behaviorist.* Philadelphia and London: J. P. Lippincott Company.

Wynne, C. D. L. 2005. "The Emperor's New Anthropomorphism." *The Behavior Analyst Today* 6(3): 151–155.

Xenophanes of Colophon. 1992. *Fragments.* Text and translation by Lesher, J. H. Toronto, ON: University of Toronto Press.

Aquaculture

Aquaculture is the farming of aquatic organisms for food consumption under managed conditions. Aquaculture farms can be found in either inland freshwater bodies or saline (salt) water bodies (also known as mariculture). It is also increasingly seen as an important strategy to combat global food insecurity in the face of unregulated overfishing around the world. Aquatic organisms include fish, crustaceans such as crabs and shrimp, and molluscs such as abalone, scallops, and oysters. The farming of molluscs and crustaceans is most often found in offshore, saline water bodies, while most fish are farmed inland.

Some believe that a primitive form of carp farming began in China as early as 4,000 years ago, but the earliest written record we have of a Chinese aquaculture manual dates to 400 BCE. However, despite its long history, aquaculture production has only gained global momentum in the past 50 years. As recently as 1950, global aquaculture production stood at less than 1 million metric tonnes (1.1 million tons), while wild capture of aquatic organisms was more than 20 million metric tonnes (2.2 million tons).

Today, aquaculture is the fastest-growing food-production sector. The Food and Agriculture Organization of the United Nations (FAO) in its most recent (2012) statistics revealed that aquaculture production is at an historic high of more than 90 million tonnes per year. In 2012, 42 percent of global fisheries output was produced from aquaculture, while since 1985 global, wild-caught fish production has peaked and stabilized at about 90 million tonnes annually. While the growth in aquaculture production is a global phenomenon, it is more evident in some parts of the world than others. In Asia, for example, aquaculture supplies more aquatic organisms for consumption than capture fisheries. There is also wide variance in the kinds of aquatic organisms that are farmed. The FAO estimates that more than 600 aquatic species are farmed commercially worldwide. Some species, like the tilapia fish, are particularly popular, with more than 140 countries farming them.

Aquaculture is an important sector of the economy. It is the livelihood of 16.6 million fish farmers worldwide, with the vast majority (97 percent) living in Asia. In poverty-stricken regions like Vietnam's Mekong Delta, local catfish farmers export their fish to affluent customers in developed countries like the United States with the aid of the national government and global institutions like the Asian Development Bank (ADB). Clearly, farmed fish is not only a subsistence food protein for the locals; it is also a highly marketable product that contributes to the general economy and an important activity that drives rural development. Besides Vietnam, a

Aquaculture, the farming of fish and aquatic organisms, is seen as a potential solution to the twin global problems of over-fishing and meeting the human food supply. Studies of the efficacy of these systems and their environmental impact are growing because the industry has expanded so quickly. (National Renewable Energy Laboratory)

host of other developing countries like Chile aim to develop the aquaculture sector to take advantage of the rising demands of global consumers. In this regard, high-value aquaculture products like abalone (a type of shellfish) are especially favored.

The aquaculture industry faces a number of challenges (although these vary significantly across locales). In inland aquaculture, a typical problem is the loss of arable land to aquaculture, as seen by the shrimp-farming industry in Southeast Asian countries like Thailand and Vietnam. Aquaculture, because of the confinement of aquatic organisms, can also suffer devastating losses of farmed stocks whenever there is an outbreak of disease. The occurrence of such outbreaks, however, can be mitigated with the use of highly controlled water tanks and sophisticated monitoring. In mariculture, where the aquatic organisms are often enclosed in bodies of saline water, changes in the salinity and temperature of the water can result in plankton blooms (sudden increases in microscopic ocean plants), which can decimate entire fish stocks due to oxygen deprivation. This also suggests the high economic risks in the aquaculture sector. Mariculture also presents risks to the surrounding environment when, for example, infectious diseases among farmed fish spread to wild populations.

Salmon farming is the standout example of a species that, though by nature migratory, has been successfully farmed. The growth cycle of wild salmon is tied to the "salmon run." Salmon runs typically occur between September and November when the fish swim from the ocean to the upper reaches of rivers in order to spawn (reproduce). The growth of wild salmon peaks in the summer prior to the annual runs, while their growth stagnates during the winter. Genetic modification of farmed salmon has successfully decoupled the growth cycle of salmon from their migratory pattern through the introduction of ocean pout genes. As the ocean pout is able to grow year round, farmed salmon genetically modified with the gene of ocean pout will similarly grow around the year, and hence grow faster to a marketable size. In addition, farmed salmon have also been genetically modified so that they can achieve improved feed conversion efficiency. In other words, in order to maximize the weight gain per unit of formula feed, salmon have to be physiologically transformed to thrive on formula feed instead of their natural diet.

Aquaculture production will increasingly be seen as vital to meeting the demand for aquatic organism food protein given that wild fish stocks have come under increasing pressure due to overfishing and inadequate regulation. To mitigate the economic and environmental risks of aquaculture, there has been growing research on closed-containment systems, which are either large, solid-wall systems that float on the water, or land-based tank systems. Such closed-containment systems will ensure better monitoring of aquatic organisms as well as minimize cross-contamination between farmed and wild species.

Harvey Neo

See also: Biotechnology; Fisheries; Fishing

Further Reading

Krause, G., Brugere, C., Diedrich, A., Ebeling, M. W., Ferse, S. C. A., Mikkelsen, E., Agúndez, J. A. P., Stead, S. M., Stybel, N., and Troell, M. 2015. "A Revolution without People? Closing the People–Policy Gap in Aquaculture Development." *Aquaculture* 447: 44–55.

Lim, G., and Neo, H. 2014. "Geographies of Aquaculture." *Geography Compass* 8(9): 665–676.

Macuiane, A. M., Hecky, R. E., and Guildford, S. J. 2015. "Changes in Fish Community Structure Associated with Cage Aquaculture in Lake Malawi, Africa." *Aquaculture* 448: 8–17.

Salgado, H., Bailey, J., Tiller, R., and Ellis, J. 2015. "Stakeholder Perceptions of the Impacts from Salmon Aquaculture in the Chilean Patagonia." *Ocean and Coastal Management* 118B: 189–204.

Aquariums

Even as water covers about 70 percent of Earth's surface and represents an important diversity from seashores to ocean floors in terms of landscape, animals, and plants, the animals and plants stay mostly unreachable to humans. The underwater

The jellyfish tank at the Monterey Bay Aquarium in California. (Jean Estebanez)

experience is accessible to very few people and then only for a very small part of the global ocean, making it one of the true limits of our Earth. Aquariums, tanks of fresh or saltwater filled with plants and animals, are organized encounters between humans and nonhumans in conditions that are unique: We may be affected, moved, and impressed by animals that are indifferent to us. They unveil a particular vision of the underwater world that we may experience firsthand and also link to wider social issues.

The practice of keeping fish alive is documented as early as 1000 BCE during the Roman and Chinese Empires. At that time, tanks had to be connected to the sea or a waterway. In 1850, the discovery by chemist R. Warington (1807–1867) and naturalist P. H. Gosse (1810–1888) of a balance between animals producing carbon dioxide (which plants need) and plants producing oxygen (which fish need) allowed aquariums to be constructed far from open waters. This discovery led to the first large-scale aquariums in Regent's Park, United Kingdom (1853); Vienna, Austria; Paris, France (1860); and Berlin, Germany (1869). *Aquarium mania*—a trend among wealthy city dwellers in England and France between 1850 and 1870—even included specialized living room furniture designed to highlight an aquarium.

We can categorize aquariums in two ways: as public venues and as the personal, home hobby. On the individual side, roughly 20 million households—6 million in the United States—own an average of 20 fishes each, fueling a home aquarium business of $5 to $6 billion a year. Aquarium fishes follow a global trade route. They are culled from productive fishing zones (Singapore, Malaysia, Brazil, Sri Lanka, and Puerto Rico) and moved to important markets (United States, United Kingdom, France, Germany, Japan, Singapore, Malaysia, China). Their financial value is almost a hundred times the value of fishes sold for food.

More than 200 million visitors pass through the doors of the roughly 240 worldwide aquariums and marine mammal parks open to the public every year. Institutions such as the Monterey Bay Aquarium (California) and the Tokyo Sea Life Park (Japan) attract 3 to 4 million visitors annually. The two main functions of public aquariums are education and entertainment. Aquariums have also evolved into scientific institutions gathering biological knowledge by studying captive species, offering suitable conditions to study rare specimens, and protecting and reproducing endangered species.

Aquariums support scientific practices, but they are also heavily based on entertainment as they try to present a spectacular, underwater show. Tanks are filled with stones, fake shipwrecks, sculptures, and/or buildings that emphasize the fishes' exoticness, wildness, and otherness (differences) in relation to humans. Contemporary aquariums tend to organize the scenery to mimic travel for visitors: We are not only looking at fish but we should be able to experience immersion in an underwater environment. The Monterey Bay Aquarium, for example, in exhibiting jellyfish uses special tanks with edges that seem to fade through the use of an optical illusion. Modern aquariums also organize animals and plants according to their places of origin. One elaborate example—the Lisbon Aquarium (Portugal)—is modeled on the divisions of the global oceans. A huge, central tank is surrounded by the Arctic World; the Indian Ocean, containing a reef barrier; the North Atlantic, with its jellyfish; and the temperate Atlantic, with giant octopi. Through their spatial organization, aquariums also reinforce the separation between humans and nonhumans, as they are located on different sides of the tanks' glass walls.

A last, important feature of large aquariums is the use of domestic versus exotic species. In the mid-19th century, fishes in aquariums were mostly local, but as transportation technologies were developed, a strong network of catchers and animal dealers began, like zoos, to gather exotic animals from abroad. Today, however, even as most contemporary aquariums present important, exotic collections, they also display local fauna (animals) and flora (plants). About a third of the Brest Océanopolis aquarium in France displays what one would encounter in a dive off the Brittany coast. There can still be a sense of otherness, however, as even local or familiar underwater animals are often very different from us—the bodies of fish, coral, and jellyfish radically differ from ours.

International trade in aquarium species and the ethics of keeping animals captive are two of the main issues public and personal aquariums face. Approximately one billion ornamental fishes are removed from the oceans every year. As opposed to freshwater aquariums that are mostly composed of captive-bred fishes, saltwater aquariums are filled with wild-caught animals. Most coral reef fishes are targeted by an unsustainable demand that is causing the disappearance of local species and a depleted local environment. Unlike mammals, marine fishes are almost exempt from international trade regulation. While zoos have faced strong criticism for caging wild animals because they overwhelmingly present mammals, aquariums are still mostly protected from disapproval, except in the case of marine mammals such as dolphins or orcas.

Jean Estebanez

See also: Aquaculture; Coral Reefs; Dolphins; Fisheries; Human-Animal Bond; Marine Mammal Parks; Non-Food Animal Products; Sharks; Taxonomy; Wildlife Rehabilitation and Rescue; Zoos

Further Reading

Brunner, B. 2005. *The Ocean at Home: An Illustrated History of Aquariums.* New York: Princeton Architectural Press.

Kisling, V., ed. 2000. *Zoo and Aquarium History: Ancient Animal Collections to Zoological Gardens.* Boca Raton, FL: CRC Press.

Taylor, L. 1993. *Aquariums: Windows to Nature.* New York: Prentice Hall.

Art, Animals in

Animals have figured prominently throughout the history of human art. In addition, animal body parts and products have been primary ingredients in artists' materials for centuries. Those working in animal advocacy have relied on art in their educational and activist campaigns. Recent examples of animal art have raised questions about the nature of art and creative endeavors.

Paintbrushes have been made with animal hair for centuries, and some pigments and dyes have historically been made from animal bodies. Tyrian purple, for example, was derived from the bodies of sea snails and was a favorite of the ancient Romans. This color was named after Tyre, a location in present-day Lebanon where it was produced. Indian Yellow (so named because it came from India) was originally made from the urine of cattle that had been fed a strict diet of mango leaves. Egg yolks have been used as a binder in tempera paints, and albumen (egg whites) was an important component of an early form of photographic prints. Later, gelatin (derived from animal byproducts) would become a standard ingredient in film and photographic paper.

Some of the earliest surviving artwork is of animals. Animals figure prominently in cave paintings found at Chauvet and Lascaux (France) and in the Bhimbetka Rock shelters (India). Representations of animals in indigenous (native peoples) rock art, dating back approximately 40,000 years, have been found in Australia. Animals have also figured prominently in art produced by indigenous cultures in North and South America. In the ancient world there are numerous examples of paintings, sculptures, and carvings from places like Egypt, in which humans are represented interacting with domesticated animals such as cattle and oxen. Bastet, a goddess in ancient Egyptian religion, was typically represented with a lion or cat head.

Bestiaries (illustrated books about animals), which provided facts about animals alongside moral and religious messages, were especially popular in Europe from the 10th to the 15th centuries. Animals have appeared frequently in the history of Christian art in Europe, but their role in this context is largely symbolic and allegorical.

Early natural history pursuits in Europe, beginning in the Renaissance period (1300–1700s), resulted in new ways of representing animals. Leonardo da Vinci (1452–1519) produced numerous studies of both human and animal anatomy that were based on direct observation and dissection. Exploration and colonization introduced new species to a curious European public, and artists were quick to produce images of these exotic-looking plants and animals. For example, in 1515, the German artist Albrecht Dürer (1471–1528) produced a woodcut of a rhinoceros that had been brought to Lisbon, Portugal, from India. In spite of the fact that Dürer worked only from a description of the animal, and therefore the image had some anatomical inaccuracies, his woodcut of the rhinoceros was popular, becoming a standard visual representation of this animal until the 18th century.

There is a long history of equestrian statues in which famous leaders are posed astride powerful-looking, majestic horses. Examples include Emperor Marcus Aurelius in Rome and King George IV in London. Throughout the history of art there are also many examples of dogs and horses appearing alongside humans in painted portraiture. The 18th- and 19th-century portraits of prized cattle by artists such as George Stubbs illustrated ideal qualities sought after by those involved in animal husbandry.

Some artists have even collaborated with animals. Olly & Suzi, a British art duo, have collaborated with many different animals by placing drawings in situations where animals such as great white sharks, lions, and crocodiles interact with them. The result is a piece that combines marks made by both the human and the nonhuman participants—pencil sketches and paint coexist with scratches, tears, and bites.

Many contemporary artists have found themselves embroiled in controversy for their use of animals in their art. Damien Hirst is best known for work in which he presents dead animals suspended in formaldehyde. Hirst draws on conventions of science and natural history while also pushing the boundaries of how art is defined. Likewise, Eduardo Kac's *Alba* (2000), a genetically engineered rabbit who turned fluorescent green in certain light conditions, has raised a number of ethical and aesthetic questions.

Today, artists like Sue Coe and Jo-Anne McArthur create images with the specific intention of raising awareness of cruelty to animals. In the 18th century, William Hogarth's series, *The Four Stages of Cruelty,* drew connections between cruelty to animals and violent behavior more generally. In the late 19th century, G. F. Watts painted *A Dedication*, a work that criticized the slaughter of animals for fashion. In the early decades of the 20th century, Morgan Dennis created imagery that was used to promote "Be Kind to Animals Week," an event started by the American Humane Association in 1915.

In recent years, there have been a number of news stories about animal artists. These examples have opened up dialogue about whether or not creativity and artistic pursuits are limited to humans. Pockets Warhol, a capuchin monkey who lives at Story Book Primate Sanctuary in Ontario, Canada, has had his art exhibited, and the sale of his paintings raises money for animal rescue efforts. A number of elephants have been celebrated as artists in sanctuaries and zoos worldwide; however, some critics have raised concerns that they are forced to learn to paint.

J. Keri Cronin

See also: Advertising, Animals in; Agency; Anthropomorphism; Ethics; Popular Media, Animals in

Further Reading

Baker, S. 2013. *Artist/Animal.* Minneapolis: University of Minnesota Press.

Baker, S. 1993. *Picturing the Beast: Animals, Identity and Representation.* Manchester: Manchester University Press.

Baker, S. 2000. *The Postmodern Animal.* London: Reaktion Books.

Donald, D. 2007. *Picturing Animals in Britain, 1750–1850.* New Haven, CT: Yale University Press.

Finlay, V. 2014. *The Brilliant History of Color in Art.* Los Angeles: J. Paul Getty Museum.

Kalof, L. 2007. *Looking at Animals in Human History.* Chicago: University of Chicago Press.

Ritvo, H. 1987. *The Animal Estate: The English and Other Creatures in the Victorian Age.* Cambridge, MA: Harvard University Press.

Thompson, N. 2005. *Becoming Animal: Contemporary Art in the Animal Kingdom.* Cambridge, MA: The MIT Press.

B

Battery Cage. *See* Factory Farming

Bees

For thousands of years and around the globe, bees have played an essential part in human lives as pollinators of food crops and producers of honey. Today, they and their products are used in many other ways, from medicine to military employment. Though habitat destruction and colony collapse disorder (CCD) have threatened bee populations on a grand scale, home bee gardens, home beekeepers, and the production of native bee nests all seek to maintain viable places for bees among modern humankind.

All bees belong to the taxonomic order (biological classification) of Hymenoptera (the order that also includes wasps and ants). There are nine families of bees: Andrenidae (sand and miner bees), Anthophoridae (carpenter bees), Apidae (honeybees; "killer bees," a more aggressive breed of *A. mellifera* [see below]; and bumblebees—all honey-producing species are found in this family), Colletidae (plasterer bees), Dasypodaidae (small African bees), Halictidae (sweat bees), Megachilidae (leaf-cutter and mason bees), Meganomiidae (small African bees), Melittidae (small African bees), and Stenotritidae (the smallest of all bees, found in Australia). Together, the nine bee families pollinate the majority of human food plants. Many wild bee species are much more efficient pollinators than are honeybees, and many areas, including Japan and Europe, make use of them for that purpose.

The first known documentation of the human-bee relationship is a cave painting in Europe from 7000 BCE: a human harvesting honey. Bees were kept in hives in Ancient Egypt between 5000 and 3000 BCE. By 2000 BCE, beekeeping was practiced in China, and Australian Aborigines made paintings from beeswax. Modern beekeeping, practiced around the globe, entails caring for colonies of bees, usually for the purpose of harvesting their products. Beekeepers provide shelter for colonies in the form of hives, combat hive illnesses and parasites, and often supplement the bees' food during the winter. Lorenzo Lorraine Langstroth (1810–1895), an apiarist (beekeeper) known as the "Father of American Beekeeping," created the most commonly used hive today, the Langstroth hive, built to utilize the concept of "bee space"—the size of the space between honeycomb cells that bees create for movement in naturally built hives.

Beekeeper Peter Hansen inspects a bee hive grid for parasitic mites in Oakdale, California, in 2005. Bees play a crucial role worldwide as crop pollinators. Several bee species are currently in danger not only from parasites but also from habitat destruction and pesticide use. (AP Photo/Ben Margot)

One domesticated species of bee, *Apis mellifera*, is most commonly used by beekeepers, but some indigenous cultures rely upon other *Apis* species. Instead of "keeping" bees, indigenous "honey hunters" harvest honey from wild colonies. Nepalese honey hunters harvest from Himalayan cliff bees (*Apis laboriosa),* which build nests on the sides of cliffs. The nomadic Boran people in central Africa are notable for their relationship with honeyguides, a bird that leads them to wild bee colonies so that it may scavenge the hive after the humans have finished.

Honey has been used for both food and medicine—current research has studied it for use in treating ailments from wounds to ulcers. Manuka honey from New Zealand is commonly accepted as an effective antibacterial by medical professionals. Other bee products used by humans include beeswax (primarily for art), bee venom (as a medicine or weapon), pollen (for food and medicine), propolis (a resin made from plant sources that is used by bees to seal holes in their hives and by humans as a medicine and adhesive), and royal jelly (a special, nutritious food produced by worker bees to feed larval [newly hatched] bees, used by humans as a dietary supplement). Bees are also interesting to other sectors of the modern Western world for their communication abilities. Scientists in the 1970s discovered that worker bees use "waggle dances" to communicate the location of food sources with respect to the position of the sun (Von Frisch 1973). Bees have gained recent employment by the American military for their ability to smell bombs and explosives (Ornes 2006).

Habitat destruction and pesticide use have devastated bee populations around the world. Many bumblebee species have disappeared completely. As a result, some of China's apple farmers now pollinate their trees by hand. In the United States, European honeybees have been imported to replace the missing native bees, as well as to support increasing food crop production for expanding human populations. However, honeybee populations currently suffer their own devastation—colony collapse disorder (CCD), identified in 2011, has been the cause of many large-scale honeybee mortality events across North America and parts of Europe during the 2000s and 2010s. So far, it has not been linked to other species. Possible explanations for this disorder include cell phone towers and cell phone emissions, agricultural pesticides, genetically modified crops, lack of genetic diversity among agriculturally produced queen bees, and bee parasites (specifically Varroa mites), among others.

CCD has raised alarm in the United States and Europe because of its effect upon food production. In 2000, honeybees pollinated over one-third and $14 billion worth of the food crops in the United States (Morse and Calderone 2000). In response to a feeling of urgency regarding CCD, it has become popular for many Americans to keep bees, even in urban settings, and many home beekeepers work to increase the genetic diversity and strength of domestic bee populations. Native bees are also available for purchase through retail stores, though they are not "kept" in hives like honeybees. Nonapiarists have joined in to provide additional support for both native bee and honeybee populations by planting bee gardens and installing native bee nests.

Heather Pospisil

See also: Communication and Language; Habitat Loss; Indicator Species; Military Use of Animals; Working Animals

Further Reading

Crane, E. 1999. *The World History of Beekeeping and Honey Hunting*. London: Gerald Duckworth & Co.

Ellis, J. 2013. *Colony Collapse Disorder (CCD) in Honey Bees*. Gainesville, FL: The Institute of Food and Agricultural Sciences, University of Florida.

Michener, C. 2000. *The Bees of the World*. Baltimore, MD: Johns Hopkins University Press.

Morse, R. A., and Calderone, N. W. 2000. "The Value of Honey Bees as Pollinators of U.S. Crops in 2000." *Bee Culture* 128: 1–15. Accessed March 31, 2015. http://www.beyondpesticides.org/pollinators/documents/ValueofHoneyBeesasPollinators-2000Report.pdf

Ornes, S. 2006. "Using Bees to Detect Bombs." *MIT Technology Review.* Accessed July 26, 2015. http://www.technologyreview.com/news/406961/using-bees-to-detect-bombs/

Root, A. I., and Root, E. R. 1910. *The ABC and XYZ of Bee Culture*. Medina, OH: The A. I. Root Company.

Von Frisch, K. 1973. "Decoding the Language of the Bee." Nobel Lecture, December 12, 1973. Accessed August 10, 2015. http://www.cs.swarthmore.edu/~meeden/cogs1/s07/frish.pdf

Bestiality

Commonly defined as sexual relations between humans and nonhuman animals, bestiality has a complex legal, social, and political history and geography. Many react to bestiality with revulsion and rejection, based on concerns ranging from animal liberation and welfare to simple disgust. However, in order to understand bestiality as *both* a type of erotic practice *and* an extreme example for debates around "proper" human-animal relations and related social issues, one must consider where it comes from and how it emerges today.

Bestiality has a long history of legal criminalization, though it has rarely been singled out and named explicitly. More often, bestiality has been criminalized as an act of "sodomy," a category that has historically included human-human and human-animal sexual "crimes against nature." Individuals engaged in bestiality have been persecuted under the same sexual norms that have criminalized same-sex acts, nonreproductive heterosexuality, and masturbation, though each act has also carried relatively distinct social meaning and legal punishment. This broad criminalization has taken hold in many Western contexts, including pre-1950s United States and Sweden and colonial-era Latin America. Other places have their own unique histories.

Criminalization and punishment for bestiality have often been applied unevenly according to a person's social position. For instance, in 17th- and 18th-century Sweden, it was most often younger men living in rural farming communities who were accused of bestiality with farmed animals. This stereotype about rural life also appears elsewhere. For instance, "sheep shagger" is a contemporary insult in the United Kingdom for Welsh people, and in Australia for New Zealanders. The slur implies negative perceptions of rural and agricultural lifestyles, socio-economic underdevelopment, and cultural backwardness. Bestiality has also often been prosecuted unevenly according to sex, more often affecting males. This is due largely to the dominant definition of sodomy as penile penetrative sex, which makes authorities largely oblivious to both female-female *and* human female-animal sex. However, in colonial New England, women were convicted and burned as witches under the charge of bestiality.

As some nations have made uneven moves to decriminalize homosexual sodomy, bestiality has been increasingly separated from other sexual practices that are considered to be outside the norm and has been established as its own category. This has at times created confusion and anxiety, leaving some places without any clarity on the legality of bestiality. For instance, in 2005, in the U.S. state of Washington, after a man died from injuries sustained while receiving anal sex from a horse, lawmakers moved quickly to recriminalize bestiality (Brown and Rasmussen 2010). They rallied public support for recriminalization by invoking feelings of disgust toward bestiality. Recriminalization has also occurred in other U.S. states. As of 2014, bestiality was a felony and/or misdemeanor in a majority of states, including Washington, even while there existed no federal law on bestiality.

Laws in Northern Europe have become similarly patchwork and complicated. Though bestiality has been banned in the United Kingdom, the Netherlands, France, and Switzerland, as of 2015 it is still legal in Belgium and Denmark, allowing for the existence of a small "bestiality tourism" circuit. Sweden passed a law in 2013 making bestiality a crime. Prior to this law, however, only instances involving evident animal abuse—at times hard to prove—could be prosecuted. Indeed, much of the criminalization of bestiality has been organized around the argument of animal cruelty. Animals *can* experience physical and psychological violence during sex with humans, though the type and degree of violence varies based on the type of sex act and the bodily structure and species of the animal. While violence against animals absolutely can and does occur, however, some who engage in sex with animals—many of whom identify as "zoophiles"—argue that their sexual relationships with animals can be nonviolent, consensual, and mutually beneficial.

This debate around consent raises questions not only about whether and how animals can consent to sex with humans—likely a question with no easy or clear answers—but also about how people understand human-animal relations more broadly. For instance, while debates around bestiality often center on questions about nonhuman animal consent, consent is rarely considered in the contexts of other human-animal relationships, such as animal farming for meat, milk, and eggs. Indeed, farmed animals like dairy cows are often artificially inseminated—a practice that involves inserting human hands and mechanical devices into a cow's vagina. However, these practices are categorized as economic production, not as sexual acts. As another example of how debates over bestiality reinforce certain social norms, in the effort to recriminalize bestiality in Washington state, supporters of the anti-bestiality bill often referred to the animals as if they were human children, relying upon assumptions about children's helplessness, instead of considering how nonhuman animals were affected *as animals*. As a final example, disgust toward bestiality often reinforces the socially constructed boundary between animals and humans. Humans are also animals, of course, but the categories of "human" and "animal" are often kept separate. The taboo against bestiality may be as much about maintaining these separate categories as it is about concern for animals. Given the complex issues around bestiality, it is important not only to consider the well-being of the animals and humans involved, but also the cultural context in which these practices of sex and sexuality emerge.

William L. McKeithen

See also: Agency; Ethics; Zoophilia

Further Reading

Beetz, A. M., and Podberscek, A. L. 2005. *Bestiality and Zoophilia: Sexual Relations with Animals*. West Lafayette, IN: Purdue University Press.

Blake, M. 2013. "Sweden Set to Ban Bestiality, Scrapping Legal Loophole That Made It Legal 'If the Animal Did Not Suffer'." *Mail Online*. Accessed June 1, 2016. http://www.dailymail.co.uk/news/article-2341789/Sweden-set-ban-bestiality-scrapping-legal-loophole-legal-animal-did-suffer.html

Boggs, C. G. 2013. *Animalia Americana: Animal Representations and Biopolitical Subjectivity*. New York: Columbia University Press.

Brown, M., and Rasmussen, C. 2010. "Bestiality and the Queering of the Human Animal." *Environment and Planning D: Society and Space* 28(1): 158–177.

Levy, N. 2003. "What (If Anything) Is Wrong with Bestiality?" *Journal of Social Philosophy* 34 (3).

Singer, P. 2001. "Heavy Petting." *Nerve*. Accessed June 1, 2016. www.utilitarian.net/singer/by/2001----.htm

Biodiversity

"Biodiversity," an abbreviation of "biological diversity," gained scientific traction in 1986 when biologist E. O. Wilson (1929–) used the term at the National Forum on BioDiversity to describe "the totality and variety of life on earth" (Goldstein 2011, 5). Since 1986, with dawning awareness that the planet is now undergoing a massive spasm of extinction, biodiversity has drawn increasing attention. Accelerating extinction rates attributable to human activity have produced among scientists an urgent call for taxonomic (classifying species) expertise and a host of proposals for slowing the loss of biodiversity and preserving or restoring the biological variety that remains.

Now widely used, "biodiversity" holds different meanings depending on the organizational level to which it refers. *Genetic biodiversity* measures genomic variation within as well as across species, while *species biodiversity* refers to the number or "richness" of species within a geographical region. *Ecosystem biodiversity* focuses on interactions between species and their environment in specific places, defining diversity as a dynamic, relational property that Fritjof Capra (1996) calls "the web of life": multiple, diverse organisms interactively performing ecological functions to maintain life in an ecosystem. Species diversity, the most common sense of the term, expresses the biological premise that the more diverse the species populations, the more stable and healthy the ecosystem. On the smaller scale of genetic diversity, the greater the genetic variety within a species, the more likely the species will withstand and adapt to environmental stressors like disease or climate change. On the grand scale of ecosystem diversity, the more varied and complex the interrelationships among organisms within and between their biomes (geographically defined habitat types such as deserts, forests, or ocean depths), the healthier and more stable the ecosystem. As a concept that cuts across organizational levels, biodiversity positions humans, plants, and nonhuman animals alike as interactive participants in the ecological systems that maintain life on Earth.

Biodiversity introduces a biological dimension to the utilitarian (usefulness), moral, and affective (emotional) terms in which human-animal relations are often discussed, ultimately placing these relations in the context of survival and extinction. Dramatically accelerating species extinction rates across the globe signal anthropogenic (human caused) threats to biodiversity today. Conservatively estimating 10 million presently existing species, an estimated current extinction rate of 1,000–10,000 species per million per year means that more than 50 percent of Earth's present biological variety may be gone by the end of the 21st century. This magnitude of biological change justifies scientists' claim that we have entered a new geological era. The label "Anthropocene" (age of humans) reflects scientific consensus that unsustainable human population growth and changes in human consumption patterns produce conditions that jeopardize biodiversity worldwide, including habitat destruction, pollution, and introduction of invasive (non-native) species.

Despite universal consensus that biodiversity is life-supporting and should be preserved, there is little agreement as yet on how to assess its value. The instrumental worth of biological diversity—its *direct usefulness* to humans—is a nurturing environment producing a wide variety of goods, such as food, medicine, and building materials, and affording recreational opportunities, such as hunting, hiking, and tourism. Humans also derive value from *indirect uses* of a biologically diverse environment—for instance, ecosystem services such as water filtration, pollination, and climate control. Economists further attribute nonuse *option* or *bequest* values to the usability of environmental resources by future generations. Apart from these instrumental and economic valuations, the *existence value* of biodiversity registers in humans' aesthetic, emotional, and spiritual responses to elements of nature. Finally, wholly apart from human interests, a diverse natural environment can be understood as having *intrinsic value*, measurable only as the self-worth of every living entity.

The current biodiversity crisis calls attention to the incomplete state of taxonomic knowledge. At present, only a small proportion of the species on Earth have been inventoried, and some species will certainly be gone before we realize they exist or understand their roles in ecosystem maintenance. Along with the call for increased taxonomic knowledge comes a demand for remedial responses to the problem of dwindling biodiversity. Conservation solutions aimed at slowing extinction rates in particular geographical areas strategically target preservation efforts to selected species. Among these are *keystone species* (e.g., pollinators) whose removal from an ecosystem would occasion the loss of 50 percent or more of other species in the system; *umbrella species* (e.g., wolves) whose habitat requirements equal or exceed those of all other species in the ecosystem; *flagship species* (e.g., polar bears), charismatic species that capture public imagination and garner support for biodiversity conservation; *indicator species* (e.g., oysters) whose prevalence is a measure of ecological vitality; and *common and widespread species* (e.g., crows), in anticipation of a time when these species will become rare and concentrated.

Another approach to the biodiversity crisis is restoration of geographical areas to previous states of diversity by eliminating invasive species—for example, cats imported to central Australia for rabbit control. Those who value the intrinsic worth of animals, however, reject this proposal on ethical grounds. While acknowledging the value of species diversity, they deny that the health of a biological community outweighs the killing of individual "invaders," often human imports such as feral donkeys and mustangs in the American West. Finally, some who recognize humans as the ultimate invasive species advocate "voluntary human extinction" through human birth control as the way to restore biodiversity to the planet, while others place their trust in biotechnology to resurrect extinct species or preserve the DNA of endangered species until their habitats are restored.

Mary Trachsel

See also: Climate Change; Endangered Species; Extinction; Flagship Species; Habitat Loss; Indicator Species; Invasive Species; Keystone Species; Species; Taxonomy

Further Reading

Capra, F. 1996. *The Web of Life: A New Scientific Understanding of Living Systems.* New York: Random House.

Gaston, K. J., and Spicer, J. I. 2004. *Biodiversity: An Introduction.* 2nd ed. Malden, MA: Blackwell.

Goldstein, N. 2011. *Biodiversity.* New York: (Facts on File) Infobase Learning Global Issues Series.

National Forum on Biodiversity. 1986. *Biodiversity,* edited by E. O. Wilson and Frances M. Peter. Washington, DC: National Academy Press.

National Forum on Biodiversity. 1997. *Biodiversity II: Understanding and Protecting Our Biological Resources,* edited by Marjorie L. Reaka-Kudla, Don E. Wilson, and Edward O. Wilson. Washington, DC: Joseph Henry Press.

Sandler, Ronald L. 2012. *The Ethics of Species: An Introduction.* Cambridge, UK: Cambridge University Press.

Wilson, E. O. 1992. *The Diversity of Life.* Cambridge, MA: Harvard University Press.

Biogeography

Biogeography is the branch of geography that studies how organisms are distributed over the surface and history of Earth. It is also the study of related patterns of variation in the numbers and kinds of living things. Biogeographers typically ask why a species is confined to its particular range, how historical and evolutionary events shape species' distributions, and the reasons for diversity being greater on continents than islands and in the tropics versus Arctic latitudes. It contributes to best practices for wildlife management and conservation, and its strong tradition of engaging with the living world renders biogeography an important subfield of animal studies.

Biogeography is divided into phytogeography, the study of plants, and zoogeography, the study of animals. Some researchers use a historical biogeographical approach, which attempts to reconstruct the origin, dispersal, and extinction records of plants and animals. This approach contrasts with ecological biogeography research, which studies present-day distributions of interactions between organisms and their environments. Another key approach is conservation biogeography, which applies biogeographic knowledge to wildlife management needs. A distinguishing feature of biogeography is that it is an observational, rather than an experimental, science because it deals with space and time at large scales, which make experimental studies impossible. Furthermore, biogeography interfaces with several traditional scientific disciplines including ecology, systematics, evolutionary biology, and the "Earth sciences" of geology, climatology, and oceanography.

Early biogeographical thought regarding plant and animal distribution was intimately tied up with travel and access to exotic specimens. Key concepts emerged in Europe during the early 19th century with the work of naturalist Alexander von Humboldt (1769–1859), who carefully calibrated distributions of plant life in Central America with physical, chemical, and environmental measurements. Naturalist Charles Darwin (1809–1882) also examined animal distributions on islands, barriers to their spread, and how they varied depending on where they were found. His contemporary, naturalist Alfred Russel Wallace (1823–1913), studied geographic variants of birds and butterflies. Wallace codified his ideas about large-scale distribution patterns by mapping points at which Pacific and Indo-Asiatic organisms met. This boundary was subsequently called Wallace's line, an important boundary between biogeographic provinces.

Biogeography as a science came of age in the 1960s with the publication of R. H. MacArthur (1930–1972) and E. O. Wilson's (1929–) book, *The Theory of Island Biogeography*. The theory attempted to predict the number of species that would exist on an island based on island size, its degree of isolation, and processes of extinction and colonization. It heralded a new paradigm within the field and beyond. Its application in conservation and influence on the science of wildlife management has been immense. Reproduced in the influential World Conservation Strategy of 1980, the theory has been deployed as a framework for devising conservation policies in many parts of the world.

More recently, biogeography has witnessed a number of exciting developments. One such development pertains to what has been called "countryside" or "matrix" biogeography, which no longer treats protected areas as islands, but as embedded in a matrix of different human-dominated habitats that act as filters for species to disperse. The Asian elephant is an interesting candidate in this regard. Land surrounding their protected areas is not unsuitable for them, but determines the extent to which they are able to move and disperse from one forest patch to another. Agricultural fields and farms are conducive, as elephants are able to utilize them, whilst heavily urbanized landscapes are not. Matrix biogeography reorients MacArthur and Wilson's model by seeking to incorporate diverse habitat corridors for elephants, rather than restricting them to forest pockets. Another important consequence, often called "reconciliation ecology," emphasizes how human-made landscapes can be made more accommodating to biodiversity. For example, efforts to make agricultural landscapes surrounding forests better suited for elephant movement can be achieved by maintaining farms that provide forage and shelter buffers, which allow elephants to move while reducing negative encounters with humans.

The field has also begun examining how biogeographies are being reconfigured in the Anthropocene, the name for the current geologic epoch of human dominance over Earth systems. For instance, novel biogeographies are evidenced in the wild elephants in south India and Sri Lanka, where centuries of trade and imports from Southeast Asia have resulted in the mixing of two genetically different populations that diverged 0.5–1.2 million years ago. Biogeographies of the Anthropocene go beyond "natural" processes to take human agency, and its planet-altering forces, seriously. It is propelling new conversations between bio- and human geographers, and is likely to be an area of research that will explode in the future.

The field of biogeography faces a number of different challenges. These include barriers to scientific development, notably in the form of biogeographic shortfalls and the need to improve the accuracy and specificity of forecasts. Two pressing shortfalls are the "Linnean shortfall" (i.e., incomplete knowledge of the number of species on Earth) and the "Wallacean shortfall" (i.e., partial information about species distribution). Both have direct bearings upon understanding the dynamics and distribution of life, and developing biogeographic theory. An associated set of challenges pertain to turning theory into practice, particularly when applying biogeographic principles for conservation practice. Generating better data; enriching links between scientists, policymakers, practitioners, and citizens; and understanding how anthropogenic (human) forces alter biogeographic processes are perhaps the most pressing arenas of future development.

Maan Barua

See also: Animal Geography; Biodiversity; Climate Change; Evolution; Geography; Species; Zoogeomorphology

Further Reading

Barua, M. 2014. "Bio-Geo-Graphy: Landscape, Dwelling and the Political Ecology of Human-Elephant Relations." *Environment and Planning D: Society and Space* 32: 915–34.

Brown, J .H., and Lomolino, M. V. 1998. *Biogeography,* 2nd ed. Sunderland, MA: Sinauer Associates.

Browne, J. 2001. "History of Biogeography." In *Encyclopedia of Life Sciences*. Wiley Online Library: John Wiley & Sons, Ltd.

Ladle, R. J., and Whittaker, R. J., eds. 2011. *Conservation Biogeography*. Oxford: Wiley-Blackwell.

Lorimer, J. 2010. "Elephants as Companion Species: The Lively Biogeographies of Asian Elephant Conservation in Sri Lanka." *Transactions of the Institute of British Geographers* 35: 491–506.

MacArthur, R. H., and Wilson, E. O. 1967. *The Theory of Island Biogeography*. Princeton, MA: Princeton University Press.

Whittaker, R. J., Araújo, M. B., Paul, J., Ladle, R. J., Watson, J. E. M., and Willis, K. J. 2005. "Conservation Biogeography: Assessment and Prospect." *Diversity and Distributions* 11: 3–23.

Wilson, E. O. 1992. *The Diversity of Life.* Cambridge, MA: Harvard University Press.

Biotechnology

Biotechnology is a rapidly growing field of academic research and commercial product development. Harnessing and manipulating the cellular structures of living organisms (both plants and animals) enables scientists to create potentially transformational products for human health and food systems, industrial needs, and the environment. The use of animals in biotechnology is controversial, however, because it raises questions of ethics, species integrity, and environmental safety.

Humans had used living organisms for thousands of years before the advent of modern technologies. For example, yeast (single-celled fungi) are used to make bread. As yeast organisms digest sugars and excrete carbon dioxide in dough, the carbon dioxide is trapped and this process causes the bread to rise. The domestication of animals is a second method of manipulating living beings. Through traditional breeding, desired traits (such as tameness or larger size) can become part of the permanent genetic code over several generations. This is the process whereby we have generated the variety of livestock and pet breeds that exist today. Modern biotechnologies take these traditional practices to an entirely new level by using targeted cellular interventions to effect changes to an organism in only one generation.

Several different processes are in use. *Cloning* is a method of making an exact duplicate of an animal by manipulating an embryo into having two sets of

Dolly, the first mammal cloned from an adult cell, in her pen at the Roslin Institute in Edinburgh, Scotland, in early December 1997. Manipulating the genetic material of animals (and possibly humans) through biotechnology remains highly controversial. Many people see these practices as "playing God" and an invasion of an animal's physical integrity, while others believe these technologies are simply an extension of traditional breeding practices. (AP Photo/John Chadwick)

chromosomes from the same parent instead of combining chromosomes from a set of male and female parents. *Genetic modification* is the targeting of specific DNA coding within an animal to either increase or decrease specific physical processes. *Transgenic animals* are the product of multiple species' DNA. They are created by taking desired DNA sequences from one animal, combining them with blastocysts (early-stage embryos) of the second animal, and then bringing those embryos to term inside the second animal.

While these processes have driven forward understandings of all manner of genetics fields through basic research, we see the wider application of animal biotechnologies today in agriculture, the biological and medical industries, as tools for conservation, and in the pet industry. Agriculture uses all three processes. Cloning provides genetic duplicates of high-value breeding animals. Genetic modifications within a breed can increase growth and lean muscle. Transgenic salmon, created by AquaBounty Technologies, have genes from both Pacific Chinook salmon and ocean pout, which helps them grow faster and year-round. The U.S. Food and Drug

Administration has recently declared that transgenic Atlantic salmon are safe for human consumption.

As with agriculture, all three processes are used in the human and animal health industries. Cloning, modifications, and transgenics have been used on species as diverse as mice, rats, goats, and pigs to produce hundreds of different animal research models to study and treat diseases. OncoMouse®, developed to have human breast cancer, is perhaps the most famous of these models because this transgenic research line was the first animal to be patented by the U.S. Patent and Trademark Office. There are two very specialized practices in these industries. Xenotransplantation is the use of nonhuman organs/parts (now mainly from pigs) in human bodies. Transgenic work on pigs has produced animals that have closer genetic matches to human systems and therefore reduce transplant rejection rates in humans. Pharming is the process of creating transgenic female animals that excrete novel materials through milk or eggs. Products developed for humans include growth hormones, tissue sealants, and blood coagulants. Biosteel®, a well-known product made by Nexia Biotechnologies, used transgenic methods to combine genes for spider silk (spiders' web-making substance) with lactation (milk-producing) genes in goats. The goats then excreted the silk proteins in their milk, which was processed into a novel fiber considered stronger than steel yet flexible and light. Although the company went bankrupt, researchers continue to explore the potential of pharming.

Cloning and transgenic methods are used for environmental management and in the pet industry. There are several companies that can clone your favorite animal companion. The successful cloning of an endangered guar Asian ox in 2001 by Advanced Cell Technologies, Inc., was heralded as a promising new way to prevent species extinction. Glofish® are pet fish that have been transgenically modified with jellyfish and sea anemone genes, making them "glow" in different colors. These types of fish have also been used to test for pollution levels and contaminants, as scientists can target the "glow" to switch on when a pollutant is encountered.

Three major concerns have emerged from a wide range of environmental, public interest, and animal advocacy groups, as well as the general public, food suppliers, and local governments. The first has to do with ethics: Many people believe that humans do not have a right to own other species and/or turn them into products for sale. The second is about species integrity and the notion that this type of science is "playing God." While traditional breeding practices can bring discomfort and harm to animals (such as bulldogs who have trouble breathing because of their short noses), biotechnologies may lead to unforeseen or invisible genetic issues that cause animal suffering. In addition, many are concerned that if we are mixing human genetic material with that of other species, how will we define what a human is? The third has to do with the long-term impact on native species or ecosystems if animals like transgenic salmon escape or are released into the environment.

Julie Urbanik

See also: Animals; Domestication; Ethics; Humans; OncoMouse; Research and Experimentation; Rights; Species; Welfare; Zoonotic Diseases

Further Reading

Biotechnology Industry Organization (BIO). n.d. "Animal Biotechnology." Accessed December 25, 2015. http://www.bio.org/category/animal-biotechnology

Emel, J., and Urbanik, J. 2005. "The New Species of Capitalism: An Ecofeminist Comment on Animal Biotechnology." In Nelson, L., and Seager, J., eds., *A Companion to Feminist Geography,* 445–457. Oxford: Blackwell.

Kolata, G. 1998. *Clone: The Road to Dolly and the Path Ahead.* New York: William Morrow.

Rifkin, J. *The Biotech Century: Harnessing the Gene and Remaking the World.* New York: Jeremy P. Tarcher/Putnam.

U.S. Patent and Trademark Office. 2000. "Class 800: Multicellular Living Organisms, June 30." Accessed December 20, 2015. http://www.uspto.gov/web/offices/ac/ido/oeip/taf/def/800.htm

Black Market Animal Trade

The black market animal trade is the illegal or illicit trade of animals for human purposes. Throughout history, humans have collected animals from their natural habitats, often threatening the welfare of individual animals and the viability of species. Humans have also taken collective action to control and regulate the collection, trade, and use of animals from the wild, ostensibly to promote a more sustainable coexistence. The black market animal trade occurs outside of the laws and regulations set forth by governments. Also known as animal trafficking, this trade involves animals both alive and dead, whole animals, and products derived from animals, such as skins, meat, feathers, and medicines.

The international black market animal trade has an estimated value in the billions of dollars. This places it among the world's top illicit markets, such as those for drugs, weapons, and humans. The illegal animal trade is growing and may threaten some species with extinction, especially when combined with other threats such as habitat loss. Generally, animals destined for the black market are collected or poached (taken illegally) in countries with high biodiversity, which also tend to be countries of the developing world such as those in Southeast Asia, Africa, and South America. These animals, and the products derived from them, are generally destined for use by people living in the developed world, in places such as the United States, countries of the European Union, and Japan. Live animals are sold as pets, status symbols, and for entertainment purposes, while dead animals are sold for food, fashion, medicine, and ornamental purposes. This trade encompasses all kinds of animal groups, including mammals, birds, reptiles, amphibians, fish, and insects.

The geography and distribution of species within the trade likely changes over time and in response to the changing tastes and preferences of consumers, the capacity of governments to enforce laws and regulations, the decline of target species' populations, and many other variables. As a result of globalization and the growth of the wealthy and middle class in parts of the developing world, specific trade routes and flows of wildlife products have shifted. For instance, China has in recent years become one of the world's major consumers of black market animal products.

Several iconic animals receive attention for the impact that the black market trade has had on their populations. African elephants have undergone a dramatic population decline, which is largely related to the ivory trade. Among other products, tigers are poached for their skins and for their bones, which are used to make tiger bone wine in China. With few, if any, wild tigers remaining within its borders, China has seen the emergence of tiger farms, or captive breeding facilities, to support the domestic demand for these products. While some of these tiger products may be legal within China, the trade in tigers and their parts across international borders remains illegal. The captive breeding of tigers may increase the demand for tiger products in other parts of the world. Other, less iconic animals are also threatened by the black market trade. Pangolins, also known as scaly anteaters, are mammals native to Africa, India, and Southeast Asia. Unique for the hardened, overlapping keratin scales covering their bodies, these animals are traded illegally perhaps more so than any other animal. Their scales are used for traditional medicinal purposes, and their meat is eaten as a delicacy. Vietnam and China are the most common markets for these animals. All eight pangolin species are considered threatened or endangered.

The international community has taken steps to regulate and control the black market animal trade. The international treaty known as CITES (Convention on International Trade in Endangered Species of Wild Fauna and Flora) came into force in the 1970s. Most of the world's countries have signed onto CITES, which requires the cooperation of its signatory countries to effectively manage the international trade of wildlife. Some countries fully abide by the regulations and have enacted even stricter domestic laws governing the trade within their own borders. A variety of issues challenge the ability of countries to manage and control the illegal trade, including, among others, corruption, weak customs enforcement, lax punishments, and the lack of political will to prosecute offenders.

Those who seek to control the black market animal trade are facing new challenges. The Internet has provided traders with new technologies to subvert existing laws and regulations. The illegal trade in wildlife may help finance organized crime, insurgency, and terrorist groups, threatening national and international security. The trade undermines the ability of some countries to effectively manage their own natural resources, creating questions about government legitimacy and accountability. The trade may cause other environmental problems such as the introduction of invasive species or the transmission of disease. Other, nontarget animals can also be impacted or endangered by the trade, such as when illegal fishing operations harvest indiscriminately.

Finally, the livelihoods of people may be disrupted by the trade. Some depend on local, natural resources for subsistence and economic benefit, and their ability to meet their basic needs may be compromised by the trade. For example, some argue that a sustainable and well-managed ivory trade could provide economic development opportunities for people in elephant range countries. At the moment, however, the elephant poaching and illegal ivory trade situation is considered too dire, and the international ban on ivory trade prohibits some opportunities for the sustainable use of products from elephants.

Gabe Wigtil

See also: Bushmeat; Elephants; Endangered Species; Exotic Pets; Extinction; Ivory Trade; Non-Food Animal Products; Poaching; Tigers; Traditional Chinese Medicine (TCM); Wildlife Forensics

Further Reading

Davies, B. 2005. *Black Market: Inside the Endangered Species Trade in Asia.* San Rafael, CA: Earth Aware Editions.

Rosen, E. G., and Smith, K. F. 2010. "Summarizing the Evidence on the International Trade in Illegal Wildlife." *EcoHealth* 7: 24–32.

Sutter, J. D. 2014. "The Most Trafficked Mammal You've Never Heard Of." Accessed March 23, 2015. http://www.cnn.com/interactive/2014/04/opinion/sutter-change-the-list-pangolin-trafficking/

Bluefin Tuna

Canned and worth pennies a half-century ago, a single bluefin tuna weighing 489 pounds (222 kilos) made international headlines in January 2013 when it sold at auction in Tokyo for a record $1.7 million. The bluefin is the most expensive fish money can buy among consumers rich enough to afford what industry insiders call "red gold" for the color of its meat (Telesca 2015). High prices have led to a black market worth billions, fueling environmentalists' concern that global demand has brought the bluefin to the brink of extinction. Despite efforts at wildlife management, overfishing renders the future of this flagship (iconic), biologically complex, historically important commercial species uncertain.

Humans today eat bluefin raw as sushi (a Japanese food combining rice, fish, and vegetables and/or fruit). It belongs to a family of over 60 species of tuna that inhabit the world's oceans. There are three varieties of bluefin that never meet: the Atlantic (*Thunnus thynnus*), the Pacific (*Thunnus orientalis*), and the Southern (*Thunnus maccoyii*). All three look alike, although the Atlantic bluefin is the largest of all tunas. Once called "giants," they grew to the size of a horse and commonly weighed well over a ton. Bluefin this size are now rare. Although overfishing contributes most to the bluefin's decline, other threats disrupt its lifecycle, including parasites, toxins such as mercury accumulating in its body from industrial pollution, and warmer, acidic oceans.

Unlike most fish, the bluefin is warm-blooded. It can heat, regulate, and stabilize its body temperature to be higher than the surrounding water. Like dolphins and orcas, it hunts in packs and cooperates by communicating with others in its school when hunting for fish. Diving to depths where the water is black and icy cold, the bluefin swims across entire oceans at speeds of 55 miles per hour using magnetic crystals that form an extrasensory internal compass. In short, the bluefin is one of the fastest, if not the most complex, fish at sea.

Unlike the white, canned meat of skipjack and albacore tuna, the bluefin is valued for the clarity of its fatty, ruby-colored flesh. Taste for the bluefin as a delicacy developed only recently in the wake of the U.S. occupation of Japan after World War II as that war-ravaged island nation looked to feed itself from the sea. In 1972, the bluefin—and sushi more generally—went global with the invention of the airplane's refrigerated box container, which drastically quickened the long-distance travel of fresh fish by air. Today, approximately half of the bluefin consumed worldwide comes from the Atlantic Ocean. Spain is the greatest exporter, and Japan the greatest importer.

The rapid globalization of a delicacy in wild fish precipitated the bluefin's decline. A commonly cited study from 2003 sounded the alarm with statistical urgency: The intensity of illegal, unreported, and unregulated (IUU) fishing led to the estimate that 90 percent of large predatory fish had been taken from the sea, eaten, or discarded as bycatch (nontarget species) (Myers and Worm 2003). Although some experts dispute such figures, scientists from the International Union for the Conservation of Nature (IUCN) currently list all three bluefin as being threatened to the following degrees: the Southern, critically endangered; the Atlantic, endangered; and the Pacific, vulnerable. However, countries participating in the Convention on International Trade in Endangered Species of Wild Fauna and Flora (CITES) have failed to list the bluefin as endangered.

The crash in fish populations, including the bluefin, is not new. Nonetheless, overfishing became particularly acute after World War II, when industrialized nations subsidized their fishing fleets and took advantage of such "efficient" technologies as sonar, steel hooks, and petroleum-powered vessels. More recently, capture aquaculture targeting the bluefin originated in Australia in the 1980s and has been adopted in Japan and some Mediterranean countries. At these tuna "ranches," wild bluefin traveling in schools are caught by "purse seine"—that is, they are encircled by a net with a drawstring that, when cinched like a purse, captures the school together. Once released into pens, the bluefin are fed other wild fish such as mackerel to fatten them for market, even though such farming practices exacerbate overfishing more generally. Other common commercial fishing methods for the bluefin include "pelagic longlines" (lines extending for miles on the ocean surface attached with a series of baited hooks) and "tuna traps" (nets erected in a maze along shorelines that entrap the bluefin, popular in the Mediterranean Basin). The latter is the oldest of methods, making the bluefin one of the first recorded fisheries in human history, stretching back before the time of Christ.

Because the bluefin crosses the high seas through zones that no nation controls directly, it has become the subject of global campaigns against overfishing. Major marine advocates such as Greenpeace, World Wildlife Fund (WWF), Oceana, and the Pew Environment Group have pressured international agencies such as the International Commission for the Conservation of Atlantic Tunas (ICCAT) to better manage the bluefin. Although harshly criticized by some conservationists, recent scholarship emphasizes that ICCAT and other regional fisheries management organizations actually perform the job asked of them. Their political mandate is to protect not the fish per se, but the export markets of the countries that sign their trade agreements.

Jennifer E. Telesca

See also: Aquaculture; Black Market Animal Trade; Endangered Species; Fisheries; Fishing; Flagship Species; Wildlife Management

Further Reading

Bestor, T. C. 2000. "How Sushi Went Global." *Foreign Policy* 121: 54–63.

Ellis, R. 2008. *Tuna: A Love Story*. New York: Alfred A. Knopf.

"The International Commission for the Conservation of Atlantic Tunas (ICCAT)." Accessed August 9, 2015. https://www.iccat.int/en/introduction.htm

Myers, R. A. and Worm, B. 2003. "Rapid Worldwide Depletion of Predatory Fish Communities." *Nature* 423: 280–283.

Safina, C. 1997. *Song for the Blue Ocean: Encounters along the World's Coasts and Beneath the Seas*. New York: Henry Holt and Company.

Telesca, J. E. 2015. "Consensus for Whom?: Gaming the Market for Atlantic Bluefin Tuna through the Empire of Bureaucracy." *The Cambridge Journal of Anthropology* 33: 49–64.

Body Modification

Body modification is usually understood as any practice that alters the form and/or function of a living body. It can be short term, long term, or permanent, and can occur either as the result of individual choice, or due to an adherence to wider cultural norms, coercion, or forced mutilation. Body modification is therefore strongly linked with power, and for nonhuman animals is largely associated with concepts of domestication and ownership. Animals are often controlled through the act of modification to their bodies in order to satisfy the social, cultural, and/or commercial demands of humans, although there are examples of animal agency through body modification as well.

The process of domestication and selective breeding has contributed dramatically to the modification of animals' bodies. A modern-day sheep, for instance, lives in a very different environment than did its wild ancestors. After shifting from mountainous terrain to the farm, sheep now have shorter legs, altered horn characteristics, and varied wool colorings, as bodily features necessary for survival in the wild have become unnecessary. Likewise, both pets and livestock have been bred to have bodies that may be smaller (such as cats), display particular physical traits (such as the long-bodied dachshund dog, or short-nosed boxer or pug dogs), be prolific breeders or egg-layers, or produce more muscle mass (meat)—all in accordance with human desires for more compliant, attractive, or productive animals.

Animals are also subject to sudden, dramatic bodily transformations over which they have little or no control. Many species have traditionally been branded to identify them as possessions of their owners. These techniques can be seen as forms of control and domination; indeed, even human beings' bodies have been modified without their consent, such as in the case of tattooing African slaves in the Americas and Holocaust victims held in concentration camps. In the present day, it is mandatory in the United Kingdom and New Zealand to implant dogs with microchips to identify and regulate ownership. Likewise, endangered species are frequently tracked using monitoring devices injected into their bodies, which transmit Global Positioning System (GPS) information back to scientists. Body modification therefore allows experts to analyze species movement or population characteristics, yet at the same time raises issues surrounding the ethics of human manipulation of wild animals, and it is a subject under some debate in the environmental management field.

Body modification is also a tool to control animal reproduction. In modern Western culture, neutering companion animals has become synonymous with responsible pet ownership, and trap-neuter-return programs (which capture, neuter, then release animals back to where they were found) are widely regarded as the most effective strategy for dealing with stray cats. Farm animal bodies are often modified to facilitate production or as products are extracted. For example, dehorning makes it less likely for cows to injure human handlers or each other, and shearing involves an involuntary (albeit temporary) modification of the body for the sheep.

Animal bodies are also modified for aesthetic purposes. Pets are groomed to ensure their comfort in hot temperatures, or simply in accordance with an owner's aesthetic preferences. Moreover, breeds of dogs, cats, rabbits, and some other species are often required to fulfill sets of physical conditions to gain entry into competitive showing or to meet what are known as "breed standards." In the United States, for example, for many dogs these standards require tail docking or ear cropping—both of which are permanent and severe body modifications that hinder canine communication, as dogs use ears and tails in their communication with each other. Similarly, cat declawing and dog "debarking" are examples of changes made for human convenience that often result in emotional distress for the animals. However, such practices are outlawed in many countries, including much of Europe, Scandinavia, Turkey, Australia, and New Zealand. On the other hand, docking the tails of lambs is accepted in these same nations, as it reduces the chance of infestations of parasitic maggots (known as "fly-strike") and is therefore performed for the good of the animal. Cultural norms therefore frequently determine whether practices are seen as beneficial or exploitative.

Finally, it is worth considering how animals might themselves perform bodily modification. It might be argued that "modification" includes only acts that are knowingly performed. This potentially excludes many animal practices such as the shedding of skins or antlers, the changing of sex (such as is done by some fish and frogs), or "housing" (as in the case of the hermit crab) as these occur as part of physical processes that animals may perform subconsciously. However, individual animals must act to augment or hasten such processes and therefore can be seen as actively participating in body modification. Deliberate acts such as sharpening teeth or claws to restrict growth, preening to attract mates, and changing color/form to either entice prey or camouflage against predators (as chameleons or octopi do) may also be seen as animals making choices to modify their bodies in response to specific situations.

Therefore, while an animal's own body modification may be understood as simply a physiological response to the external environment, it remains that it is a way in which animals exert control over themselves and their social/physical environment, which humans do as well. Conversely, forced body modification of nonhuman animals can be addressed as part of broader social systems that regard animals as products or possessions, altering them to fit human needs and desires.

Linda Madden

See also: Agency; Domestication; Evolution; Factory Farming; Non-Food Animal Products; Pets; Spay and Neuter; Tracking

Further Reading

Anderson, K. 1997. "A Walk on the Wild Side: A Critical Geography of Domestication." *Progress in Human Geography* 21(4): 463–485.

Bennett, P., and Perini, E. 2003. "Tail Docking in Dogs: A Review of the Issues." *Australian Veterinary Journal* 81(4): 208–218.

Cazaux, G. 2007. "Labelling Animals: Non-Speciesist Criminology and Techniques to Identify Other Animals." In Beirne, P., and South, N. eds., *Issues in Green Criminology:*

Confronting Harms against the Environment, Humanity and Other Animals. Portland, OR: Willan Publishing.

Clutton-Brock, J. 1992. "The Process of Domestication." *Mammal Review* 22(2): 79–85.

Holloway, L. 2007. "Subjecting Cows to Robots: Farming Technologies and the Making of Animal Subjects." *Environment and Planning D* 25(6): 1041–1060.

Jacoby, K. 1994. "Slaves by Nature? Domestic Animals and Human Slaves." *Slavery and Abolition* 15(1): 89–99.

Wilson, R. P., and McMahon, C. R. 2006. "Measuring Devices on Wild Animals: What Constitutes Acceptable Practice?" *Frontiers in Ecology and the Environment* 4(3): 147–154.

Bovine Growth Hormone

Bovine growth hormone (BGH), also known as bovine somatotropin (bST or BST), is a naturally occurring hormone in cows (bovines) that plays a role in metabolism. BGH is also produced synthetically in laboratories and is known as recombinant bovine growth hormone (rBGH) or recombinant bovine somatotropin (rBST). The use of rBGH is a feature of industrial, large-scale dairy production. The synthetic form has been used in dairy industries to boost cows' milk production, but its use is controversial because of links to poor animal welfare and potentially to cancer in humans.

All animals produce growth hormones that help regulate the living organism's bodily processes throughout life. When a normal female mammal has offspring, her mammary glands will produce milk that will nourish the baby or babies for the period of time before they can eat solid food. A typical mammalian milk-production cycle increases for a period of time after the birth of offspring until production reaches a peak, then slowly decreases until stopping altogether. In the 1930s, Russian scientists discovered that administering BGH (extracted from dead animals' pituitary glands) to cows would cause a slower decrease of milk-producing cells in the udder, allowing the cows to remain at peak milk production longer. Research in the 1970s–1980s led to the discovery of how to manufacture BGH in a laboratory, allowing for more efficient production of the substance.

The efforts of the companies Genentech and Monsanto led to the synthetic production of rBGH. In the 1980s, the U.S. Food and Drug Administration (FDA) deemed food produced from cows who had received rBGH safe for human consumption. In the early 1990s, rBGH (typically given by injection) was deemed safe for animals by the FDA's Center for Veterinary Medicine. Although pursued by other chemical and pharmaceutical companies, Monsanto's version of rBGH was the first to receive FDA approval, and the company began to market its product in early 1994 under the name Posilac®. Since 2008, the drug company Eli Lilly has sold this animal drug through its Elanco division, having bought the Monsanto division that produced it. Corporate-sponsored research claims that U.S. cows' milk production

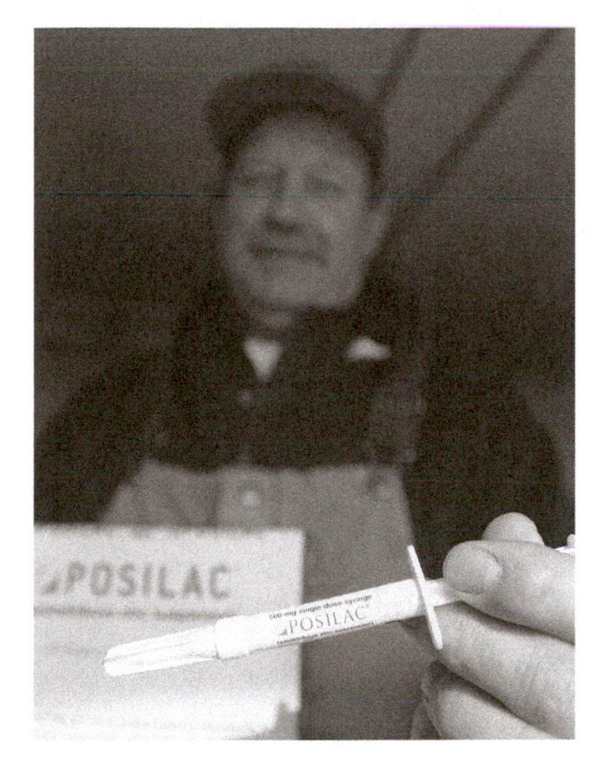

Dairy farmer Darrell Reece holds a dosage of Posilac (the brand name of rBGH produced by Monsanto) at his farm in Knob Lick, Kentucky, in 2004. The use of rBGH is controversial for both human health and animal welfare reasons. It has been banned in such locations as Canada and the European Union but is still widely used in the United States. (AP Photo/Patti Longmire)

increases by approximately 15 percent when they receive rBGH (Raymond et al. 2010). However, the increased milk production does require that the animals be fed more.

The use of rBGH has been and is controversial, however, and its use is banned or limited in a number of locations—currently Australia, Canada, the European Union, Israel, Japan, and New Zealand. In the United States, the product is currently available in all 50 states, and it has been sold since its approval in 1994. The percentage of dairy farms using rBGH ranged from approximately 9 percent for small operations to almost 43 percent for large farms in 2007 (Sechen 2013).

There are two main reasons rBGH is controversial, the first being its potential negative effects on human health. Although no studies have shown conclusively that rBGH negatively impacts human health, concern has arisen from the uncertainty. The locations that have banned it cite this uncertainty as a major reason for the ban. Despite a lack of conclusive proof of harm to humans, data have been gathered that indicate a potential link to diseases such as breast and prostate cancers. This link arises primarily from what is known as "insulin-like growth factor," or IGF-1, which has been associated with an increased risk of certain cancers if found at increased levels in humans. According to the American Cancer Society, testing has shown at least slightly higher levels of IGF-1 in rBGH-treated cows' milk, but the impact on cancer risk from drinking this milk is still unknown, with more research being needed. Although not considered to be the biggest contributor, rBHG is also considered a potential cause of early-onset ("precocious") puberty in girls in the United States.

Another primary reason for controversy over rBGH is its effect on animal health and welfare. A study published in the *Canadian Journal of Veterinary Research* found that cows treated with rBGH were 25 percent more likely to develop mastitis (inflammation of mammary glands), 40 percent less likely to conceive, and 55 percent more likely to become lame (unable to walk/walking with difficulty) (Dohoo et al. 2003). An earlier study from the European Union Scientific Committee on Animal Health and Animal Welfare (1999) showed a 14–79 percent increased risk of mastitis, causing one veterinarian to say that the estimates were "not only statistically significant but also biologically relevant and of considerable welfare concern" (Kronfeld 2000, 1719–1720). Indeed, Canada's and the European Union's initial bans of rBGH were motivated in large part by animal health and welfare concerns.

Limiting the sale or use of rBGH in the United States has been much more difficult. Groups such as the Consumers Union and the Organic Trade Association have pushed, unsuccessfully, for nationwide labeling requirements indicating milk/milk products from treated cows. There has also been legal controversy over whether it was misleading for companies to label their products as coming from *nontreated* cows. However, this type of labeling is legal in all 50 states as of 2010. Rather than legal restrictions, many limits on the sale of milk from treated cows have come from decisions by large retailers such as the grocery store chain Safeway and Starbucks Coffee to cease selling it. Research has shown that these various forms of pressure are likely having an effect, however, as usage of the drug has been in decline in recent years.

Connie L. Johnston

See also: Advocacy; Biotechnology; Concentrated Animal Feeding Operation (CAFO); Factory Farming; Humane Farming; Husbandry; Livestock

Further Reading

American Cancer Society. 2014. Recombinant Bovine Growth Hormone. Accessed September 17, 2015. http://www.cancer.org/cancer/cancercauses/othercarcinogens/athome/recombinant-bovine-growth-hormone

Dohoo, I. R., DesCôteaux, L., Leslie, K., Fredeen, A., Shewfelt, W., Preston, A., and Dowling, P. 2003. "A Meta-analysis Review of the Effects of Recombinant Bovine Somatotropin 2:

Effects on Animal Health, Reproductive Performance, and Culling." *The Canadian Journal of Veterinary Research* 67: 252–264.

Donohoe, M., Epstein, S., Hansen, M., North, R., and Wallinga, D. n.d. *A Public Health Response to Elanco's "Recombinant Bovine Somatotropin (rbST): A Safety Assessment."* Accessed September 15, 2015. http://phsj.org/wp-content/uploads/2007/10/Response-to-Elanco-rBST-Safety-Assessment2.doc

Kronfeld, D. 2000. "Recombinant Bovine Somatotropin and Animal Welfare." *Journal of the American Veterinary Association* 216(11): 1719–1720.

Raymond, R., Bales, C. W., Bauman, D. E., Clemmons, D., Kleinman, R., Lanna, D., Nickerson, S., and Sejrsen, K. 2010. *Recombinant Bovine Somatotropin (rbST): A Safety Assessment*. Greenfield, IN: Elanco.

Scientific Committee on Animal Health and Animal Welfare. 1999. *Report on the Animal Welfare Aspects of the Use of Bovine Somatotropin.* Brussels: European Commission.

Sechen, S. J. 2013. *Bovine Somatotropin (bST)—Possible Increased Use of Antibiotics to Treat Mastitis in Cows*. Rockville, MD: United States Food and Drug Administration, Center for Veterinary Medicine.

Breed Specific Legislation

Breed specific legislation (BSL) is a legal method used in many parts of the world for controlling which dogs can be present in certain areas. BSL is most often associated with pit bulls but can also apply to any dog breed considered dangerous, such as rottweilers or Doberman pinschers. These laws are controversial because they target an entire breed, or group, of animals rather than focusing on individual dogs or the humans who are in charge of them. These laws provide an excellent example of the complexity of legal conflicts over domestic animals.

With the increasing numbers of dogs as pets (about 70 million in the United States alone) and feral, or free-roaming, dogs (numbers unknown) in countries like the United States, Canada, and across Europe, there has been an increase in the numbers of dog bites and even deadly attacks on humans. In the United States, about 4.5 million dog bites are reported each year, with about 27,000 people needing some type of reconstructive surgery (AVMA n.d.). Most serious dog bites (those requiring medical attention) happen to children by unneutered (i.e., intact testes) male dogs normally known to the child. In the past several decades in the United States, data has shown that more than 25 different breeds have been involved in attacks in which a person was killed (Hussain 2006). The total annual costs for these bites/attacks in the United States alone is over $450 million between insurance payouts, medical care, legal fees, and veterinary care.

The public health problem of dog bites/attacks has been addressed by legal systems in two ways. "Dangerous dog" laws are those that take effect *after* a dog of any breed has bitten/attacked someone. These laws include such actions as evaluating the dog for temperament issues, confiscating the dog, serving the owner with a dangerous dog notice/summons, fining the owner, requiring the dog to wear a muzzle or be contained, requiring the owner to carry extra insurance and/or place "vicious" or "dangerous" dog signs in their yard, and sometimes euthanizing the dog (usually in cases where a fatal attack has occurred). BSL are laws designed to *prevent* bites/attacks by banning an entire breed of dogs from being in a certain area—regardless of the temperament of any individuals of a banned breed. Local governments may use either or both of these methods; however, it is the BSL method that has become highly controversial. Proponents of BSL argue that these laws are needed to adequately protect people—especially children—from being exposed to animals that can cause such harm and that it is better to simply ban a breed than try to control or regulate the owners. Opponents of BSL argue that these laws stigmatize breeds without actually allowing for individual behaviors of dogs or requiring accountability for dog owners.

One of the breeds most often targeted by BSL is the pit bull, which is technically not one breed, but a group of breeds—similar to the spaniel or retriever breeds. Pit bulls include the American Staffordshire terrier, the Staffordshire bull terrier, and the American pit bull terrier. Historically bred for both fighting prowess and loyalty to humans, these dogs had a positive image throughout much of the 20th century as a family dog. A series of fatal attacks in the 1980s, on both adults and children, and the increasing use of pit bulls as a symbol of aggressive masculinity culminated in a 1987 *Sports Illustrated* magazine cover depicting them as terrifying. In addition, the same study that showed a variety of breeds were involved in fatal attacks in the United States revealed that nearly one-third of those were committed by pit bull breeds. While many breeds, especially smaller breeds like Chihuahuas, statistically bite more often, the strength and tenacity bred into pit bull breeds can allow an attack to turn fatal. These disproportionate statistics, along with negative media publicity have resulted in a public backlash against them. In the United States, however, at least 24 other breeds, including Chihuahuas, have been targets of BSL legislation (NCRC 2014).

According to the American Veterinary Medical Association (AVMA), however, there is no direct evidence showing that BSL helps to reduce dog bites/attacks (AVMA 2015) by pit bulls or any other breed. They state that evidence shows that the most impact comes from both educating owners to be more responsible for the proper socialization and treatment of dogs, and teaching the general public how to better understand the "language" of dogs (e.g., raised fur, growling). Furthermore, targeting entire breeds for exclusion does not take into account the individual nature of dogs, including their upbringing, socialization, training, neutering status, and location and manner in which the dog is kept (e.g., in the house, on a chain, behind a fence).

Organizations such as the National Animal Control Association (NACA) believe that a better method for reducing bites/attacks is to assess dogs based on their individual actions and to have more comprehensive, enforceable, and funded "Dangerous dog" laws (NACA 2014). Notably, other reputable national organizations such as the American Bar Association, the American Kennel Club, the U.S. Centers for Disease Control, and the International Association of Canine Professionals are among a long list of those whose public position is that BSL is an ineffective method for reducing dog bites/attacks because it does not address the fundamental need for public education and owner responsibility.

Julie Urbanik

See also: Animal Law; Dogs; Fighting for Human Entertainment; Pets; Pit Bulls

Further Reading

Animal Farm Foundation. 2015. "Breed Specific Legislation Map." Accessed January 13, 2016. http://www.animalfarmfoundation.org/pages/BSL-Map

AVMA. 2015. "Dog Bite Risk and Prevention: The Role of Breed." Accessed January 12, 2016. https://www.avma.org/KB/Resources/LiteratureReviews/Pages/The-Role-of-Breed-in-Dog-Bite-Risk-and-Prevention.aspx

AVMA. n.d. "Infographic: Dog Bites by the Numbers." Accessed January 12, 2016. https://www.avma.org/Events/pethealth/Pages/Infographic-Dog-Bites-Numbers.aspx

Dogs Bite. 2016. "Legislating Dogs." Accessed January 13, 2016. http://www.dogsbite.org/

Hussain, S. 2006. "Attacking the Dog-Bite Epidemic: Why Breed-Specific Legislation Won't Solve the Dangerous-Dog Dilemma." Accessed January 5, 2016. http://ir.lawnet.fordham.edu/flr/vol74/iss5/7

NACA. 2014. "NACA Guidelines." Accessed January 13, 2016. http://c.ymcdn.com/sites/www.nacanet.org/resource/resmgr/Docs/NACA_Guidelines.pdf

NCRC. 2014. "Breed Specific Legislation (BSL) FAQ." Accessed January 15, 2016. http://nationalcanineresearchcouncil.com/dog-legislation/breed-specific-legislation-bsl-faq/

Swift, E. M. 1987. "The Pit Bull: Friend and Killer." *Sports Illustrated,* July 27: 74.

Broiler Chicken. *See* Factory Farming

Buddhism. *See* Eastern Religions, Animals in

Bullfighting

Bullfighting is on the decline, yet events still take place in eight countries: Spain, France, Portugal, Colombia, Venezuela, Peru, Ecuador, and Mexico. Every year, approximately 250,000 bulls are stabbed and speared multiple times before typically suffering a slow death.

Bullfighting (known in Spain as *corridas*) is one of the varieties of *fiestas* (celebrations) that include the regional *Toro de la Vega,* the South American *corralejas*, and many others. In Spain alone there are about 16,000 *fiestas* in which bulls are

An assistant bullfighter stabs a bull during a Novillero bullfight at the Las Ventas bullring in Madrid in 2008. Although bullfighting's popularity is declining, it is still practiced in a number of countries such as France, Mexico, and Spain (as in this image from Madrid). In 2013, Spain's Congress declared the increasingly controversial practice an official part of the country's cultural heritage. A number of other countries, such as Argentina, Cuba, and the United Kingdom, have made bullfighting illegal. (AP Photo/Paul White)

killed. Bullfighting is often depicted as an artistic and symbolic fight between "man and beast," perpetuating the idea that it is ethically acceptable to inflict pain on beings that are outside of human society. Some scientists and sociologists view bullfighting as a way for the fans to identify with the grand persona of the bullfighter and/or to satisfy sadistic impulses.

In 2013, amid increasing controversy, Spain's Congress voted to officially declare bullfighting part of the country's cultural heritage. Although its precise origins are unknown, forms of bullfighting or contests involving humans and bulls in the Mediterranean region are thought to date from before the Christian era. The Spanish national hero El Cid (ca. 1043–1099) is alleged to have been one of the first to fight a bull from horseback. Bullfights became supported by Spanish royalty and elites and came to be held on days of local and religious significance.

A typical bullfight is divided into three *tercios* (thirds), with two 20-minute bullfights in each. In the first *tercio*, the bullfighter's assistants (*banderilleros*) provoke the bull using large colorful capes (*capotes*). The presiding official of the *corrida* signals the entry of the *picadors*, whose job is to injure the bull's neck using a lance. This is done to break down the neck muscles and make it difficult for the bull to raise his head, thereby impeding the ability to make sudden and abrupt movements. If the bull wants to charge, he has to exert himself more, and this is considered "more beautiful" because the movement is exaggerated. The second *tercio* is called "el tercio de banderillas." The purpose of this *tercio* is to use *banderillas* (wooden sticks with spiked ends) to tear muscles, nerves, and blood vessels. The bullfighter must stab at least four *banderillas* into the bull before the next and final act can take place. In the third *tercio*, the bullfighter has 10 minutes (with the possibility of being allowed 5 extra minutes) to kill the bull by inserting a sword in the cervical (neck) vertebrae immediately below the skull and cutting the animal's spinal cord. Sometimes this procedure is not performed well and the bull is still alive but paralyzed, having his tail and ears cut off while conscious. A myth that has been scientifically refuted by the Association of Veterinarians against Bullfighting and Animal Abuse (AVATMA) is that adrenaline and endorphins released during the *corrida* block the ability of the bull to feel pain.

A number of countries—Argentina, Canada, Cuba, Denmark, Germany, Italy, the Netherlands, New Zealand, and the United Kingdom—have banned bullfighting. There are even certain regions within the bullfighting countries that have banned it, including the Canary Islands, Catalonia and the Balearic Islands in Spain, and most of France. Towns in Spain, Ecuador, Venezuela, France, Portugal, and Colombia have also declared themselves to be anti-bullfighting towns. However, bullfighting and other *fiestas* that involve the use of bulls are heavily subsidized in Spain by the national government. Britain's League against Cruel Sports estimates that over 550 million euros ($600 million) from European sources go annually to the bullfighting industry (League 2016). In October 2015, the European Parliament overwhelmingly voted against financing this industry with public subsidies.

Child protection is also a concern within bullfighting. There are two significant issues: child bullfighters and the exposure of children to extreme violence. In Spain, children can enroll in bullfighting school, but not until age 16. Such schools are also on the rise in Mexico, where there is no age limit. Several scientists argue that witnessing a bullfight can lead to negative psychological effects on children, affecting their moral judgement and empathy, and desensitizing them to violence. In 2013, 140 scientists and academics signed a letter arguing that bullfighting desensitizes young people to violence. According to the UN's Committee on the Rights of the Child, the violence of bullfights is harmful to youngsters: "The Committee is concerned about the physical and mental well-being of children involved in the training of bullfighting, and performances associated with it, as well as the mental and emotional well-being of child spectators who are exposed to the violence of bullfighting" (United Nations 2014, 10). For this reason, the Madrid Assembly decided to restrict content showing violence toward people and/or animals (including bullfighting) on television during times when children can be expected to be watching. The Madrid City Council also decided to withdraw public subsidies to its bullfighting school.

The International Anti-Bullfighting Network, made up of some 100 animal advocacy organizations from more than 20 countries, has been established recently to coordinate actions internationally in order to educate people about the suffering of bulls and horses used in bullfighting, protect children from exposure to animal cruelty, and provide rigorous scientific data to help campaigners lobbying to ban these bloody *fiestas*.

Núria Querol i Viñas

See also: Advocacy; Art, Animals in; Cruelty; Fighting

Further Reading

La Tortura No Es la Cultura. Accessed January 8, 2016. http://www.latorturanoescultura.org/de/news-blog/88-informacion-general-sobre-la-tortura-no-es-cultura/157-140-scientists-and-academics-address-letter-to-spanish-congress

League against Cruel Sports. 2016. Accessed January 7, 2016. http://www.league.org.uk/our-campaigns/bullfighting/what-are-we-doing-about-bullfighting

Paniagua C. 1994. "Bullfight: The Afición." *Psychoanal Q* 63(1): 84–100.

Querol, N. 2013. "Minors and Bullfighting, Scientific Data and Recommendations." Accessed January 8, 2016. http://www.gevha.com/analisis/articulos/parlament-de-catalunya/1371-minors-and-bullfighting-scientific-data-and-recommendations

United Nations. 2014. *Convention on the Rights of the Child.* Accessed January 7, 2016. http://tbinternet.ohchr.org/Treaties/CRC/Shared percent20Documents/PRT/CRC_C_PRT_CO_3-4_16303_E.pdf

Zaldivar, J. E. "El sufrimiento del toro en la lidia: Lesiones anatómicas, alteraciones metabólicas y neuroendocrinas." Accessed January 8, 2016. http://www.gevha.com/analisis/articulos/parlament-de-catalunya/1028-el-sufrimiento-del-toro-en-la-lidia-lesiones-anatomicas-alteraciones-metabolicas-y-neuroendocrinas

Bushmeat

The term "bushmeat" refers to meat from nondomesticated amphibians, birds, reptiles, and mammals (e.g., the giant forest hog, *Hylocheoerus meinertzhageni*) in tropical forest and savanna regions. It is a contentious topic within conversations about both environmental conservation and economic development. Whereas meat from animals hunted in industrialized countries such as Canada or the United States is called "game," "bushmeat" invokes controversy associated with hunting animals in tropical Africa, Latin America, and Asia. While hunting affords a crucial food source and monetary income for millions of people, bushmeat is frequently targeted by wildlife conservation initiatives, which depict its consumption as a threat to biodiversity. Negotiations about resource management in less-industrialized countries frequently emphasize bushmeat, making it a pivotal concept at the interface of biodiversity conservation and socioeconomic development. On one hand, overhunting does pose a serious risk to certain species; on the other hand, hunting bans can restrict subsistence livelihoods that rely on wild meat, thereby marginalizing certain groups of people and creating a justice problem that generates conflict between (local) resource users and (global) resource conservation efforts.

Wild-sourced meat is often significantly more accessible than farm-raised meat in many parts of tropical Africa, Latin America, and Asia. This is particularly true in Central Africa's Congo Basin forests, where wildlife is abundant while environmental, political, economic, and technological constraints make raising livestock difficult and cost prohibitive. Often the only viable source of protein for many people in these regions, bushmeat consumption is estimated at 51 kg (112 lbs.) annually per person in forested areas in the Congo Basin, and 63 kg (139 lbs.) in Latin America (Nasi et al. 2011). In many places, the importance of bushmeat is such that complex social systems often develop to manage hunting grounds to ensure that certain species are not overhunted. Wild meat is also valued in urban areas in the tropics and beyond—for example in cities such as New York, London, and Shanghai—where it is marketed as an expensive delicacy to middle- and upper-class consumers.

Practiced for millennia throughout the tropics, subsistence hunting (that is, hunting for basic food needs) and low levels of commercial hunting have generally entailed little risk of species loss. However, with increased demand for meat (primarily in urban areas) and access to technologies that enable more efficient hunting (such as guns and metal snare traps), some scholars anticipate unsustainable levels of commercial hunting in many tropical and subtropical forest and grassland regions. Resource extraction such as industrial logging often exacerbates the risk of overhunting by extending road networks that facilitate both increased access to remote areas of tropical forest and ease of meat transport from hunting sites to markets. Some contend that overhunting has reached a critical threshold denoting a "bushmeat crisis," particularly in West and Central African forests, and that the potential loss of species is a concern to global conservation goals and local livelihoods alike (Bennett et al. 2002). To address the risk of overhunting, governments and wildlife conservation NGOs (nongovernmental organizations) act together to enact and enforce laws prohibiting some combination of the following: hunting certain species, hunting with particular technologies (such as wire snares and automatic weapons), hunting in specific locations (for example, national parks and other protected areas), and selling or buying the meat of certain animals or from certain locales.

Bushmeat management strategies tend to emphasize *production* (i.e., processes associated with meat supply, including hunting and transporting) without attending to its *consumption* (i.e., networks of demand that extend to urban and international markets). Some suggest that the resulting emphasis on criminalization of hunting unfairly targets the rural poor, who often have few options besides wild meat. By the same token, local hunting bans are seen as failing to address root issues of the bushmeat crisis because commercial hunting operations are frequently orchestrated through networks of powerful individuals that often can be traced to urban areas far from hunting grounds. Recognizing that commercial hunting is a dynamic process that threatens local livelihoods in addition to wildlife populations, the Bushmeat Crisis Task Force (or BCTF, a collaboration of individuals and organizations that coordinated activities regarding the commercial bushmeat trade from 1999 to 2009, primarily in Africa) advocated a holistic approach to research, policy, education, and monitoring that includes attention to broader social, political, and economic contexts.

As an intersection of biodiversity conservation and socioeconomic development, the bushmeat issue exemplifies the society-environment interaction in which humans and nonhumans are deeply entangled. Along these lines, recent scholarship in anthropology has helped create a more nuanced understanding of the issue by considering hunter-hunted relationships as dynamic encounters between various groups of humans and nonhumans. For example, in their respective recent papers, Hardin and Remis (2009) and Robinson and Remis (2014) suggest viewing the bushmeat trade as composed of intimate relationships between hunters and animals, and discuss the multiple ways that animal species are valued economically, symbolically, and ecologically by various human societies. They demonstrate, for example, that hunted species have become more nocturnal in response to the presence of firearms and also that hunters' understanding of certain species' behavior and population dynamics has resulted in less overhunting than might be expected. Viewing human-animal interaction as coexistence (where humans and animals are interdependent) contrasts with the typical depiction as conflict (where hunters degrade the natural world). This more nuanced perception of hunters as part of a dynamic system suggests the value of working with hunters on programs that aim to conserve biodiversity without impinging on local livelihoods.

Nathan Clay

See also: Biodiversity; Human-Wildlife Conflict; Hunting; Indigenous Rights; Poaching; Wildlife Management

Further Reading

Bennett, E. L., Milner-Gulland, E. J., Bakarr, M., Eves, H. E., Robinson, J. G., and Wilkie, D. S. 2002. "Hunting the World's Wildlife to Extinction." *Oryx* 36(4): 328–329.

Davies, G. and Brown, D., eds. 2007. *Bushmeat and Livelihoods: Wildlife Management and Poverty Reduction.* Oxford: Blackwell Publishing Ltd.

Hardin, R., and Remis, M. 2009. "Transvalued Species in an African Forest." *Conservation Biology* 23(6): 1588–1596.

Milner-Gulland, E. J., Bennett, E. L., and the SCB 2002 Annual Meeting Wild Meat Group. 2003. "Wild Meat: The Bigger Picture." *Trends in Ecology and Evolution* 18(7): 351–357.

Nasi, R., Taber, A., and Van Vliet, N. 2011. "Empty Forests, Empty Stomachs? Bushmeat and Livelihoods in the Congo and Amazon Basins." *International Forestry Review* 13(3): 355–368.

Robinson, C. A. J., and Remis, M. J. 2014. "Entangled Realms: Hunters and Hunted in the Dzangha-Sangha Dense Forest Reserve (APDS), Central African Republic." *Anthropological Quarterly* 87(3): 613–633.

Wilkie, D. S., and Carpenter, J. F. 1999. "Bushmeat Hunting in the Congo Basin: An Assessment of Impacts and Options for Mitigation." *Biodiversity and Conservation* 8: 927–955.

C

Canadian Seal Hunt

Seal hunting (variously known as "harvesting," "slaughter," and "killing"), or sealing, is currently practiced in eight countries, with most of the world's hunting taking place in Canada. The seal hunt is surrounded by controversy due to the clash between animal rights and environmentalist concerns and economic interests. In the 1960s, protesters pressured the Canadian government to pass legislation limiting the killing. As a result, killing quotas were introduced. The hunt has led to widespread protest by animal-rights activists, as well as other concerned groups, and by some international governmental institutions. Conservationists have demanded reduced rates of killing, arguing that the hunt is cruel as well as threatening to the very survival of seals as a species.

There are two main reasons for the Canadian harp seal hunt: the seal products and the hunters' desire to keep the seals from eating the fish stocks on the eastern seaboard. Most sealing in Canada occurs in late March in the Gulf of St. Lawrence and in the northeast of Newfoundland in April. Harp seals—the main species hunted, called "hair seals"—depend on their blubber as their defense against the cold, as their pelts (skin with fur) have no underfur. Most seals killed are those under four months old that have just grown out of their white-coat stage. Canada sells pelts to eleven countries, with Norway, Germany, Greenland, and China purchasing the largest quantities. Economic, cultural, and environmental factors, as well as climate change, affect seal hunting.

Indigenous inhabitants (now called First Nation peoples, or Inuit) of Northern Canada traditionally regarded the animals they hunted as sentient (beings able to feel pain), intelligent beings that shared their environment and deserved their respect. Traditionally, the Inuit diet was rich in fish, whale, and seal, with seal meat being an important source of fat, protein, and vitamins, and the pelts vital for providing warmth. According to an analysis by the anthropologist Ann McElroy, Inuit believed that food security depended on observances of taboos, involving respect and humility toward seals. Inuit used to believe that if taboos were broken, Sedna, a goddess who controls sea mammals, would withhold animals from hunters.

When the first European settlers landed on the east coast of Canada in the 18th century, there were an estimated 30 million harp, hood, and gray seals. Commercial trading companies such as the Hudson's Bay Company (HBC) engaged in large-scale sealing, trapping, and exports of seal skins in the mid-19th to early 20th centuries.

The Canadian seal hunt has been a very controversial activity that has pitted commercial sealers and subsistence hunters against animal advocates and much of the general public. Public outcry has been strong, especially because many of the seals killed are babies ("pups"), such as the one pictured here in 2015. (Engelbert Fellinger)

The HBC and other commercial traders competed for the labor of the Inuit, and the native Inuit chiefs controlled the goods and rations exchanged for animal skins. In the 1840s, 546,000 seals were killed annually. This rate of killing continued well into the 20th century.

During the second half of the 20th century, the seal populations declined to approximately 2 million. Presently, the aquaculture (marine agriculture) industry is regulated by 17 federal departments and agencies, with Fisheries and Oceans Canada in the lead role. It manages fisheries, regulates the seal hunt, sets quotas, and works with the Canadian Sealers' Association, which promotes sealing.

The seal hunt provides an important source of revenue for the island of Newfoundland's economy (as well as for the Canadian economy more generally), with global demand for pelts, leather, oil, and meat providing part-time employment for up to 6,000 people. The value of the Canadian seal hunt was estimated to be roughly 40 million Canadian dollars in 2014. In 2012, sealers killed over 325,000 seals with an additional 10,000 seal quota allowed for the traditional hunt by the First Nation peoples. Both the quotas and the number of seals actually killed every year fluctuate, due to shifts in public opinion, media coverage, political decisions, the accuracy of statistics, and illegal hunting. The seal population was estimated at 505,000 in 2014.

At present, there is little evidence that a significantly expanded population of Inuit still hunt for subsistence (basic needs/survival) but rather for sport and commercial profit, using modern equipment such as rifles, outboard motors, and snowmobiles. Due to large discrepancies in numbers, as well as definitions of "traditional subsistence," there is controversy about which seals or seal parts are used for local consumption and which are sold commercially. This highlights political dimensions to what is considered to be hunting in traditional ways. Comments by the Sea Shepherd Conservation Society (a nonprofit marine conservation organization) imply that the seal hunt is an easy way for the Canadian government to solve structural social problems like unemployment and mask large-scale commercial exploitation of the sea, which has led to drastic depletion of fish stocks.

Conservationists have been relatively successful in mobilizing public opinion against sealing, leading to boycotts, changes in export regulations, and eventually bans on international trade of marine mammal products, prohibiting sales of items such as seal leather and pelts. Many activists, as well as scientists who are experts on the subject, have pointed out the inhumane nature of seal killing (with seal pups being killed with clubs and spikes), and thus have argued in favor of more humane methods. The International Fund for Animal Welfare began to campaign against *all* forms of sealing in the 1970s, with limited success. Many ethics scholars question the very act of killing, "humanely" or not, asking whether commercial seal hunts can be morally justified in the 21st century.

Helen Kopnina

See also: Advocacy; Cruelty; Ethics; Human-Wildlife Conflict; Indigenous Religions, Animals in; Indigenous Rights; Rights

Further Reading

Fisheries and Oceans Canada. 2015. http://www.dfo-mpo.gc.ca/index-eng.htm

Lavigne, D. M., and Lynn, W. S. 2011. "Canada's Commercial Seal Hunt: It's More than a Question of Humane Killing." *J. Animal Ethics* 1(1): 1–5.

McElroy, A. 2013. "Sedna's Children: Inuit Elders' Perceptions of Climate Change and Food Security." In Kopnina, H., and Shoreman-Ouimet, E., eds., *Environmental Anthropology: Future Directions*. New York: Routledge.

Sea Shepherd Conservation Society. 2012. "Seal Hunt Facts." http://www.seashepherd.org/seals/seal-hunt-facts.html

Canned Hunting

Canned hunting, also known as captive or high-fenced hunting, is a type of hunting where hunters pay a fee to shoot animals in captivity. These hunts are a type of trophy hunting, where hunters keep parts of slain animals as trophies or souvenirs.

In canned hunts, hunters often pay very high prices to shoot wild, sometimes endangered, animals that are trapped behind fences. Canned hunting is very controversial and is criticized by both animal welfare groups and many hunting organizations. Although canned hunting is increasing in popularity in many African countries, canned hunts also exist in the United States.

The animals involved in canned hunts often come from breeders, zoos, or circuses, or are captured from the wild as babies. They are usually hand-fed so that they become accustomed to being around humans, making them easier to hunt. As the animals grow older, they are fed at regular intervals in specific locations within the enclosures in order to increase the chances that the hunters will be able to more easily find the animals. In other words, the animals are raised in such a way that they become tame so that they guarantee a "kill" for hunters. Because semi-tame animals are easier targets for hunters, the shooting preserves where canned hunts take place can offer a "no kill, no pay" deal, ensuring that the animals will not be able to escape.

In the United States, there is no federal law that bans canned hunting. However, about half of all U.S. states have laws that prohibit the practice. The Humane Society of the United States estimates that there are about 1,000 canned hunt facilities in the United States, about half of which are in Texas. Although the Endangered Species Act (ESA) was created in 1973 in the United States to protect endangered or threatened animals, captive-bred wildlife are not covered by the Act if state law permits a hunt. Moreover, the 1966 federal Animal Welfare Act does not apply to animals that are kept in game preserves, leaving very little protection for animals that are would-be trophies in canned hunts.

While canned hunting practices exist in many countries, it is becoming increasingly popular in South Africa in particular. African game animals, most often lions and rhinos, are bred in captivity to supply the animals for canned hunts. Lion cubs are often taken from their mothers quickly after birth to be bottle-fed and become more accustomed to humans. This separation also brings the mother back into a reproductive cycle to ensure a continually refillable stock of animals for the facilities. Often, the big cats used in canned hunts are bred and raised on game farms whose owners tell visitors and tourists that the cubs are orphaned and are being prepared for reintroduction into the wild. Visitors can feed lion cubs by bottle under the impression that they are helping orphaned cats get a second chance at life. Once the cubs are grown, however, they are transferred to enclosures where they become the targets for captive hunts.

Trophy hunters are attracted to canned hunts because of the guarantee of a successful kill. A wild lion that is shot on a safari may cost 10 times as much as a captive-bred lion in a shooting preserve. Canned hunting is a big business that creates large revenue for those owning the facilities. For example, by one estimate, the canned hunting industry in South Africa brought in about $70 million in profit in 2012 (Bennett-Smith 2013). A hunter can pay up to $38,000 to hunt a lion in a controlled environment.

Over the last decade, the rate of canned hunting has been increasing exponentially. A contributing factor is the increasing demand for trophy hunting from foreigners visiting South Africa to participate in hunts. Demand for lion parts for traditional Asian medicine has also contributed to the rise in canned hunting and creates a rapidly growing source of revenue for canned breeding facilities. Although many canned hunting facilities claim to only offer nonendangered exotic animals, there is evidence to suggest that endangered animals are victims of these practices. The breeders might argue that shooting a captive-bred animal rather than a wild one is better for conservation of wild populations. However, many conservationists, hunters, and animal welfare groups argue that this is not the case.

Interestingly, canned hunting is one area where many hunting organizations and animal welfare groups find common ground. Animal welfare groups such as the Humane Society of the United States argue that it is cruel to remove babies from their mothers to be raised by humans for the sole purpose of making them easier to hunt. Similarly, both animal welfare groups and hunting organizations argue that it is wrong to hunt animals that have no way of escaping the hunter. For instance, in 2007, the cable news network CNN showcased a video that showed a lion in a canned hunting facility being shot while it was pressed up against a fence trying to find a way to escape. This alarming episode raised a lot of attention about the unfair nature of canned hunts. Hunting groups like Boone & Crockett, Pope & Young, and the Izaak Walton League raise concerns about the unethical and unsportsmanlike nature of canned hunting, as it removes the concept of the "fair chase," where an animal has a fair chance of escaping the hunter.

Stefanie Georgakis Abbott

See also: Endangered Species; Ethics; Hunting; Trophy Hunting

Further Reading

Bennett-Smith, M. 2013. "Canned Lion Hunting Report Suggests South African Business Booming after Regulations Lifted." *Huffington Post*, June 5. Accessed July 1, 2015. http://www.huffingtonpost.com/2013/06/05/canned-lion-hunting-video-south-africa_n_3386878.html

Humane Society of the United States. n.d. "Captive Hunts." Accessed July 1, 2015. http://www.humanesociety.org/issues/captive_hunts/

Oke, F. 2007. "Shooting Lions in South Africa." *CNN*, June 8. Accessed July 1, 2015. http://www.cnn.com/exchange/blogs/in.the.field/2007/06/shooting-lions.html

Times Media Reporters. 2015. *Rise of Canned Lion Hunting in South Africa: Reducing the King of Beasts to Easy Prey*. Johannesburg, South Africa: Times Media Books.

Cats

Felis catus, or the domestic cat, has long had significance in human society. A deep social interest in felines has inspired depiction of cats in art since early domestication, and they are the subject of countless photographs, books, films, and songs. Cats are the most popular subject on the Internet with websites and social media pages dedicated to individual cats. They are one of today's most common pets. Cats are widely acknowledged for their sociability and worth as companions or service animals, and they are respected for keeping vermin populations in check. But cats also kill birds and spread disease, often through feral populations. The cat's place in society has never been stable: They have been revered and feared, adored and despised, and worshiped and persecuted throughout history.

It is commonly thought that cats were domesticated in Egypt, but genetic studies and archaeological finds indicate that cats may have been domesticated in the Near East as many as 12,000 years ago. Cats, having only recently split from wild relatives, retain close genetic and social ties to the wild. Despite these wild ties, a 2007 study revealed that domestic cats carry genes linked to memory, fear conditioning, and stimulus-reward learning, which lead to tameness. The presence of these "domestication" genes prompted researchers to propose that cats actually domesticated themselves (Montague et al. 2014).

Cats were worshiped as gods in Egypt between 2500–945 BCE. Sekhmet was a human-lion hybrid deity that destroyed enemies and protected against evil and plague. Her sister, cat-headed Bastet, often depicted as the left eye of the sun god, was linked to mystical powers in the moon, intuition, and the unconscious. As a fertility goddess she was specifically associated with women. During this period, cats were mummified and ritually buried with royalty and others of high stature. Women emulated the cat's physical beauty with make up by drawing cat eyes with kohl eyeliner. Many cat figurines exist from this time period, indicating the popularity of the cat in Egyptian society.

Christianity pushed cats from their high social place. In Medieval Europe, cats were considered heretical and associated with witches. Historian Irina Metzler speculated that the cat's independence was unsettling for humans because the cat was not subservient to the humans God chose to rule over them (2009). Cats were associated with pagan (non-Christian) religions and cults, and they were demonized along with pagan women. The historical linking of women and cat persecution provided a foundation for the modern gendering of cats in western society as the female complement to man's best friend, the dog.

Fear and superstition of cats carried into 18th-century France, where cats were tried for witchcraft and burned at the stake. People still harbor superstition of black cats today, which are among the least adopted and most frequently euthanized cats at animal shelters. The mass killing of cats in Europe led to an abundance of vermin, which later spread the bubonic plague (or Black Death) that devastated Europe in the 14th century.

Cat breeding emerged in the mid-19th century. The world's largest pedigree cat registry, Cat Fanciers' Association (CFA), now recognizes 41 cat breeds, of which 16 are "natural breeds" that emerged without human intervention. The other breeds were developed over the last 50 years through the practice of pedigree breeding for distinct physical characteristics and personality traits. For example, Scottish Folds are bred for their turned down ears, and Ragdolls are bred for their large size (they average 20 pounds). The nearly hairless and highly energetic Sphynx is perhaps the most unique breed, often described as a suede hot water bottle with a mischievous personality (CFA 2015). The oldest (and fastest, clocked at 30 miles per hour) cat breed is the Egyptian Mau, descended from ancient Egyptian cats.

While cats are appreciated for their vermin hunting, they are not appreciated for their bird hunting. Problems with cats hunting and killing large numbers of birds are particularly pronounced on islands such as Tasmania, Australia, where cats are not native. Pet cats can wear bells or brightly colored collars that warn birds, but uncollared feral cats more successfully kill and potentially damage native bird populations.

The wild propensity of cats results in large feral (free roaming) cat populations, particularly in cities, where cats, like humans, live in close quarters. Feral cats are distinct from stray cats in that feral cats are typically born wild, and strays are typically lost or abandoned former pets. Individual feral females can birth 120 or more kittens in a 10-year period, leading to large feral cat populations that often conflict with people due to noise, concern about zoonotic diseases (those which can be transferred from cats to humans), and potential feral cat contact with pets. Management agencies and cat advocacy groups generally agree that it is advantageous to reduce the feral cat population, but they are divided on the ways to do so. Agencies often trap and euthanize due to limited resources, while many cat rescue groups advocate for the trap-neuter-return (TNR) approach in which captured feral cats are neutered and returned to their free-roaming lives.

The cat's wild nature has been highlighted through centuries of literature and art. Cats are often depicted as clever, aloof, independent, arrogant, and wild at heart. Rudyard Kipling (1835–1936) wrote of the cat who refused the lure of domestication: "He is the Cat that walks by himself, and all places are alike to him" (Kipling, 1907, 214).

Anita Hagy Ferguson

See also: Advertising, Animals in; Art, Animals in; Domestication; Eastern Religions, Animals in; Feral Animals; Hoarding, Animal; Human-Animal Bond; Intelligence; Literature, Animals in; Pets; Popular Media, Animals in; Service Animals; Tigers; Zoonotic Diseases

Further Reading

Appleby, M., Weary, D., and Sandøe, P., eds. 2014. *Dilemmas in Animal Welfare*. Boston: CABI International.

Cat Fanciers Association. "Sphynx." Accessed March 15, 2015. http://www.cfainc.org/Breeds/BreedsSthruT/Sphynx.aspx

Darnton, R. 1985. *The Great Cat Massacre and Other Episodes in French Cultural History*. New York: Random House.

De Mello, M. 2012. *Animals and Society: An Introduction to Human-Animal Studies.* New York: Columbia University Press.

Helgren, J. A. 2013. *Encyclopedia of Cat Breeds*. Hauppauge, NY: Barron's Educational Series, Inc.

King, B. J. 2010. *Being with Animals: Why We Are Obsessed with the Furry, Scaly, Feathered Creatures Who Populate Our World*. New York: Doubleday.

Kipling, R. 1907. "The Cat Who Walked by Himself." *Just So Stories*. New York: Doubleday.

Lipinski, M. J., Froenicke, L., Baysac, K. C., Billings, N. C., Leutenegger, C. M., Levy, A. M., Longeri, M., Niini, T., Ozpinar, H., Slater, M. R., Pedersen, N. C., and Lyons, L. A. 2008. "The Ascent of Cat Breeds: Genetic Evaluations of Breeds and Worldwide Random Bred Populations." *Genomics* 91(1): 12–21.

Metsler, I. 2009. "Heretical Cats: Animal Symbolism in Religious Discourse." *Medium Aevum Quotidianum* 59: 16–32.

Montague, M. J., Li, G., Gandolfi, B., Khan, R., Aken, B. L., Searle, S. M. J., Minx, P., Hillier, L. W., Koboldt, D. C., Davis, B.W., Driscoll, C. A., Barr, C. S., Blackistone, K., Quilez, J., Lorente-Galdos, B., Marques-Bonet, T., Alkan, C., Thomas, G. W. C., Hahn, M. W., Menotti-Raymond, M., O'Brien, S. J., Wilson, R. K., Lyons, L. A., Murphy, W. J., and Warren, W. C. 2014. "Comparative Analysis of the Domestic Cat Genome Reveals Genetic Signatures Underlying Feline Biology and Domestication." *PNAS* 111(48): 17230–17235.

Saunders, N. 1991. *The Cult of the Cat*. London: Thames and Hudson.

Thomas, E. M. 1994. *The Tribe of the Tiger: Cats and Their Culture*. New York: Simon and Schuster.

Chimpanzees

Chimpanzees are great apes, a specific type of primate known for intelligence and genetic similarity to humans. Chimpanzees are found in 21 African countries, where habitat loss, hunting, disease, and illegal trade contribute to the fact that they are endangered. While there are no official figures, chimpanzees are found in captivity all over the world, including in zoos, laboratories, sanctuaries, and as pets. A range of laws govern the conservation and captive care of chimpanzees globally.

Chimpanzees are considered the closest living primate relatives of human beings, and scientists place both species within the same biological group known as *Hominidae*, or great apes. The scientific name for chimpanzees is *Pan troglodytes*. All four chimpanzee subspecies are considered to be endangered and at risk of extinction (WWF 2015).

Chimpanzees eat a varied diet, primarily consisting of fruits and vegetation, but they are also known to hunt and eat meat. Chimpanzees live in groups called "communities" that range from about 15 to 150 individuals, with groups often breaking into smaller parties for periods of time, then later pooling together again. Male chimpanzees remain in the group where they were born for their entire lives, while females typically emigrate to a new community during adolescence.

Chimpanzees live in 21 countries across equatorial Africa, where experts estimate there are 150,000 to 250,000 remaining (WWF 2015). The number and geographic distribution of chimpanzees has declined sharply in the last 50 years, largely as a result of human activities such as land conversion from forest by industries such as logging, mining, and agriculture. Increased contact and conflict with people, arising from habitat loss and degradation, have also played a role. For example, crop-raiding is a primary harm caused by chimpanzees and a problem for some communities. Disease and hunting are also threats to chimpanzee populations in some areas (Arcus Foundation 2014).

People keep chimpanzees in captivity in many countries. The welfare of chimpanzees in captivity presents challenges owing to their high intelligence and therefore complex needs and abilities, as well as long lifespans of up to 50 years or more. Experts concur that captive care for chimpanzees must go beyond adequate space, sanitation, and nutrition to provide for social, cognitive, and other needs, with an emphasis on allowing for natural behaviors.

In chimpanzee habitat countries, lawful captivity is generally limited to zoos and sanctuaries. Biomedical testing on chimpanzees has been limited in Africa historically, especially since exportation for research was stopped under international treaties. Illegal trade, which includes pets, exhibition, and trade in body parts, is estimated to affect about 1,900 chimpanzees per year. Hubs of such illegal trade include Guinea and Democratic Republic of Congo (Stiles et al. 2013). Across Africa, a network of sanctuaries cares for approximately 1,000 chimpanzees rescued from illegal trade and other dangers (Arcus Foundation 2014).

Common types of captivity in nonhabitat countries include zoos, entertainment, biomedical testing, and sanctuaries. Evidence traces trade and exhibition of chimpanzees back for centuries. The quality of modern zoos varies globally, with some providing specialized environments and care, while others do not meet international care standards or provide for basic needs. Some studies suggest that people who see chimpanzees in zoos or in other unnatural settings like advertisements often express support for commercial use or private ownership and fail to identify chimpanzees as endangered.

Many countries banned or abandoned research and testing on chimpanzees in the 1980s and 1990s, leaving Gabon and the United States among the last countries to allow such practices. In the United States, large-scale experiments on chimpanzees began by the 1920s and continued until 2013, when the government ruled in favor of new standards and a plan to phase out its experiments (Arcus Foundation 2014). A 2015 ruling under the U.S. Endangered Species Act further restricted biomedical testing and other commercial uses of chimpanzees. As a result of these two rules, testing on all chimpanzees is strictly limited, and chimpanzees may be eligible for "retirement" to sanctuaries in the United States.

Research studies have revealed a number of similarities between chimpanzees and humans. The most extensive studies began approximately 55 years ago. Studies of behavior and biology have provided knowledge of chimpanzees—for example, documenting evidence of culture, personality, play, and sophisticated communication—as well as human and primate evolution. The publication of the chimpanzee genome increased public interest in genetic similarities—estimated to range from 96 to 99 percent—between chimpanzees and humans (The Chimpanzee Sequencing and Analysis Consortium 2005).

Owing to chimpanzees' similarities to humans, their needs and capabilities, their extinction risk and other factors, there are a number of laws ranging from international to local for their protection. Indeed, some experts argue that chimpanzees have inherent rights similar to those of humans.

Debra Durham

See also: Advertising, Animals in; Animal Cultures; Animal Law; Black Market Animal Trade; Circuses; Deforestation; Endangered Species; Ethology; Extinction; Great Apes; Human-Wildlife Conflict; Research and Experimentation; Rights; Shelters and Sanctuaries; Species; Taxonomy; Zoonotic Diseases; Zoos

Further Reading

Arcus Foundation. 2014. *Extractive Industries and Ape Conservation,* 1st ed. State of the Apes. New York: Cambridge University Press.

The Chimpanzee Sequencing and Analysis Consortium. 2005. "Initial Sequence of the Chimpanzee Genome and Comparison with the Human Genome." *Nature* 437 (7055): 69–87. doi:10.1038/nature04072

Nonhuman Rights Project. 2015. "The Nonhuman Rights Project." Accessed October 21, 2015. http://www.nonhumanrightsproject.org/

Stiles, D., Redmond I., Cress, D., Nellemann, C., and Formo, R. K., eds. 2013. *Stolen Apes—The Illicit Trade in Chimpanzees, Gorillas, Bonobos and Orangutans. A Rapid Response Assessment.* United Nations Environment Programme. Accessed October 21, 2015. http://www.un-grasp.org/news/121-download

WWF. 2015. "Chimpanzees." Accessed October 21, 2015. http://wwf.panda.org/what_we_do/endangered_species/great_apes/chimpanzees/#distribution

Christianity. *See* Western Religions, Animals in

Circuses

The circus is an entertainment enterprise that combines performances of acrobats, clowns, and trained animals. Originating in equestrian shows, the presentation of wild animals became a central feature during the golden age of Western circus entertainment at the end of the 19th century. Questionable training practices and husbandry, however, have led to critiques of animal performances.

The origin of the modern circus is usually associated with Philip Astley (1742–1814), a former English cavalryman, although traveling acrobatic and animal acts like dancing bears had been known before. In the mid-18th century, horsemanship in Europe began to move away from its aristocratic and military roots, and former cavalrymen worked as riding instructors and trick riders. In London in 1768, Astley opened a horse riding school that presented horse shows in the afternoons. Although his was not the only horse show, he became the first to combine it with acrobatic acts, clowns, and magic shows. He also introduced the circular shaped arena, which later became a typical sign for the circus. The first to use the term "circus" for his enterprise, however, was Charles Hughes (1747–1797), a former employee of Astley who opened his Royal Circus, Equestrian and Philharmonic Academy in 1782 near Astley's amphitheater.

Most of the early circuses in Europe had stationary buildings. The oldest and one of the few still operating today is the *Cirque d'Hiver*, founded in Paris in 1852. In the United States, however, the circus had to adapt and become mobile to reach its audiences. The circus tent—first used by Joshua Purdy Brown (ca. 1802–1834) in 1825—and the train as a mode of transportation for circuses—started by William Cameron Coup (1836–1895) in 1871—were both U.S. innovations. Coup and his partner, Phineas Taylor Barnum (1810–1891), introduced in their *P.T. Barnum's Museum, Menagerie & Circus* two further specialties of the American circus: the addition of multiple rings to increase visitor capacities, and sideshows in which they presented human oddities, such as the African American albino twins "Eko and Iko," and animal oddities such as a 7-foot-tall horse and a 32-inch-tall cow.

While the early circuses predominantly featured horse performances, wild-animal acts featuring lions, elephants, or even crocodiles were established later. Parallel to the first circuses in the late 18th century, traveling menageries became popular after restrictions on the ownership of certain animal species such as lions were eased in the Netherlands and Britain. It was in these menageries that the first lion tamers like Henri Martin (1793–1882) appeared. Although Isaac A. Van Amburgh (1811–1865) is often recognized as being the first to combine a menagerie and circus by performing with his lions at Astley's in 1838, most circuses at the time did not have the necessary funds to keep wild animals. The change started with Carl Hagenbeck (1844–1913), a German animal trader and, in 1907, founder of a zoo in Hamburg. His company also encompassed a circus, traveling menageries, and human zoos exhibiting indigenous ethnic groups such as the Somali (from Africa), Lapps (from Northern Europe), or Bella Coola (from North America), considered exotic by the European and North American audiences. From 1872 on, leading American circus entrepreneurs like James A. Bailey (1847–1906) and Barnum began buying animals for menageries accompanying their circuses as sideshows. By the 1890s, animal performances were routinely included in the main show. Early lion-tamers like Van Amburgh, who was infamous for his brutal treatment of the animals, often presented themselves as "human masters taming the beast" and as an allegory on the biblical sovereignty of humanity over nature. Influenced by Charles Darwin's

Responding to animal advocacy critiques and public pressure, in March 2015 *Ringling Bros. and Barnum & Bailey* announced that it will retire its elephant herd by 2018. Training elephants with "bullhooks" has been one of the primary focal points of critique. (AP Photo/ Bill Sikes)

work *On the Origin of Species* (1859) a generation later, Hagenbeck became ambassador for a more gentle form of animal training, working with the natural behavior of the animals and presenting them as peaceful and playful. Other trainers, though, kept training their animals with whip, chair, and gun blanks.

Critiques of animal well-being in circuses led to the founding of some of the first animal welfare organizations like the *American Humane Association* in 1877 and the passing of early legislation like the *Wild Animal in Captivity Protection Act 1900* in Britain. Even today, animal advocacy organizations question the training methods, husbandry, and transport conditions in circuses, emphasizing that the animals did not volunteer to perform and demanding a general prohibition of animal performances. The use of steel-tipped bullhooks to train elephants and the long periods in transport especially are criticized. A new generation of circuses has reacted to the public critique. Whereas the circus genre of Cirque Nouveau, which is best known through the Canadian-based *Cirque du Soleil*, founded in 1984, completely abandoned animal acts, circuses like the German *Circus Roncalli*, founded in 1976, returned to the origins of the circus by only featuring horse performances. At the same time, a new generation of animal trainers in Europe and the United States, like lion and tiger trainer Alexander Lacey, supported by circus enthusiast organizations, defends the possibility of humane animal training. Although individual cities and states worldwide had already prohibited the use of animals in the circus, in 2009 Bolivia became the first country to ban it, followed by Peru, Greece, Cyprus, Paraguay, Columbia, the Netherlands, and Slovenia. In March 2015, *Ringling Bros. and Barnum & Bailey,* one of the biggest traditional circus enterprises in North America, announced that it will retire its elephant herd by 2018. Still, however, many circuses continue to feature elephants, tigers, lions, and other animals.

Jan-Erik Steinkrüger

See also: Animal Law; Elephants; Ethics; Horses; Husbandry; Zoos

Further Reading

Kotar S. L. and Gessler, J. E. 2011. *The Rise of the American Circus, 1716–1899*. Jefferson, NC and London: McFarland and Company.

Nance, S. 2013. *Entertaining Elephants: Animal Agency and the Business of the American Circus*. Baltimore: The Johns Hopkins University Press.

Simon, L. 2014. *The Greatest Show on Earth: A History of the Circus*. London: Reaktion Books.

Tait, P. 2012. *Wild and Dangerous Performances: Animals, Emotions, Circus*. Basinstoke and New York: Palgrave Macmillan.

CITES. *See* Wildlife Forensics; Primary Documents

Climate Change

Climate change refers to shifts in the statistical distribution of weather patterns worldwide over an extended period, irrespective of cause. Such changes may be measured in terms of changing average weather conditions or by weather events or trends at the extremes of historical distribution patterns. In public policy discussion, climate change is synonymous with global warming, the documented century-long rise in the average temperature of Earth's climate system. The anticipated effects of climate change in the 21st century include warming temperatures, rising sea levels, altered patterns of precipitation, and increased spread of deserts. Its impacts upon nonhuman animals are the subject of growing attention.

There is disagreement over whether or not climate change is caused by humans. Some argue that the burning of fossil fuels and other human activities are causing warming temperatures, loss of sea ice, rise in sea levels, fiercer storms, and increasing drought worldwide. Others argue that human-caused greenhouse gas emissions are not substantial enough to alter Earth's climate and that warming is a natural process—an aspect of long-term, fluctuating climatic patterns. A 2013 analysis of more than 11,000 peer-reviewed studies concluded that 97 percent of those studies taking a position adopted the view that climate change is human caused.

The Intergovernmental Panel on Climate Change (IPCC) uses the term "climate change" to refer to any change in climate, over time, whether due to human activity or to natural variability, while the UN Framework Convention on Climate Change (UNFCCC) uses the term only to designate change in climate that is directly or indirectly attributable to human activity that alters the global atmosphere and occurs in addition to natural climate variability over time.

In combination with continuing human population growth, intensifying development, and increases in consumption, climate change—with its potential to affect the entire biosphere (all of Earth's ecosystems as a whole) and to alter terrestrial (land) and marine (water) habitats—has implications for the welfare and survival of billions of companion animals, farm animals, and wildlife worldwide.

When it comes to wildlife, human dominance and exploitation of the biosphere, including direct loss of habitat and extensive extraction and use of natural resources, has been decisive in its impact over the last 200 years. Still, many view climate change as a threat of a different order. The *Fourth Assessment Report* of the IPCC concluded that climate change has resulted in noticeable species redistribution and range shifts, adaptations (such as altered bird migratory pathways), diminished genetic diversity, population decline, and biodiversity loss on all continents and in most oceans. The report concluded that 20–30 percent of species assessed would be at increased risk of extinction if global average warming increases exceed 1.5–2.5°C (2.7–4.5°F), relative to late-20th-century temperatures.

To a great extent, in the public imagination, the polar bear is the iconic animal at risk from climate change because its small worldwide population and slow reproduction rate make rapid evolutionary adaptation improbable. But there is increasing attention to the presumed threats of climate change—including diminishing ice and other habitat, and declining or disappearing food sources—to numerous other species, including penguins, sea turtles, whales, wolverines, seals, lobsters, frogs, coral, and cod.

Several reports emphasize the significance of animal agriculture as a source of greenhouse gas emissions responsible for climate change, including emissions from animal digestion and the decay of manure, the production and transportation of animal feed, energy use in agricultural facilities, postslaughter transportation, and refrigeration and packaging of animal products. A 2013 UN review put the animal agriculture industry's emissions at 7.1 gigatons per year, nearly 15 percent of all greenhouse gases associated with human activity. With an estimated 70 billion animals raised annually for food worldwide, environmental and animal organizations have advocated reduction in meat and animal product consumption to mitigate greenhouse gas emissions and key externalities (external costs to the whole of society) associated with most contemporary meat production, such as deforestation for grazing and feed crop cultivation, and air, soil, and water pollution.

The implications of climate change for companion animals could include hazards and effects relating to food security, water purity, general health, longer feline reproductive seasons that lead to increased breeding, and abandonment. Increasing temperatures could expose household pets and other companion animals to new vector-borne diseases spread by fleas, ticks, and mosquitoes.

In the early 21st century, the U.S. federal government and a number of states began to require that emergency and disaster evacuation plans accommodate animals. The Pets Evacuation and Transportation Standards (PETS) Act, passed after Hurricane Katrina hit the city of New Orleans in 2005, recognized the failures of prior evacuation planning that did not take account of evacuees' desire to see their pets rescued. These measures will become more important if climate change results in the displacement of people and their companion animals. Much less attention has been paid to preparedness planning for animals held in institutional settings like industrial and small-scale farms, laboratories, zoos, and other environments.

Bernard Unti

See also: Biodiversity; Endangered Species; Extinction; Human-Animal Bond; Wildlife

Further Reading

Beugnet, F. C., and Marié, J-L. 2009. "Emerging Arthropod-Borne Diseases of Companion Animals in Europe." *Veterinary Parasitology* 163: 298–305.

Cook, J., Nuccitelli, D., Green, S., Richardson, M., Winkler, B., Painting, R., Way, R., Jacobs, P., and Skuce, A. 2013. "Quantifying the Consensus on Anthropogenic Global Warming in the Scientific Literature." *Environmental Research Letters* 8.

Gerber, P., Steinfeld, H., Henderson, B., Mottet, A., Opio, C., Dijkman, J., Falcucci, A., and Tempio, G. 2013. *Tackling Climate Change through Livestock—A Global Assessment of Emissions and Mitigation Opportunities.* Rome: Food and Agriculture Organization of the United Nations. http://www.fao.org/docrep/018/i3437e/i3437e.pdf

Hannah, L., ed. 2012. *Saving a Million Species: Extinction Risks from Climate Change.* Washington, DC: Island Press.

Mann, M., and Kump, L. 2015. *Dire Predictions: Understanding Climate Change.* New York: DK Publishing.

Reisinger, A., and Pachauri, R. K., eds. 2008. *Climate Change 2007: Synthesis Report. Contribution of Working Groups I, II and III to the Fourth Assessment Report of the Intergovernmental Panel on Climate Change.* Geneva: Intergovernmental Panel on Climate Change. https://www.ipcc.ch/pdf/assessment-report/ar4/syr/ar4_syr_full_report.pdf

Cockroaches

Virtually unchanged in form and function at least since the Carboniferous Age (often referred to as the Age of Cockroaches) 400 million years ago, the cockroach—in fact all arthropods (beings with outer skeletons) and therefore all insects—evolved from a segmented worm (*onychopheron*) when Earth consisted of a single supercontinent (Pangea) in one vast ocean. The cockroach is, indeed, a living fossil. When Pangea split and drifted apart to become Eurasia/Africa and the Americas, related species of cockroach began journeys that carried them to every corner of the globe. Most of the 4,000-plus species of cockroach now live in habitats far removed from humans: boreal and rain forests, caves, burrows and hives, brush and beach. Although some 40 species do live close to man, only four (the American, German, Asian, and Australian cockroaches, all of tropical origin), have chosen to share our warm dwellings, becoming—in our minds—pest species. Actually, these cockroaches serve the same function in our homes and cities as their relatives do in the wild: "a food-waste disposal ecosystem service," as Steve Minsky put it recently in "Insect Aside" (Minsky 2015, 85). The cockroach's diet of decaying organic matter both in the wild and in human habitations traps considerable nitrogen which when released in their feces, enriches the soil, providing food for plant life, and may well have been a crucial factor in the rise of plants on the planet.

Ancestors of the social insects (bees, ants, termites), cockroaches are communal and exhibit complex social behavior—caring for offspring and sharing reserves and

Throughout the world, there are over 4,000 species of cockroach. However, only four of these species live in human dwellings. Widely reviled as pests, cockroaches exhibit complex social behaviors and care for their offspring. (iStockPhoto.com)

habitats—much as humans do. Recent studies show that cockroaches find both pleasure and security in tight, dark, enclosed spaces, and in touch. In fact, "a living touch triggers cockroaches to make babies faster" (Nuwer 2014). Like all insects, cockroaches have three body segments (head, thorax, and abdomen), six legs, and compound eyes. As T. H. Huxley claimed in 1869, cockroaches are the archetypal insect; more recently, Bernd Heinrich called them "the quintessential insect" (Heinrich 1996, 21). They breathe through a system of tubes (called trachea) branching throughout their bodies. Recent studies reveal that these tubes serve as a complex chemical laboratory, providing the cockroach with pheromones (chemical stimulants) that trigger defense, procreation, and communication. While retaining the arthropod's outer skeleton, cockroaches developed a soft, slippery, outer skin or cuticle that prevents dehydration and lubricates their speedy escapes into tight spaces. A study at the University of Oulu in Finland revealed that cockroaches also have evolved the ability to pool light signals similar to time-lapse photography, allowing them to see quite well in what seem to humans totally dark places (Nuwer 2015). Many species have wings or the remnants of wings that fold tightly against their bodies when not in use, thus preserving the aerodynamic form that allows them to move quickly and slip into even the narrowest of cracks and fissures in wall or floor, rock or soil.

Each of these ancient characteristics of form and function has helped and continues to help the cockroach survive conditions that have proved lethal to other species. Through each of the planet's great extinction events, the cockroach survived the die-offs in the late Permian and Paleozoic Eras (245 million years ago), the extinction of the dinosaurs in the late Triassic (208 million years ago) and Jurassic (114 million years ago) eras, and the deaths of the great mammals—mammoths, giant sloths, saber-toothed cats—in the Eocene (37 million years ago) and Pleistocene (10,000 years ago) eras. Whether present-day species of cockroach will survive the current, or sixth, extinction (now referred to as the Anthropocene because humans seem to be the main cause of the changes that threaten Earth today) remains to be seen. However, a recent study revealed that cockroaches possess distinct personalities as well as the ability to reach a consensus based on the responses of numbers of individuals—a behavior usually attributed to so-called advanced species like humans. This suggests to researchers that cockroaches will indeed survive global climate change.

Despite their contributions to the planet, their extraordinary survival skills, and their contributions to human cultures worldwide, cockroaches remain little studied and little understood. Most of scientific research surrounding them has been aimed at exterminating the four pest species that continue to share human habitations. David George Gordon, in *The Complete Cockroach,* suggests that knowing this insect's natural history should lead us to "regard the cockroach in a new light." Instead of seeing the cockroach as "an accursed nuisance," the cockroach can now be recognized as "a wizened old soul—one whose ancestors were around when the continents were formed, and witnessed the emergence and disappearance of the dinosaurs, and who watched an agile chimpanzee-like primate become *Homo sapiens*" (Gordon 1996, xiii–xiv).

Marion W. Copeland

See also: Agency; Animals; Climate Change; Communication and Language; Evolution; Intelligence; Nuisance Species; Sentience

Further Reading

Copeland, M. W. 2003. *Cockroach*. London: Reaktion Books.

Gordon, D. G. 1996. *The Complete Cockroach: A Comprehensive Guide to the Most Despised (and Least Understood) Creature on Earth*. New York: Ten Speed Press.

Heinrich, B. 1996. *The Thermal Warriors: Strategies of Insect Survival*. Cambridge, MA and London: Harvard University Press.

Lauck, J. 1998. *The Voice of the Infinite in the Small: Revisioning the Insect-Human Connection*. Mill Spring, NC: Shambala.

Minsky, S. 2015. "Insect Aside." *Scientific American*, February: 85.

Nuwer, R. 2014. "A Loving Touch Triggers Cockroaches to Make Babies Faster." *Salon*, April 5. Accessed May 22, 2015. http://www.salon.com/2014/04/05/a_loving_touch_triggers_cockroaches_to_make_babies_faster_partner/Smithsonian commentaries

Nuwer, R. 2015. "Household Pest Sees the Light." *Scientific American*, March: 20.

Schweid, R. 1999. *The Cockroach Papers: A Compendium of History and Lore*. New York and London: University of Chicago Press.

Communication and Language

Are a dog's ears up or down? Is a cat's tail swishing or at rest? The question of whether animals communicate and/or have their own languages is one that may seem self-evident to people with pets, but these have actually been major controversial topics for scholarly fields such as ecology, ethology, and evolutionary biology because animal communication is different than animal language. This topic is important to explore in the context of human-animal relations because, if humans understand animals as beings who communicate and have their own languages, it might shift how we treat them as a collective society.

We can trace the roots of the scientific study of animal communication back to naturalist Charles Darwin (1809–1882) and his 1872 publication *The Expression of the Emotions in Man and Animals*. This book was a continuation of his work on natural selection and the theory of evolution. In it he argued that if one studies emotional expressions (such as fear, happiness, surprise, and anger) of humans and animals in terms of how they use these expressions to visually communicate states of being, there is great similarity across species—thereby providing evidence of evolutionary connections as well.

Today, many scholarly fields are working on animal communication and language studies. These include biology, ecology, physical anthropology, ethology, and linguistics. The goal of these studies is to 1) further our understanding of how animals communicate, 2) understand the evolution of communication and language across species, and 3) help with issues such as wildlife management, pet care, and pest control. Animal communication can be understood as the cues or signals given by a "sender" organism to a "receiver" organism, resulting in specific behavioral actions. These signals and cues occur in a variety of ways depending on the species and the situation. Common methods include the use of vocalizations (to locate young, alert for predators, mating) or other sounds (such as the chirping of crickets rubbing their wings together), visual signals such as changing colors to show sexual status or defensive warnings (hair raising), chemical signals such as releasing smells (a skunk's defensive release, dogs marking with urine), and even touch (grooming, licking, fighting). All animal species communicate in some way. Animal communication can be quite complex. For example, vervet monkeys are known to produce alarm calls for specific predators such as eagles, snakes, or leopards. These different calls allow the group to take the appropriate defensive actions for each predator.

Human language, on the other hand, is seen as being different from nonhuman forms of communication. We can understand language as a system that uses sounds and/or images to convey abstract or symbolic meaning in an organized manner through the use of grammar and sentence structure. For example, saying or writing the words "She is a citizen" conveys meaning about the abstract concept of someone's political status. In addition, language is typically fluid and adaptable, and therefore different from straightforward communication such as alarm calls.

Scholars have long thought that human language is one of the traits that sets humans apart from other animals, but individual animals illustrate the complexities of differentiating between animal communication and language. Alex (1976–2007), an African grey parrot bought from a pet store, was the subject of a long-term study by American chemist and animal behavioralist Dr. Irene Pepperberg (1949–). Alex learned more than 100 human words, could count up to eight, and understood abstract concepts such as none and same versus different. He even recombined human words in ways that made sense to him. Koko the Gorilla (1971–), born in the San Francisco zoo, has worked since she was an infant with American animal psychologist Francine Patterson (1947–), who taught her to "speak" using modified American Sign Language. Koko understands spoken human English and uses upwards of 2,000 signs. Both animals have used human language to the extent of a human toddler of above average intelligence. Alex and Koko are clearly examples of the intelligence and adaptability of other species, yet there is

Koko, a four-and-a-half-year-old gorilla who has been taught sign language, communicating with her trainers. Animals such as Koko have helped demonstrate that other species are capable of communication and even language, breaking a barrier that was once thought to be the main difference between humans and other animals. (Bettmann)

controversy as to whether or not they truly speak a language comparable to the full extent of human language. Those studying language capabilities in other species argue that defining language using only the standard of human language may be doing a disservice to other species and blinding us to deeper understandings of them, and that perhaps we need to recognize human language as, like other evolutionary processes and forms of communication, part of a continuum across species and not an either/or proposition.

The idea of animal communication for many people also includes certain humans who feel they have a special ability to communicate with other species. In popular culture these people are often called "whisperers" (The Edge 2010). Interestingly, the term itself is much older and dates back to the early 1900s and Irishman Daniel Sullivan, who rehabilitated abused and/or vicious horses. Cesar Millan is the official "dog whisperer" of the modern era, with his own television show and training product line. He encourages people to gain a better understanding of both how to communicate with their dogs and how their dogs are communicating with them. Finally, Sonya Fitzpatrick calls herself an animal psychic and hosts her own television show *The Pet Psychic*. She claims she actually experiences what other animals are experiencing, including pain, and helps humans with various issues with their dogs.

Julie Urbanik

See also: Animal Cultures; Emotions, Animal; Ethology; Evolution; Intelligence; Sentience

Further Reading

Bradbury, J., and Vehrencamp, S. 2011. *Principles of Animal Communication,* 2nd ed. Sunderland, MA: Sinauer Associates, Inc.

Darwin, C. 2009. *The Expression of the Emotions in Man and Animals.* London and New York: Penguin Classics.

The Edge. 2010. "A Brief History of Animal Communication." Accessed January 8, 2016. http://www.edgemagazine.net/2010/01/a-brief-history/

Kim, M. 2014. "Chirps, Whistles, Clicks: Do Any Animals Have a True 'Language'." Accessed January 15, 2016. https://www.washingtonpost.com/news/speaking-of-science/wp/2014/08/22/chirps-whistles-clicks-do-any-animals-have-a-true-language/

Patterson, F., and Linden, E. 1981. *The Education of Koko.* Austin, TX: Holt, Rinehart and Winston.

Pepperberg, I. 2008. *Alex and Me.* New York: HarperCollins Publisher.

Roitblat, H., Herman, L., and Nachtigall, P., eds. 1993. *Language and Communication: Comparative Perspectives.* Hillsdale, NJ: Lawrence Erlbaum Associates, Publishers.

Wilson, E. O. 1975. *Sociobiology: The New Synthesis.* Cambridge, MA: Harvard University Press.

Companion Animals. *See* Pets

Compassion. *See* Empathy; Ethics

Compassion Fatigue. *See* Shelters and Sanctuaries

Concentrated Animal Feeding Operation (CAFO)

CAFO is the U.S. Environmental Protection Agency's (EPA) acronym for "concentrated animal feeding operation," although it is also often used popularly to stand for "confined animal feeding operation." In its broader, non-EPA usage, it is associated with *any* large-scale, industrial animal agriculture, not only within the United States. Although CAFO is a specific EPA classification, it is often used interchangeably with "factory farm." CAFOs have developed a negative reputation in recent decades because of increased public knowledge of the animals' living conditions, considered to contribute to poor animal welfare, as well as the facilities' environmental impacts.

The EPA uses the acronym AFO (animal feeding operation) to indicate an operation that "congregate[s] animals, feed, manure and urine, dead animals, and production operations on a small land area" (EPA 2015). In these operations, animals are solely fed processed food, rather than obtaining their own food through grazing or foraging over a larger area. AFOs that exceed certain number thresholds (variable depending on the species) are automatically classified as CAFOs. For example, an AFO with 700 or more mature dairy cows is a large CAFO. AFOs with lower numbers of animals will be classified as CAFOs if the facilities meet certain regulatory criteria related to their contribution of pollutants, especially with respect to water, into the surrounding environment. The EPA states that out of all AFOs, approximately 15 percent are CAFOs.

The AFO form of animal agriculture came into being in the mid-20th century in North America and Western Europe. Prior to this time, animals were commonly raised for food on farms that had far fewer animals but utilized more space per animal. Although frequently fed by the farmers, the animals also obtained food by grazing on extensive pastureland, foraging in nearby fields and forests for roots and above-ground vegetation, and scratching around barnyards for insects. In terms of animals' bodily conversion of food into flesh or milk, this form of feeding is inefficient in two primary ways. First, the animal must expend energy to locate food, thereby using some of the calories gained from it. Second, sufficient land is required for these food resources. For example, a herd of cattle needs enough pasture to provide fresh grass while that eaten in one area is regrowing.

On AFOs, in contrast, the animals have all their food provided to them, thereby reducing both the energy spent in finding food and also the amount of necessary land. Mid-20th-century developments in animal nutrition science allowed for the production of feed that more efficiently met the animals' needs (and additionally promoted more weight gain) than they could do on their own through grazing or foraging. As an example, in the case of cows, this feed is largely corn- and/or soy-based and is very different from the natural roughage they would consume while grazing.

Because of the animals' close confinement, AFOs can allow for easy spread of disease, and therefore antibiotics are routinely used as a preventive. Such antibiotics were originally developed during World War II to help stem the spread of disease among soldiers in close quarters. This frequent antibiotic use in AFOs is widely considered to be the primary cause of growing bacterial resistance to antibiotics.

The AFO form can easily become a polluting CAFO because harmful byproducts of production are concentrated in a small area. Pollution threats include elevated levels of nitrogen and phosphorus in water, causing fish die-offs; runoff/seepage of animal waste into water resources, potentially contaminating drinking water with both bacteria and residual antibiotics; and ammonia and hydrogen sulfide gases (from animal waste) in the air, causing/triggering respiratory problems (e.g., asthma) in humans. In addition to health hazards, many residents who live near CAFOs/ AFOs complain of significantly bad odors. One notable example of the potential negative impact of CAFOs is the aftermath of 1999's Hurricane Floyd in the U.S. state of North Carolina. This storm caused significant flooding in the eastern part of the state, the location of numerous CAFOs. Hog waste storage ponds (known as "lagoons") overflowed into surrounding water bodies and an estimated 30,000 hogs, 700,000 turkeys, and over 2 million chickens drowned, with many of their carcasses ending up as water pollutants (National Weather Service 2009).

Routine animal welfare concerns resulting from CAFOs/AFOs include the overcrowding of animals into confined areas. In the case of pigs, chickens, and turkeys, this confinement is usually indoors, where the animals have no access to fresh air and receive little natural light. They also frequently spend long periods of time in their own waste. Dairy cows may be housed indoors or out; beef cattle are typically housed outside. Although cattle may have access to light and air, they also frequently have no alternative but to stand or lie in waste. Severe confinement also limits normal movement and behaviors.

Environmental protection/justice and animal advocacy groups are vocal critics of CAFOs/AFOs. For example, the Sierra Club in the U.S. state of Michigan has several pages of advice for citizens who want to challenge existing/proposed CAFOs in their communities. Organizations such as the North Carolina Environmental Justice Network highlight that CAFOs/AFOs, because of their locations and workforce, disproportionately affect the poor and people of color. A number of animal advocacy organizations, such as the Humane Society of the United States and Farm Sanctuary, include discussions of CAFOs in their efforts to educate the public about industrial agriculture's abuse of animals.

Connie L. Johnston

See also: Advocacy; Cruelty; Ethics; Factory Farming; Humane Farming; Husbandry; Livestock; Meat Eating; Welfare

Further Reading

EPA (Environmental Protection Agency). 2015. "What Is a CAFO?" Accessed September 18, 2015. http://www.epa.gov/region07/water/cafo/

Farm Sanctuary. n.d. "Factory Farming: Destroying the Environment." Accessed September 19, 2015. http://www.farmsanctuary.org/wp-content/uploads/2013/01/Environment-Brochure-FINAL-3-24-09.pdf

Humane Society of the United States. "Farm Animal Welfare." Accessed September 19, 2015. http://www.humanesociety.org/news/publications/whitepapers/farm_animal_welfare.html

National Weather Service. 2009. "Hurricane Floyd Impacts." Accessed September 19, 2015. http://www.erh.noaa.gov/mhx/Floyd/Impacts.php

Nierenberg, D. 2005. *Happier Meals: Rethinking the Global Meat Industry.* Danvers, MA: Worldwatch Institute.

North Carolina Environmental Justice Network. 2014. "NCEJN Spotlights CAFO Organizer." Accessed September 19, 2015. http://www.ncejn.org/

Sierra Club. 2015. "How to Stop Approval of a New CAFO." Accessed September 19, 2015. http://www.sierraclub.org/michigan/how-stop-approval-new-cafo

Consciousness. *See* Sentience; Primary Documents

Conservation. *See* Biodiversity; Tracking; Wildlife Management

Coral Reefs

Coral reefs are diverse saltwater ecosystems built by colonies of marine invertebrates. Although at first glance reefs may appear to be rock, they are comprised of calcium carbonate ($CaCO_3$) structures secreted by corals. Found throughout the world's oceans, coral reef ecosystems are among the most biologically rich and economically valuable marine resources in the world. Coral reefs are under immense threat from anthropogenic (human-caused) activities, and their survival depends on changes to the ways in which human communities interact with reef ecosystems.

Corals belong to the exclusively aquatic phylum *Cnidaria* and are related to jellyfish and sea anemones. Corals are divided into two groups: hard and soft coral. Hard corals, or *Scleractinia,* produce a rigid skeleton made of calcium carbonate in crystal form called aragonite. Also known as stony coral, they are the world's primary reef-builders. Soft corals, or *Alcyonacea*, do not produce calcium carbonate or form reefs, although they may be present in reef ecosystems.

Most corals obtain the majority of their nutrients and energy from *Symbiodinium* (colloquially called *zooxanthellae*), photosynthetic algae that live symbiotically (in an interdependent relationship) within coral tissue cells. These corals require sunlight and grow at shallow depths in clear, warm water ranging from 70–85°F (Hoegh-Guldberg 1999). For this reason, most of the world's reefs are located between the Tropics of Cancer and Capricorn in the Pacific Ocean, Indian Ocean, Caribbean Sea, Red Sea, and Persian Gulf. Corals without *Symbiodinium* can live in darker, colder waters, including the genus *Lophelia,* a reef-building coral found at depths between 260 feet and over 9,800 feet, within an average temperature range of 39–54°F (NOAA 2015).

Coral reefs are among the most biodiverse ecosystems in the world and have been characterized as the "rainforests of the sea." Although coral reefs cover less than 1 percent of the world's ocean floor, they are home to at least 25 percent of all marine life, including 4,000 species of fish, worms, molluscs (e.g., clams, oysters), crustaceans (e.g., crabs, shrimp), echinoderms (e.g., starfish, sea urchins), sponges, and cnidarians (NOAA 2015).

Throughout the world, people depend on reef systems for food and livelihood. Coral reefs support abundant fisheries and are a significant food source for over a billion people worldwide (NOAA 2015). The annual global economic value of coral reefs is estimated at $9.9 trillion, generated by fisheries, aquaculture, aquarium trade, recreational fishing, scuba diving, snorkeling, and other tourism activities (Costanza et al. 2014).

Many human activities have direct negative impacts on coral ecosystems. Overfishing (the taking of fish beyond sustainable levels) is a problem common to many reef systems. This can disrupt balance in reef ecosystems, facilitating the excessive population growth of corallivores (coral predators) including the crown-of-thorns starfish (*Acanthaster planci*). Destructive fishing practices threaten the survival of coral reefs throughout the world. Blast fishing (fishing with explosives like dynamite or hand grenades) indiscriminately kills fish, reef animals, and corals within the blast area. Cyanide fishing (using poison to stun and capture fish for the aquarium trade) harms the target species as well as surrounding coral and many other reef dwelling animals. Fishing gear—including fishing line, gill nets, traps, trawls, and anchors—abrade, fracture, and entangle corals through direct physical contact with the reef structure. Reefs are also intentionally damaged through coral mining and destroyed for the creation of shipping channels and canals.

Reef systems are also damaged by pollutants, including runoff nutrients and pesticides from agriculture, wastewater, industrial effluent, and trash, particularly plastics. The introduction of pollutants like fertilizers lead to eutrophication (enrichment with chemical nutrients), upsetting the balance of the reef by enhancing algal growth and crowding out and smothering corals. Air pollution has also been directly linked to stunted growth and interferes with reproduction in corals.

Global warming has been linked to many harmful effects on coral reefs. Rising sea levels can "drown" coral, as they cannot stay close enough to the surface for *Symbiodinium* to photosynthesize. Coral reefs are also very sensitive to water temperature changes of even one to two degrees (Hoegh-Guldberg 1999). Increasing sea temperatures have triggered major coral-bleaching events, as the corals expel the algae within their tissues, revealing the white of their skeletons. Warming seawater also changes migration patterns in fish populations, potentially exposing reefs to dangers of invasive species who could disrupt entire ecosystems. Pollution, warming sea temperatures, and coral bleaching all weaken corals, also leaving them vulnerable to an array of different coral diseases.

Another impact of global climate change, ocean acidification, results from increases in atmospheric carbon dioxide, which dissolves with water to form carbonic acid, acidifying the ocean. With changes to the pH of ocean waters, corals experience reduced calcification and cannot absorb the calcium carbonate they need to maintain their skeletons, leaving them weak and vulnerable to damage and dissolution.

Facing such stresses from direct and indirect human activity, it is estimated that at least 19 percent of the world's coral reefs are dead and about 60 percent are under threat (Burke et al. 2011). Conservation measures, such as education programs, improved fishery management, regulation of recreational activities on reefs, pollution reduction measures, habitat protection, and the establishment of Marine Protected Areas (MPAs), biosphere reserves, national monuments, and marine parks offer some hope of protecting reef systems from further damage, although the ongoing impacts associated with global warming remain a significant threat to these delicate ecosystems.

Sharon Wilcox

See also: Biodiversity; Climate Change; Fisheries; Fishing; Wildlife

Further Reading

Burke, L., Reytar, K., Spalding, M., and Perry, A. 2011. *Reefs at Risk Revisited.* Washington, DC: World Resources Institute.

Costanza, R., de Groot, R., Sutton, P., van der Ploeg, S., Anderson, S. J., Kubiszewski, I., Farber, S., and Turner, R. K. 2014. "Changes in the Global Value of Ecosystem Services." *Global Environmental Change* 26: 152–158.

Dubinsky, Z., and Stambler, N., eds. 2010. *Coral Reefs: An Ecosystem in Transition.* New York: Springer.

Hoegh-Guldberg, O. 1999. "Climate Change, Coral Bleaching and the Future of the World's Coral Reefs." *Marine and Freshwater Research* 50(8): 839–866.

International Coral Reef Initiative. 2015. Accessed April 1, 2015. http://www.icriforum.org

Sheppard, C. 2014. *Coral Reefs: A Very Short Introduction.* Oxford, UK: Oxford University Press.

United States Coral Reef Task Force. 2014. Accessed April 1, 2015. http://www.coralreef.gov

United States National Oceanic and Atmospheric Administration (NOAA): Coral Reef Conservation Program. 2015. Accessed April 1, 2015. http://coralreef.noaa.gov

Wilkinson, C. 2008. *Status of the Coral Reefs of the World: 2008.* Townsville, Australia: Global Coral Reef Monitoring Network and Reef and Rainforest Research Centre.

Cruelty

Many animals' encounters with humans involve experiences of cruelty—through negligence related to care, sadistic pleasure in causing pain or suffering, or everyday practices that are widely culturally accepted (e.g., bullfighting) or about which there is little public knowledge (e.g., painful laboratory experiments). Emerging concern for animals frequently manifests itself as protection against cruelty. The concept of cruelty may be viewed differently depending on the circumstances, with differences arising both culturally and legally.

Cruelty is typically broadly defined as causing pain and/or suffering, either intentionally or unnecessarily. For example, many anticruelty laws define an act as cruel if it inflicts *unnecessary* pain and suffering. For example, a veterinarian causing a dog pain while giving a vaccination would not be cruel, as that act would be deemed necessary for the animal's health, but a person sticking a needle into a dog for fun would be considered cruel because that person's enjoyment would be viewed as unnecessary in relation to the dog's pain.

Historically in Western society, many appeals against animal cruelty have been made in a religious context, connecting mercy to animals with divine mercy to humans. One example is *A Dissertation on the Duty of Mercy and Sin of Cruelty to Brute Animals* (1776), by the British Reverend Humphrey Primatt (1736–1799). Concerns have also been voiced in the contexts of moral philosophy and the law, as with the British philosopher and attorney Jeremy Bentham (1748–1832), who stated that animals' capacity to suffer made them worthy of both moral consideration and legal protection. Some individuals have been more concerned with cruelty to animals affecting humans' behavior to each other.

Many legislators have been moved by religious and moral beliefs and helped enact anticruelty laws in a number of governments at all levels. Early laws were primarily concerned with livestock. For example, the United Kingdom's first animal-related law was the 1822 "An Act to Prevent the Cruel and Improper Treatment of Cattle," which imposed fines for unnecessarily beating or abusing horses, sheep, or cattle. In the United States, the first substantive legislation, passed by the state of New York in 1829, made it a misdemeanor to kill, maim, beat, or torture one's own or another's horses, cattle, or sheep.

Whether an act is legally considered cruel frequently depends on the type of animal, the location of the act, and the context. In the United States, the pain and suffering of animals in laboratories is regulated by the Animal Welfare Act, but a major point of contention for animal advocates is the subjective definition of "unnecessary," as the necessity of practices that may cause pain and suffering is left largely to the discretion of those who design experiments that use animals. Also currently in the United States, all anticruelty laws related to companion animals have been passed at the state or local level and many, if not most, phrase the law in terms of *intentionally* causing pain and suffering. For farmed animals, all treatment other than that related to transport or slaughter is covered by state or local law and, although many argue that industrial farming practices (e.g., hot-iron branding of cattle) today can be considered cruel, many states exempt those practices because they are considered to be usual and customary. In contrast to the inconsistent state- and local-level laws in the United States, European Union regulations typically provide more (and consistent) protection to animals from cruel treatment.

In terms of non-Western countries, India passed the Prevention of Cruelty to Animals Act in 1960. Among other things, the law prohibits beating, kicking, cruelly killing, overworking, providing insufficient food and water, or allowing a disabled animal to die in the street. In 2009, the National People's Congress in China received proposed anticruelty legislation drafted by a group of experts, but as of early 2016, it had yet to be passed.

Early anticruelty organizations such as the United Kingdom's Royal Society for the Prevention of Cruelty to Animals (RSPCA, established 1824) and the American Society for the Prevention of Cruelty to Animals (ASPCA, established 1866) are still active today. The early organizations were, like the legislation, focused on preventing cruelty to livestock and, additionally, carriage horses. It was not until decades later, when the keeping of companion animals became more commonplace, that they began to also include these animals. Today, animal companions are a major focus of these two organizations, as is *promoting* animal well-being and not just *preventing* cruelty.

For many in Western society, the farming and eating of dogs in parts of Asia—practiced by some but by no means all people—is seen as intolerably cruel. Also in the West, the treatment of many farmed animals (e.g., severe confinement of pregnant pigs) would be judged cruel if the practice were used on dogs. Cows, while eaten and treated with little regard throughout much of the world, are revered and protected throughout India. These views about different species and cruelty rest on deep cultural ideas about animals' utility and relationship to humans. If animals are considered sacred, or viewed as being companions within a culture, they will be more likely to be protected from unnecessary pain and suffering.

Finally, in using the terms "cruel" and "cruelty" to indicate wrongdoing, perpetrators are almost universally assumed to be human (at least from a present-day, Euro-American cultural perspective). Whereas we can feel anguish for a mouse caught and played with by a cat, and may even call the actions cruel, we rarely if ever ascribe malicious intention to the cat.

Connie L. Johnston

See also: Advocacy; Animal Law; Body Modification; Bullfighting; Empathy; Ethics; Factory Farming; Fighting for Human Entertainment; Husbandry; Institutional Animal Care and Use Committees (IACUCs); Research and Experimentation; Rights; Sentience; Social Construction; Vivisection; Welfare

Further Reading

Bentham, J. 1781. *An Introduction to the Principles of Morals and Legislation.* Oxford: Clarendon Press.

Favre, D., and Tsang, V. 1993. "The Development of Anti-Cruelty Laws During the 1800s." *Detroit College of Law Review* 1: 2–35.

Martin, R. 1822. "An Act to Prevent the Cruel and Improper Treatment of Cattle." *Statues of the United Kingdom of Great Britain and Ireland.* Accessed November 15, 2012. www.animalrightshistory.org/animal-rightslaw/romantic-Legislation/1822-uk-act-ill-treatment-cattle.htm

Prevention of Cruelty to Animals Act. 1960. Accessed September 23, 2015. https://www.animallaw.info/statute/cruelty-prevention-cruelty-animals-act-1960

Primatt, H. 1776. *A Dissertation on the Duty of Mercy and Sin of Cruelty to Brute Animals.* London: R. Hett.

Tomaselli, P. M. 2003. "Overview of International Comparative Animal Cruelty Laws." *Animal Legal and Historical Center.* East Lansing, MI: Michigan State University College of Law. Accessed September 21, 2015. https://www.animallaw.info/article/overview-international-comparative-animal-cruelty-laws

Wang, Q. 2009. "Draft Law to Punish Animal Cruelty." *China Daily.* Accessed September 21, 2015. http://www.chinadaily.com.cn/china/2009-06/19/content_8300745.htm

D

Deforestation

Deforestation refers to forest loss resulting in marked changes to the tree landscape. The process of forest loss affects deciduous forests, mountain alpine forests, humid tropical forests, and dry xerophytic (desert) forests. Animals that live in forest environments, such as bears, jaguars, pine martens, and orangutans, to name a few, are threatened with habitat depletion that results from forest loss. As species numbers decline, so too does species diversity as more and more species become extinct as their forest homelands go out of existence. Deforestation impacts animals as well as the trees and forests in which they live.

The deforestation process may range from a relatively small number of trees selected for felling to the other extreme, where a significant portion of trees no longer exists and the forest cover is decreased significantly. Deforestation may be incremental and slow, or it may be catastrophic and quick. Forest loss is a process occurring over time, though it might appear as a discrete event when viewed as, for example, a photograph of a deforested area showing a line on one side of which there are trees, and on the other side none. Reforestation or afforestation is the reverse of deforestation, signaling forest gain through either natural regeneration of trees or concerted efforts on the part of humans to replant trees.

Among the factors in the process leading to a reduction of forest cover on Earth's surface, deforestation is the most controversial, especially when human agents are implicated as expanding or accelerating deforestation. Trees have long served as means to human ends because of their exchange value: the ability of humans to take a tree and exchange it for something else (such as food, medicine, shelter, fuel, and even symbols of the life process among our species *Homo sapiens*), thereby creating value in the human realm. Problems arise when the exchange benefits one side of the relationship but harms the other side.

There are numerous causes of deforestation: some human, some nonhuman, some both. Lightning strikes, for example, can burn large tracts of forest in a relatively short period of time. Cyclones and hurricanes, tornadoes, floods, insect infestations, and drought can also have drastic effects upon forest cover. Natural events such as the abrupt climate change that occurred approximately 40,000 years ago and marked the last ice age—a significant cooling of Earth and expansion southward of glaciers—have impacted Earth's forest cover and led to a reduction of humid tropical forests. With the present population of our species, *Homo sapiens*, nearing the double-digit billion mark, and with our reliance on carbon-based fuels sending emissions into Earth's atmosphere, we are entangled in deforestation, even when a catastrophic event seems to be nonhuman in nature, such as a forest fire burning hotter because of forestry practices.

Of the world's remaining intact forest canopy—the boreal (northern) forests of Canada, Alaska, and Russia; the tropical (equatorial) rain forests of the Amazon basin in South America and the Congo basin in central Africa; and the island tropical rain forests of Madagascar, Borneo, and New Guinea—most concern is over the fate of tropical rain forests. These forests are valued over boreal forests because they host a vast biological diversity, including an astonishing 50 percent of Earth's animal species (Nature Conservancy 2015). For example, as deforestation expands in rain forests, the welfare of numerous animal species is threatened: jaguars are on the decline in the Amazon, orangutans are threatened with extinction in southeast Asia, and some lemur species in Madagascar are on the brink of extinction.

Rain forests may also contain the key to cures of human diseases (for example, the Madagascar Periwinkle, *Catharanthus roseus* (L.), which is native to Madagascar, has provided alkaloids used to treat cancer). Moreover, humid tropical forests act as large heat sinks, taking in large amounts of carbon from the atmosphere and producing clean air in return. Cutting rain forests releases their stored energy into the atmosphere. Having industrialized countries pay carbon credits to, for example, the Amazonia rain forest, which contains over half of the Earth's remaining rain forests, for helping to clean Earth's dirty air, raises the value of rain forests. Carbon credits may in turn slow pressures on the Amazonia rain forest from the globalized soy and beef industries, which turn deforested land into a commodity. Cutting trees to raise cattle impacts negatively animal species that live in the trees. It harms the forest animals, warms Earth's climate, and reduces biological diversity.

Jeffrey C. Kaufmann

See also: Biodiversity; Climate Change; Endangered Species

Further Reading

Chazdon, R. L. 2014. *Second Growth: The Promise of Tropical Forest Regeneration in an Age of Deforestation*. Chicago: University of Chicago Press.

Hecht, S. B., Morrison, K. D., and Padoch, C., eds. 2013. *The Social Lives of Forests: Past, Present, and Future of Woodland Resurgence*. Chicago: University of Chicago Press.

Hecht, S. B., and Cockburn, A. 2010. *The Fate of the Forest: Developers, Destroyers, and Defenders of the Amazon*. Updated edition. Chicago: University of Chicago Press.

Nature Conservancy. 2015. "Rainforests: Facts about Rainforests." Accessed June 25, 2015. http://www.nature.org/ourinitiatives/urgentissues/rainforests/rainforests-facts.xml

WWF. 2015. "Overview." Accessed June 25, 2015. http://www.worldwildlife.org/habitats/forest-habitat

Designer Breeds

Designer breeds are mixes of two different domestic, purebred animals (those whose parents are both of the same breed) or two different wild species. The resulting name of the cross breed usually combines the names of the two wild species or the two domesticated purebreds. While breeding selectively has long been a major part of the process of human domestication of other species, the move to so-called designer breeds of today signifies to many a new level of the commodification of animals (seeing animals as products or objects that can be sold, rather than as individual subjects in their own right).

Designer breeds are not true biological classifications of different subspecies but are more accurately described as cultural classifications. For domestic purebreds, these cultural classifications originally related to utility (e.g., herding dogs) and now are based mostly upon physical conformation (e.g., color and size). Many people are of the opinion that these new breeds have been created simply to generate demand for a new "product." Examples of designer breeds with wild species include ligon or liger (lion/tiger), zonkey (zebra/donkey), cama (camel/llama), and beefalo (buffalo/cow). Examples of designer domestic breeds of dogs, which are the most popular and diverse group, include labradoodle (Labrador retriever/poodle), goldendoodle (golden retriever/poodle), chihchon (chihuahua/bichons frise), puggle (pug/beagle), cockapoo (cocker spaniel/poodle), border shepherd (border collie/German shepherd), and border collie terrier (border collie/Jack Russell terrier). Designer cat breeds include Bengal (wild Asian leopard cat/domesticated cat), oriental shorthair (Siamese/American and British shorthair/Abyssinian/Russian blue), and the ocicat (Siamese/Abyssinian/American shorthair). The "designer" designation properly applies only to first-generation hybrids (i.e., offspring of parents from the two original breeds or species), as subsequent generations usually do not conform to the desired characteristics of the designer breed.

Designer breeds differ little in terms of genetics from the breed of either parent. In this manner, designer breeds are as much an artificial label as any of the dog breeds that first came into being during the late-Victorian period (late 1800s) in England and became outward symbols of middle-class status. In effect, nonworking breeds were developed to show that the owner had the disposable income necessary to purchase and maintain something of sentimental value but with no inherent utility, as a working dog would possess. This exercise of human control over the conditions of animal reproduction, population, and classification are much the same now as in 1900.

In the 1980s, Wally Conron, a former manager at the Royal Guide Dog Association of Australia, bred a dog he called a labradoodle in an attempt to combine a Labrador-type dog's temperament with the minimal allergic reaction that people sometimes experience with the single-layer coat of poodles. Conron admitted the name was a gimmick designed to generate interest because the demand for mixes was minimal while demand for purebreds was high. Humans have been manipulating canine genetics for thousands of years, and therefore the labradoodle presented nothing truly new. However, Conron referred to his creation as a "Frankenstein," indicating that he viewed the mix as something new and unsettling. At about this same time, dog breeder Wallace Havens created the puggle in the United States, which is the most popular designer breed there.

Designer dogs are not recognized by dog registries such as the American Kennel Club (AKC) and therefore cannot be shown at official dog shows like the Westminster Dog Show in New York City. However, the American Canine Hybrid Club (ACHC) has arisen in order to collect fees for meeting many designer dog owners' desire to validate his or her purchase via registration. The original purpose of registration was to ensure the value of stud service, the breeding fee paid to the owner of a male dog for mating to another owner's female dog. The owner of the female dog would then sell the puppies at a profit based upon the lineage (ancestry) and conformity to the breed's standards (size, color, temperament). However, because designer dogs are not reproducible across generations (except for labradoodles) to an identifiable standard, the point of a designer breed registry is divorced from the original intent of a purebred registry.

Many humans erroneously assume that designer dogs are necessarily healthier and freer of congenital (inherited) defects found in purebred dogs because there has been less linebreeding (the mating of offspring to their parents in order to produce successive generations of offspring who have the desirable or marketable breed traits of their parents, grandparents, etc.), such as occurs with many show dogs. However, although breed-specific genetic diseases are fewer in mixed breeds, mutations that occurred in ancestral dogs prior to the establishment of the breed are also found in designer dogs. For example, the genomic sequence responsible for diseases such as retinal atrophy (a group of genetic diseases in canines involving degeneration of the retina and the ability to see clearly and in color) is found in both established breeds and subsequent mixes who have retained this genetic sequence during breed standardization.

John T. Maher

See also: Cats; Dogs; Domestication; Exotic Pets; Pets; Welfare

Further Reading

Associated Press. 2014. "Breeding Blunder: Labradoodle Creator Laments Designer Dog Craze." February 7. Accessed June 12, 2014. http://www.today.com/pets/breeding-blunder-labradoodle-creator-laments-designer-dog-craze-2D12072744

Haraway, D. 2003. *The Companion Species Manifesto: Dogs, People, and Significant Otherness*. Chicago: Prickly Paradigm Press.

Haraway, D. 2007. *When Species Meet (Posthumanities)*. Minneapolis: University of Minnesota Press.

Mooallem, J. 2007. "The Modern Kennel Conundrum." February 4. Accessed June 12, 2014. http://www.nytimes.com/2007/02/04/magazine/04dogs.t.html?pagewanted=all

Ritvo, H. 1987. *The Animal Estate: The English and Other Creatures in the Victorian Age*. Cambridge, MA: Harvard University Press.

Thurston, M. E. 1996. *The Lost History of the Canine Race: Our 15,000-Year Love Affair with Dogs*. Kansas City, MO: Andrews and McMeel.

Wentworth, L. 1911. *Toy Dogs and Their Ancestors: Including the History and Management of Toy Spaniels, Pekingese, Japanese and Pomeranians*. London: Forgotten Books.

Dissection

Dissection is the close examination of the parts and systems that make up a deceased organism. In the case of animal dissection, it is the close examination of the internal and external features of an animal's structures and systems. Ever since the time of the Italian philosopher Alcmaeon of Croton (fifth century BCE), often considered to be the father of anatomy, dissection has been used in the study of many aspects of animal biology including, but not limited to, physiology (the study of the function of animals and their parts), morphology (the study of the shape and form of animals and their parts), taxonomy (the classification of animals), and anatomy. In the 2,500 years of dissection, there have been improvements in the instruments used for dissection, and the way in which animals are obtained has changed; however, the basics remain very much the same. More recently, the science of dissection has been faced with a number of issues and technological advancements that leave the future of dissection as a teaching tool uncertain in some areas of science.

The tools used for any dissection have not changed dramatically since the time of Alcmaeon, and any textbook on the subject will provide a list that includes scissors, forceps, scalpels, probes, needles, and safety equipment. Variations in size, shape, and uses of these tools exist, but the essential components of a dissection kit vary little. For example, scissor types can vary from microscissors used for cutting small blood vessels as well as dissection of tiny invertebrates, to large bone cutters used to access the chest cavities of mammals and birds. Dissection of small animals in dishes containing water allows for the internal organs to be teased apart without causing undue damage because the tissue floats, resulting in a more holistic view of the internal structure of the animal. As a tool, dissection is still providing significant contributions to scientific fields. For example, advancements in neuroscience (the study of the brain) based on dissection of animal models have provided important breakthroughs in restoring hearing, sight, and damaged nerves in humans. Advances in diabetes research have also benefited from the dissection and anatomical analysis of kidneys in pigs and rabbits.

A recent change to dissection as a learning tool is the limitation of species that can be used. For a species to be used as an example in teaching biology, it needs to

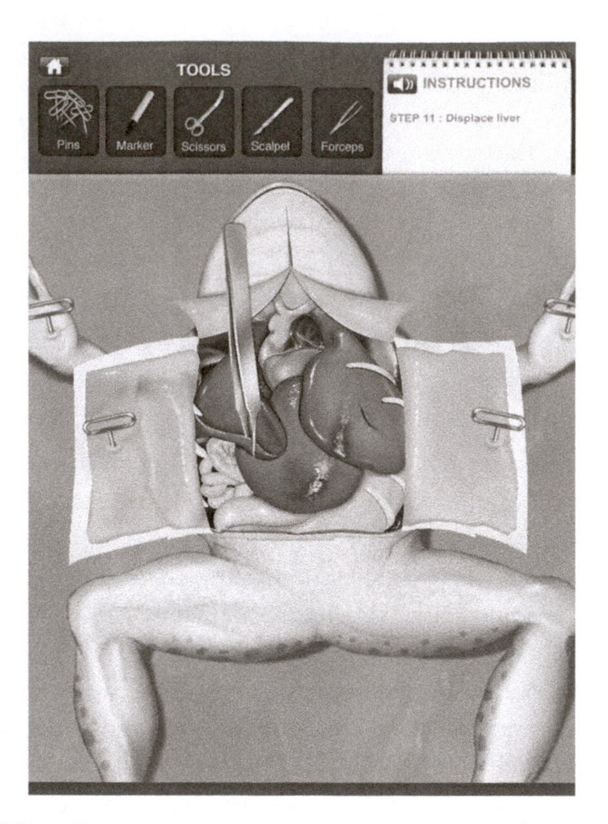

A screen view of Frog Dissection, an interactive computer simulation program designed for use by K–12 students on computers or tablets. Programs that provide a "hands-on," virtual dissection experience are promoted by animal welfare groups as humane alternatives to traditional classroom exploration. (Courtesy of mLab Emantras/Punflay)

be representative of its class (species group), abundant enough to use in large quantities, and easily available. Approval based on ethical considerations is required for animal use in almost all countries; however, the application of these approvals varies from country to country. Generally, the dissection of any vertebrate or invertebrate animal that has a high level of brain function and is able to perceive pain (e.g., the octopus) requires the oversight of an ethics committee; however, "lesser" animals such as worms do not. Some species are used routinely and widely for dissection in teaching. Rats and fetal (unborn) pigs are often used as models of mammalian dissection because they are easily accessible from farms and abattoirs (slaughterhouses) and provide good models for comparisons to human anatomy. Chickens from local farms provide an easy and reliable source for bird dissections. Of the lower-order vertebrates (amphibians, fish, sharks, and reptiles), fish and frogs are preferred, as other groups can be difficult to source in large quantities.

As a primary tool for demonstrating the internal anatomy of animals, dissection has been a requirement in the teaching of biology. Recently, the right of students to opt out of dissections on moral or religious grounds has created a change in attitude toward the necessity of dissections. Examinations of U.S. high school students' opinions of animal use in dissections showed that over 70 percent of students thought it wrong to breed animals for dissections, and almost 40 percent said they would object to any animal tissue being used. At the undergraduate university level, dissection has been banned in India for ethical and religious reasons. As a result of these changing views, the United Kingdom has not required whole animal dissection in A-level biology (high school) since 1990. In California and Florida (United States), legislation has been passed protecting the rights of students who wish to opt out of animal dissections. It is common in Western cultures to now provide alternatives to dissection at the high school level. In most higher education institutions around the world, however, dissection is still part of the curriculum.

The push to find other ways to study the internal anatomy of animals without dissection has resulted in a number of alternative demonstration techniques. These include demonstration videos, 3-D models, and digital dissection software, all of which provide a different method of "dissecting" an animal without using an actual creature. Comparative studies into the application of these techniques have indicated that they can be as effective as live animal dissection in teaching students the internal anatomy of animals. While these replacements result in an equal declarative knowledge, combining them with actual dissections can provide the tactile experience that works to reinforce the learning. Dissection still provides the practical experience that is required for some levels of biological study; however, it is becoming less of a requirement for overall animal biology.

Peter Derbyshire

See also: Ethics; Research and Experimentation; Vivisection

Further Reading

De Iuliis, G., and Pulerà, D. 2006. *The Dissection of Vertebrates: A Laboratory Manual.* Cambridge, MA: Academic Press.

De Villiers, R., and Monk, M. 2005. "The First Cut Is the Deepest: Reflections on the State of Animal Dissection in Biology Education." *Journal of Curriculum Studies* 37(5): 583–600.

Dogs

The relationship between humans and dogs (*Canis familiaris*) arguably offers the ultimate example of coexistence. Wolves have been identified as the earliest animal to establish a relationship with humans, beginning a process of domestication that has led to the contemporary dog. Based on archaeological evidence, the human-dog relation has existed for between 12,000 and 17,000 years. However,

DNA evidence suggests that the domestication of wolves may have begun 100,000 years ago.

Since the beginning of the human-dog/wolf relationship, we have seen the emergence of a wide array of dog breeds (the Kennel Club in the United Kingdom currently recognizes 215 separate breeds), all of them created by humans based on human needs and desires. As these needs and desires have altered over time, the scale of the population of specific dog breeds and the number of breeds has fluctuated. At the same time, the nature of specific breeds has altered, as their dominant role in human society has changed. This is exemplified by the border collie, which emerged as a working sheep dog on farms in the United Kingdom in the 1800s. These dogs were bred by humans for their intelligence, obedience, and athleticism. Today, this breed is often to be found in dog shows, bred purely for their physical appearance rather than their ability to herd sheep.

The relationship between humans and dogs, and the roles the latter play in human society, is a complex and constantly evolving one that is specific to a given place and culture. Such roles include acting as a source of food—a prevalent practice in many parts of Asia, where dog meat is a cuisine. There is also a history of eating dog meat in Western countries such as Greece and Ireland in the medieval era (5th to the 15th centuries), though this seems to have been associated with the avoidance of starvation. However, a French cookbook from 1870 includes a variety of dishes that include dog meat. There is also a strong history of dog meat being eaten in Latin America (where the Aztecs bred a hairless dog for human consumption), Africa, and Polynesia, with reports of its continuation into the contemporary era in parts of the latter two (e.g., the Congo Basin and the Philippines). In addition, dogs often act as fashion accessories and indicators of social status in the same way that ties and handbags are used by humans to construct a personal identity. While the contemporary image of such dogs may be associated with celebrities, the practice can be traced back to medieval Europe. Some dogs are also utilized in a variety of ways to assist humans. These include guide dogs, bomb and drug detection dogs, and guard dogs, among others. Dogs also play a prominent role in human entertainment through sports such as sled and greyhound racing, both of which are popular sports throughout the world in countries as diverse as the United Kingdom, the United States, Australia, and India. Finally, dogs play a role in the physical, emotional, and social well-being of humans through their position as family pets. The royalty of medieval Europe utilized dogs for such purposes yet only recently has scientific research identified how dogs can aid human well-being.

Keeping dogs as pets as we do today has its origins in Western society, with dogs being one of the most popular pets alongside the cat in countries such as Canada, the United Kingdom, Australia, and New Zealand. However, the practice of keeping dogs as pets is rapidly becoming a global phenomenon, with an increasing number of households outside of the West, in countries as diverse as China, Iran, India, and Brazil, for example, having at least one pet dog. Within many of these households, the dog is viewed as possessing at least human-like sentience (i.e., consciousness, which relates to an ability to feel and perceive subjectively and to empathize). Many humans are happy not just to share their house with a dog but also their bed, food, and leisure time.

There has been a recent trend to identify pet dogs as *companions* of humans as an attempt to break away from the perceived, traditionally subservient, position of dogs, characterized by the use of the term "pet." The distinction is that a pet is seen to be owned by a human whereas a companion is exactly that: a friend, an equal (or near-equal), and therefore not something that can be owned. Yet just because a dog is a pet or the owner of a dog prefers to refer to the animal as a companion does not automatically mean that the dog becomes an equal to the human, with rights enshrined in law and no longer owned as an object. Rather, the dog remains, as it has done for millennia, in a subordinate position, as an animal in a human society where humans set the rules and expectations.

The relationship between humans and dogs is not governed by how we label dogs, though it may be reflected by it. Rather, it is governed by how we perceive dogs and particularly their sentience. If we see dogs as not being sentient, then we can suggest they have no rights or welfare needs. However, we live in an era when more and more people are arguing that dogs have sentience, are capable of interacting with us as active agents, experience emotions in ways that—while not necessarily the same as humans—are not totally dissimilar, and therefore have welfare concerns and should have rights.

Neil Carr

See also: Body Modification; Designer Breeds; Domestication; Fighting for Human Entertainment; Fox Hunting; Greyhound Racing; Human-Animal Bond; Pets; Pit Bulls; Puppy Mills; Service Animals; Working Animals

Further Reading

Bradshaw, J. 2011. *In Defence of Dogs: Why Dogs Need Our Understanding*. London: Allen Lane.

Carr, N. 2014. *Dogs in the Leisure Experience.* Wallingford, UK: CABI.

Sanders, C. 1999. *Understanding Dogs: Living and Working with Canine Companions.* Philadelphia: Temple University Press.

Dolphins

Dolphins and porpoises, along with whales, belong to the biological order of cetaceans. There are presently 40 known species of dolphins. (Due to the closeness in physical appearance between porpoises and dolphins, they are often wrongly thought to be the same species.) Dolphins have had long, enduring relationships with human societies, due in part to their intelligence and sociability. Research has shown that dolphins possess the abilities of self-recognition, both visual and vocal learning, and long-term memory that can span decades. Dolphins are also known to have highly evolved culture within their social groups (or pods). Dolphins have developed a sophisticated sonar sensing ability that enables them to send out sounds that will bounce off other objects, allowing them to distinguish objects in murky waters as far as 50 feet away. These attributes arguably demonstrate that dolphins possess sophisticated behavior and thinking abilities. In popular culture, dolphins (and marine mammals in general) have always been portrayed as highly intelligent and friendly beings.

Besides the capturing of dolphins for entertainment and for consumption, some species of dolphins are increasingly threatened due to the loss of their habitat or the general degradation of their environment. For example, in the Yangtze River in China, the longest river in Asia, the *Baiji* dolphin was declared extinct in 2006. Other river dolphins faced with extinction include the Ganges River dolphin and Indus River dolphin, both of which are found in India. At present, Maui's dolphin, found in the coastal waters of New Zealand, is the rarest dolphin in the world; estimated to number less than 60, it faces imminent extinction.

Films such as *Free Willy* (1993) and *Whale Rider* (2002), as well as the television series *Flipper* (first telecast in the United States in 1964 and remade in 1995 for four seasons) highlight the deep bonds that cetaceans can form with humans. Due largely to such positive portrayals of dolphins, the use and abuse of these animals have attracted much controversy. Since the 1960s, the U.S. Navy has maintained a Marine Mammal Program to train bottlenose dolphins and sea lions for specific military tasks like underwater mine detection and clearance, as well as ship and harbor protection. From its inception there have been perennial criticisms centered on animal welfare concerns over the use of marine mammals for military purposes. Russia also maintains a similar military program.

Cetaceans, particularly bottlenose dolphins and orcas, are popular in marine mammal parks, where they are held captive to perform for paying audiences. Sometimes referred to as killer whales, orcas are the largest member of the dolphin family. The plight of captive cetaceans was in the public consciousness with the release of the 2013 documentary *Blackfish*. The documentary made a series of claims over the dismal welfare of orcas kept in captivity by the U.S. marine park SeaWorld. While SeaWorld has steadfastly denied claims of mistreatment of orcas, since the documentary's release it has seen continual declines in visitor numbers, profits, and stock prices. Nonetheless, as of 2015, there are still 58 captive orcas in 14 marine mammal parks, spread across eight countries.

Opposition to captive cetaceans also occurs outside North America. Since 2013, there has been persistent campaigning against the captivity of wild-caught bottlenose dolphins in a marine park in the city-state of Singapore. Taking a slightly different approach, the campaign centered on the fact that it is unnatural for wild dolphins to be kept in captivity, regardless of whether their physical welfare is compromised or not.

Given the allure of dolphins and the controversies surrounding captivity, dolphin watching has been suggested as an ethical way to sustain human fascination with dolphins. Although ecotourism seeks to connect visitors to nature in ways that are not harmful to the environment, humans, or animals, research suggests that cetacean behavior does change as a result of being part of ecotourism activities. Such changes include becoming used to food handouts from humans, as well as having less time for socializing and feeding. Others have argued, however, that dolphin–human interactions are natural and point to numerous historical accounts throughout the millennia. However, it can be argued that the scale and frequency of deliberate marine mammal encounters that one sees in ecotourism is significantly different, in terms of its impact and artificiality, from the more random interactions of the past.

Perhaps the most controversial direct "use" of dolphins is the human consumption of dolphins for meat. While the consumption of cetaceans for meat occurs globally, the issue gained prominence with the release of *The Cove* in 2009. This award-winning documentary highlighted the practice of dolphin hunting in Japan. In particular, it is a biting critique of the practice of "drive hunting," whereby dolphin pods are herded into a cove by manipulating their sonar systems. The latter is achieved by using metal rods to make loud clanking noises in the water to disorient the dolphins. The documentary also exposed the cruelty with which the dolphins are killed, via spears and knives and in full view of one another.

The proponents of such dolphin hunting and the consumption of the meat (among them, the Prime Minister of Japan) defended these acts on two main grounds. First, they argue that these activities are not illegal. Second, they see these acts as part of their traditional culture. The documentary and the issues it highlighted have become so highly politicized in Japan that efforts to screen it in Japan to a wider audience have been met with protests by Japanese conservatives.

Harvey Neo

See also: Ecotourism; Intelligence; Marine Mammal Parks; Sentience; Sharks

Further Reading

Chang, K. H. K. 1978. "Case of the Japanese Dolphin Kill." *Human Nature* 1: 84–87.

Kessler, M., Harcourt, R., and Heller, G. 2013. "Swimming with Whales in Tonga: Sustainable Use or Threatening Process?" *Marine Policy* 39: 314–316.

Lusseau, D., and Higham, J. E. S. 2004. "Managing the Impacts of Dolphin-Based Tourism through the Definition of Critical Habitats: The Case of Bottlenose Dolphins in Doubtful Sound, New Zealand." *Tourism Management* 25: 657–667.

Neo, H., and Ngiam, J. Z. 2014. "Contesting Captive Cetaceans: (Il)legal Spaces and the Nature of Dolphins in Urban Singapore." *Social and Cultural Geography* 15(3): 235–254.

Rauch, A. 2013. *Dolphins*. London: Reaktion Books.

Domestication

Derived from the Latin *domesticus* ("of the household"), "domestication" in the biological sense describes human control of other species through artificial selection (deliberate breeding to produce desired physical and behavioral traits). Animal domestication can also be understood sociologically as cultivating interspecies social relationships. As a significant human cultural development, animal domestication rivals tool-making, writing, and mathematics, having dramatically magnified human capacity for work, travel, survival, and warfare. Like plant domestication, it also entails selective human allocation of biological resources (e.g., grazing land) to domesticates at other species' expense. Beyond human culture, ants' "farming" of aphids introduces the possibility of nonhuman models of domestication.

The earliest domesticated animals were wolves, who likely began self-domesticating in various locations between 20,000 and 30,000 years ago. While initially wolves may have affiliated with humans for food and comfort, they eventually forged hunting and companion partnerships. In time, humans managed to selectively breed wolves into the many dog breeds of today.

Subsequent animal domestication events arose independently in diverse locales (Central America, the Andes, Southeast Asia, the Tigris and Euphrates Valley, the Arabian Peninsula, and Northeast Africa) between 10,000 and 12,000 years ago. In these locations, the presence of useful wild animals (e.g., pigs and cattle) near agricultural populations escalated the development of geographically rooted human cultures with increasingly complex social structures. From each site, domesticated animals spread geographically with human trade.

Behavioral qualities that pre-adapt animals for domestication include comfort-seeking dispositions, tolerance of humans, submission to captive feeding and breeding, and modifiable social structures that allow humans to control or adopt key individuals or assume their social roles. Species possessing these behavioral traits in varying degrees can be genetically modified by artificial selection for physical and behavioral traits. Traditional "barnyard domestication" uses natural breeding practices to deliberately select for one trait, but another may incidentally occur, because behavioral and physical traits can be genetically linked; for instance, livestock may be deliberately bred to produce more offspring, a physical trait that is genetically linked to docility (calmness), a behavioral trait that will incidentally occur because of this link. Domestication practices of modern agribusiness employ biotechnology to select for genetic markers statistically linked to physical and/or behavioral traits (e.g., chickens bred for large breast muscles and tolerance of close confinement).

Though often confused with "taming," domestication affects the genome of the species as well as the behavior of individuals. The ancient practice of elephant keeping that originated in present-day Pakistan illustrates both the difference and the close connection between taming and domesticating (Clutton-Brock 2012). Traditional keeping of Asian elephants for transportation, hauling, and other heavy labor now persists only residually in the profession of the mahout (elephant handler). Currently, 16,000 Asian elephants live in captivity in 11 Asian countries; captured and tamed but not necessarily bred or permanently maintained as captives, they are legally defined as domesticates, though they are genetically indistinguishable from wild elephants. Over time, however, the genetics of Asian elephant populations have been partly managed through human relocation of elephants accidentally or intentionally, permanently or temporarily released into the wild. As the number of elephants born, bred, and kept in captivity increases, and as "wild" elephant populations dwindle, the genetic effects of captivity on the elephant species will also grow.

Stephen Budiansky (1992) has argued that even domesticated predator-prey relationships, such as relationships between humans and livestock, are mutually beneficial in terms of long-term species survival, observing that no domesticated species has ever gone extinct (though domesticated breeds have disappeared). Donna Haraway (2008) maintains that domestication should also be framed as interpersonal relationships experienced on a human timescale. Viewed this way, domestication is a social process jointly negotiated by individual human and nonhuman animals. To illustrate, Haraway describes working with her dog in canine agility training. Beyond genetic transmission of physical and behavioral traits, she argues, human/canine co-evolution is an ongoing cultural accumulation, preservation, and transmission of countless such interspecies experiences.

Whatever its dynamics, domestication is both celebrated as a lynchpin of human progress and denounced as human interference with nature. At issue are whether and how to weigh human benefits against costs to domesticated animals or nature as a whole. Genetic manipulation and confinement conditions imposed on billions of domestic animals raise concern for the structural stability of animals physically modified to human desires—for example, turkeys whose skeletal health is compromised by their oversized breast muscles, or bulldogs whose shortened noses inhibit breathing. The moral charge that confinement conditions further prevent animals from engaging in natural behaviors often accompanies environmental concern when natural resources are allocated to domesticates, thereby limiting other species' access to those resources, reducing biodiversity and weakening the stability of the planet's ecosystems (e.g., cutting rainforest for pastureland).

Finally, although animal domestication is usually regarded as an exclusively human activity, some ant species keep "herds" of aphids, which they move among feeding sites and train to excrete a sweet, nutritious substance in response to physical stimulation. When other food is scarce, ants may also consume aphids from their herds, suggesting parallels with human agriculture (Wilson 2012). Direct comparisons between ants' "farming" of aphids and human domestication of other animals through selective breeding are impossible, however, as it does not appear that the ants could intentionally manipulate the aphids' reproduction. At the very least, this example indicates that domestication occurs in ecological contexts outside of human society and calls into question the simplistic, unidirectional model used to explain human domestication of other animals.

Mary Trachsel

See also: Biodiversity; Biotechnology; Cats; Deforestation; Dogs; Horses; Husbandry; Livestock

Further Reading

Budiansky, S. 1992. *The Covenant of the Wild: Why Animals Choose Domestication.* New York: William Morrow.

Clutton-Brock, J. 2012. *Animals as Domesticates: A World View through History.* East Lansing, MI: Michigan State University Press.

Darwin, C. 1988. *The Works of Charles Darwin. Vols. 19–20. Variation of Plants and Animals under Domestication.* London: Wm. Pickering.

Driscoll, C. 2014. "Animal Domestication." *Henry Stewart Talks Online Collection.* Accessed August 24, 2015. http://hstalks.com/main/view_talk.php?t=2851

Haraway, D. 2008. *When Species Meet.* Minneapolis: University of Minnesota Press.

Wilson, E. O. 2012. *The Social Conquest of Earth.* New York: Norton.

Domestic Violence and Animal Cruelty

"Domestic violence is defined as a pattern of abusive behavior in any relationship by one partner to gain or maintain power and control over another intimate partner through the use of physical, sexual, emotional, economic or psychological threats or actions" (Department of Justice 2014). Any family member can be a victim or perpetrator of domestic violence; however, women are more often the victims, when compared to men. One victim seldom discussed in incidents of domestic violence is the family companion animal. Companion animals have been redefined as a part of the continuum of domestic violence and are considered an indicator of other problems in dysfunctional and violent families. They are vulnerable to abuse as a way to manipulate and control victims of domestic violence, and victims will often remain in a violent relationship to avoid leaving their companion animal behind. Awareness about the link between domestic violence and animal cruelty is imperative for the safety of companion animals and victims of abuse.

Batterers abuse companion animals to maintain control over their victims, to keep the family isolated, to control the family with fear, and to punish the victim for leaving or showing independence. Some methods used to manipulate and control victims include "threats to physically abuse companion animals, denying access to veterinary care, [and] threatening to 'get rid' of the companion animals" (Hardesty 2013, 9). Common types of physical abuse include "pinching, hitting, choking, drowning, shooting, stabbing, and throwing the animal against a wall or down stairs" (Carlisle-Frank 2004, 39). In some instances batterers deny a companion animal food or water. The treatment of the family pet by the batterer is indicative of future behavior in the relationship. Batterers who abuse companion animals are more likely to exhibit violent and controlling behaviors than those who do not abuse pets.

Children who live in violent families are often exposed to animal cruelty. When children witness abuse of a companion animal, it increases the likelihood of future animal cruelty by copying the violence that they have encountered. Children may also harm the family companion animal to demonstrate their own dominance, or in an attempt to save the animal from what they believe to be imminent harm by the abuser. Early studies reveal that over one-third of domestic violence victims reported that a companion animal had been hurt or killed by their child (Ascione 1998).

There is a strong bond between companion animals and victims of domestic violence, resulting in some victims staying in the relationship for fear of an animal's safety. Women entering domestic violence shelters are 11 times more likely to report abuse of a companion animal compared to women who said they did not experience abuse (Ascione 2007). Over half of the victims of domestic violence in one study report their partners threatened abuse of the family pet (Volant et al. 2008). Threats or harm directed at the companion animals can result in strong, negative emotional responses on the part of the victim. When victims cannot remain connected with their companion animals, it is one more trauma they must endure. Likewise, it is very stressful for companion animals to be uprooted into a strange environment without their human guardian. Currently, few domestic violence shelters are equipped to accommodate companion animals due to the health and safety of other sheltered domestic violence victims, the cost of housing and veterinary care, space, and program resources.

The movement to shelter companion animals with domestic violence victims is primarily a grassroots effort led by community organizations. Two examples include Ahisma House and RedRover. Ahisma House, located in the U.S. state of Georgia, is one of the first community programs to house victims and companion animals together. Run primarily by volunteers, and financially supported by grants and donations, Ahisma House maintains a central place for victims of domestic violence and their companion animals to seek refuge and access community resources. RedRover is a program that offers grants to assist animals in crisis, including domestic violence situations. Two types of grants are available: Safe Escape and Safe Housing. Safe Escape grants are provided for emergency housing and veterinary visits, providing some financial relief for victims of domestic violence leaving their abusers. Safe Housing grants are made available to agencies that wish to alter their housing so that companion animals can stay with their human guardian.

There is substantial concern for the welfare of companion animals, and the U.S. federal and state governments have responded to public demands for nonhuman animal safety. In 2015, the Federal Bureau of Investigation (FBI) included a category for the commission of animal cruelty in their National Incident Based Reporting System (NIBRS), a data-collection instrument that reveals national crime trends in the United States. Similarly, all U.S. states have increased the penalty from a misdemeanor to a felony in cases of animal cruelty. To specifically address companion animal victimization in violent homes, states have enacted legislation placing companion animals on protection orders. Maine was the first state to allow judges to include companion animals on protection orders in 2006. At present, 29 states and the District of Columbia have passed legislation that includes pets on protection orders.

Dawna Komorosky

See also: Advocacy; Empathy; Human-Animal Bond; Pets

Further Reading

Ascione, F. R. 1998. "Battered Women's Reports of Their Partners' and Their Children's Cruelty to Animals." *Journal of Emotional Abuse* 1(1): 119–133.

Ascione, F. R., Weber, C. V., Thompson, T. M., Heath, J., Maruyama, M., and Kanna, H. 2007. "Battered Pets of Domestic Violence: Animal Abuse Reported by Women Experiencing Intimate Violence and by Nonabused Women." *Violence Against Women* 13(4): 354–373.

Carlisle-Frank, P., Joshua, F. M., and Nielson, L. 2004. "Selective Battering of the Family Pet." *Anthrozoos* 17: 26–42.

DeGue, S., and DiLillo, D. 2009. "Is Animal Cruelty a 'Red Flag' for Family Violence? Investigating Co-Occurring Violence Toward Children, Partners, and Pets." *Journal of Interpersonal Violence* 24: 1036–1056.

Hardesty, J. L., Khaw, L., Ridgway, M. D., Weber, C., and Miles, T. 2013. "Coercive Control and Abused Women's Decisions about Their Pets When Seeking Shelter." *Journal of Interpersonal Violence* 28(13): 2617–2639.

U.S. Department of Justice. 2014. "Domestic Violence." Accessed April 3, 2015. http://www.justice.gov/ovw/domestic-violence

Volant, A. M., Johnson, J. A., Gullone, E., and Coleman, G. J. 2008. "The Relationship between Domestic Violence and Animal Abuse: An Australian Study." *Journal of Interpersonal Violence* 23: 1277–1295.

E

Ear Cropping. *See* Body Modification

Earthworms

Earthworms (*Oligochaeta*) are a biological class of segmented worm (Annelid) related to marine worms (*Polychaeta*) and leeches (*Hirudinea*). Of the 6,000 species of earthworm, the appearance and behavior of the species known as the common earthworm (*Lumbricus terrestris* or *L. terrestris*) has ensured they are a familiar sight on European and American soils. Although valued for aerating and fertilizing soil, and their use as fishing bait, common earthworms have also been considered an invasive species that can have a negative impact on biodiversity.

Living in underground burrows, at depths reaching 1.8 meters (6 feet), earthworms consist of over 100 segments, each containing muscles and outer bristles called setae. This allows movement. With no eyes, ears, or lungs, they rely on light receptors and vibrations to sense surroundings, and they breathe through their skin. Five heart-like structures, close to the head end (prostomium), pump blood around the body. Without these the creature will not survive. Contrary to a popular myth, cutting a worm in half does not create two living worms, just a shorter worm which will regenerate a new rear end.

As largely invisible subterranean (underground) animals, their surfacing above ground is usually to feed and mate. Although they are hermaphrodites (individuals with both male and female reproductive organs), earthworms reproduce sexually. Mating, which can take up to three hours, involves inverted earthworms locking at the clitellum (the thickened glandular section of body), wrapping together in a thick tube of mucous, and the mutual exchange of sperm. Long after separation (up to 12 months later), eggs are injected and fertilized in an egg cocoon, where they incubate for up to 90 days before hatching. Remarkably, an acre (4,046.86 square meters/43,560 square feet) can hide up to six million earthworms (Chaline 2011). Made up of 70 percent protein, they are nutritious food for birds and other predators.

Indigenous (native) to Western Europe, *L. terrestris* now has a global distribution, found in most temperate to mild climates, including parts of Canada, India, South Africa, and Australia. For over 2,000 years, humans have transported them over this invasive range—most notably in relation to the movement of soil and plants in the horticulture trade, and importation and sale of earthworms as fishing bait. In

With no eyes, ears, or lungs, earthworms rely on light receptors and vibrations to sense surroundings and breathe through their skin. Although earthworms' activity can benefit soils for agriculture, they can also have a negative impact on forest biodiversity. (Jlmcloughlin/Dreamstime.com)

the United States, for example, earthworms (also known as night crawlers or vitalis) can be bought as bait in all of the lower 48 states.

Earthworms are, however, more than fishing bait. Often overlooked and undervalued, they deserve far more respect. The Greek philosopher Aristotle (384–322 BCE) described them as "intestines of the earth." Gilbert White (1720–1793), the English parson-naturalist, applauded their role as "great promoters of vegetation." In *The Natural History of Selborne* (1789, 181), White maintained that "[t]he earth without worms would soon become cold, hard-bound and void of fermentation, and consequently sterile." Charles Darwin (1809–1882) made no secret of his passion for earthworms, conducting over 40 years of research and devoting an entire book to them. In his 1881 publication, *The formation of vegetable mould, through the action of worms, with observations on their habits*, Darwin concluded: "It may be doubted whether there are many other animals which have played so important a part in the history of the world, as have these lowly organised creatures" (313).

According to nature writers Marren and Mabey (2010, 41), with this research "Darwin effectively changed their image from slithering nullities, even pests, to benevolent animals of great benefit to mankind." In 1942, Thomas J. Barrett (1884–1975), a pioneer of vermiculture (worm growing or worm farming) wrote *Harness-*

ing the Earthworm. The aim of his manual was to aid understanding of what he thought to be the most important animal in the world, ultimately showing their practical benefits to soil building, soil conditioning, and plant nutrition.

Earthworms are considered keystone detritivores (an organism that feeds on dead plant or animal matter) and "ecosystem engineers." Their presence changes soil in three ways. First, their movement creates burrows that open the soil, allow water and air to penetrate, and prevent compaction. Second, organic matter from the surface is taken underground as they feed, thus increasing humus (organic substance made up of decayed plant or animal matter) and soil fertility. Third, soil particles are digested and excreted as feces known as casts. These are rich in minerals such as nitrogen and phosphates.

Whereas these processes are highly beneficial to agriculture and horticulture, they can have adverse effects on forest biodiversity, and their presence can be detrimental to ecosystems previously without native earthworm populations. According to the Great Lakes Worm Watch, set up by the University of Minnesota, invasive earthworms can cause a host of negative changes that "affect small mammal, bird and amphibian populations, increase the impacts of herbivores like white-tailed deer, and facilitate invasions of other exotic species such as European slugs and exotic plants like buckthorn and garlic mustard."

Daniel Allen

See also: Biodiversity; Invasive Species; Keystone Species

Further Reading

Allen, D. 2013. *The Nature Magpie.* London: Icon Books.

Barrett, T. J. 1942. *Harnessing the Earthworm.* Boston: Bruce Humphries, Inc.

Chaline, E. 2011. *Fifty Animals That Changed the Course of History.* Hove, UK: David & Charles.

Darwin, C. 1881. *The formation of vegetable mould, through the action of worms, with observations on their habits.* London: John Murray.

Great Lakes Worm Watch. Accessed January 27, 2016. http://www.nrri.umn.edu/worms/

Invasive Species Compendium. 2014. "Datasheet: *Lumbricus terrestris.*" Accessed October 15, 2015. http://www.cabi.org/isc/datasheet/109385

Marren, P. and Mabey, R. 2010. *Bugs Britannica.* London: Chatto & Windus.

White, G. 1789. *The Natural History of Selborne.* London: Arrowsmith.

Eastern Religions, Animals in

Eastern religions, including the South Asian-origin religions of Hinduism and Buddhism, and the East Asian-origin religions and philosophies of Daoism, Confucianism, and Shintoism, often incorporate elements of nonhuman nature, including landscapes, plants, and animals, as objects of worship. This is also in keeping with polytheism (belief in many gods and many forms of the divine) in Eastern religions, especially in the case of Hinduism, Daoism, and Shintoism. Continuities between human and animal worlds in Eastern religions are reflected in the notion of the soul, which is believed to be in both humans and animals. While traditions of care toward animals in Eastern religions have been cited as important for transforming human-animal relations across the world, some attitudes toward animals are also matters of controversy. Bans on cattle slaughter in parts of South Asia can be attributed to Hinduism but also discriminate against Hindu and non-Hindu communities that depend on cattle meat and skins for their livelihoods. Reverence toward whales in Japan is linked to Buddhist and Shinto traditions, but also coexists with Japanese commercial whaling, which has been the target of international opposition.

In Eastern religions, the human and nonhuman worlds are viewed as linked to one another. In Hinduism, souls can migrate from animals to humans and vice-versa in the process of reincarnation or rebirth. Buddhism also connects humans and animals through its notion of the continuity of life through rebirth, though there is no belief in the existence of an eternal soul. The Confucian notion of ren (being humane) can include caring for animals, and in Shintoism, kami (spirits or souls) are associated with humans and nonhumans. Because Eastern religions have been influenced by one another—for instance, by the spread of Hindu and Buddhist beliefs across Asia—the status accorded to animals in a specific context often reflects a combination of various religious traditions.

Specific forms of relating to animals are an important aspect of expressing religious and cultural identities. Cows are sacred in Hinduism and represent fertility and sustenance, with milk and milk products being important in ritual practices. The Vedas—the earliest Hindu texts, which may have been composed in oral form as early as 1500 BCE—mention the value of cows and milk. The god Krishna, in his young avatar, is depicted as a cowherd inhabiting a rural idyll where milk is plentiful. Cattle have also been, and continue to be, important in agricultural activities across Asia. The coexistence of economic and cultural practices is displayed in the uses of cattle dung in India, important for fertilizing agricultural fields, in cooking as fuel, in surfacing floors of traditional homes, and creating artwork for cow-related festivals.

In Hinduism, animals are depicted as vahana (or carriers) of deities. Vahana not only serve as animals on which deities travel, but often signify the deity itself and augment the power attributed to deities. An example would be Nandi the bull, the vahana of Shiva, god of destruction and dance. Nandi's sculptural representation is found at the entrance to Shiva temples, where it acts as gatekeeper and protector. The goddess Durga, a manifestation of shakti (power or energy) and a consort of Shiva, rides on a lion, and her vahana signifies and enhances her ability to destroy evil. A number of Hindu deities combine human and animal features, such as Ganesh, the elephant-headed god, and Hanuman, the monkey god. The strict separation between human and animal worlds is bridged though such more-than-human deities.

Reverence for animals can lead to either their killing, when animals become linked to good fortune and become important in ritual practices, or a ban on their killing, when protection of specific animals becomes part of religious norms. The rise of an ethic of ahimsa, or nonviolence, in South Asian religions can partly be traced to concerns about the treatment of animals and how kindness to animals was viewed as part of living a moral life. The ban against killing of cows in Hinduism is linked to the rise of Jainism and Buddhism around the sixth century BCE. Nonviolence toward all living creatures, including animals, is a central tenet of Jainism (Jains are vegetarians). Jainism and Buddhism also challenged the practice of animal sacrifices, which was prevalent in Hinduism. Thus, changes were initiated in terms of the species of animal sacrificed—for instance, buffaloes and goats being sacrificed in ritual practices rather than cattle. In Daoism, not killing or causing harm is an important principle and resembles Buddhist precepts.

The slaughter of cattle, especially cows, and the consumption of beef are matters of controversy in India, and are often targeted to be controlled or banned, especially by Hindu right-wing political parties and groups. Such bans discriminate against Hindu and non-Hindu communities who utilize beef as food and are traditionally involved in the selling of cattle meat and the disposal of slaughtered cattle. The notion of cattle slaughter as illegal therefore must be viewed not only through the lens of religious ethics, but also through the lens of the social inequalities it perpetuates.

Demand for rare and endangered animal parts—such as tiger skins, elephant tusks, and rhinoceros horns—and the illegal hunting that seeks to satisfy this demand threatens the loss of species in East and South Asia. Commercial whaling has been justified by Japan as part of a long tradition of small-scale whaling and associated practices of reverence for beached whales. The Shinto god, Ebisu, the god of fishers and good fortune, is sometimes depicted as a floating whale. Opponents of whaling, however, argue that large-scale commercial whaling does not fit these traditional Japanese beliefs, and whaling may soon be discontinued in Japan.

Pratyusha Basu

See also: Ethics; Indigenous Religions, Animals in; Traditional Chinese Medicine (TCM); Western Religions, Animals in; Whaling

Further Reading

Blakely, D. 2003. "Listening to the Animals: The Confucian View of Animal Welfare." *Journal of Chinese Philosophy* 30(2): 137–157.

Dalal, N., and Taylor, C., eds. 2014. *Asian Perspectives on Animal Ethics: Rethinking the Nonhuman.* New York: Routledge.

Jha, D. N. 2002. *The Myth of the Holy Cow.* London: Verso.

Kemmerer, L. 2012. *Animals and World Religions.* New York: Oxford University Press.

Kemmerer, L., and Nocella, A., eds. 2011. *Call to Compassion: Reflections on Animal Advocacy from the World's Religions.* Brooklyn, NY: Lantern Books.

Korom, F. 2000. "Holy Cow! The Apotheosis of Zebu, or Why the Cow Is Sacred in Hinduism." *Asian Folklore Studies* 59(2): 181–203.

Naumann, N. 1974. "Whale and Fish Cult in Japan: A Basic Feature of Ebisu Worship." *Asian Folklore Studies* 33(1): 1–15.

Regan T., ed. 2010. *Animal Sacrifices.* Philadelphia: Temple University Press.

Ecotourism

The practice of ecotourism is largely focused on nature and wild animals. In a world that is quickly urbanizing and full of environmental catastrophes, many people seek out an ecotourism experience because they are able to encounter wild animals without harming local environments or peoples. In fact, ecotourism's focus on protection means it is an important element in the preservation of (endangered) animal species.

The International Ecotourism Society (TIES) defines ecotourism as "[r]esponsible travel to natural areas that conserves the environment, sustains the well-being of the local people, and involves interpretation and education" (TIES 2015). The term "ecotourism" can be traced back to the late 1960s and early 1970s, when it arose as a consequence of dissatisfaction with mass tourism. Mass tourism favors a strictly profit-centered approach, often ignoring the social and ecological impacts of people visiting a particular destination. In contrast, ecotourism aims for tourism that impacts the environment minimally, protects animals, and respects and benefits host cultures, while giving tourists an educational experience and maximum recreational satisfaction. Altogether, the ecotourism model is meant to be ecologically and socially responsible and sustainable. With a very strong focus on the tourists' interaction with animals that live in nature, ecotourism is often built on human-animal relations, as the following example will show.

Creating a national park attracts tourists who want to see wild animals and are willing to pay to enter the park. Because of this financial return, the animals create an income for poor local populations, which leads to a different type of engagement between various local, indigenous groups of people and the animals. Protected parks such as the world-famous ecotourism destination Masai Mara National Reserve in Kenya and the adjoining Serengeti in Tanzania have been set aside for tourist use, which has resulted in many tourists passing through the area to watch for large predators and other African mammals. Although most of the Maasai people have never been into these parks themselves, some of them are able to get a job at a park and profit financially from the wild animals. Before ecotourism became the focus of these parks, the local communities were often excluded from tourism initiatives and were further disadvantaged because they had to leave their lands so the national park could be created. Today, when Maasai speak to tourists, they explain that wildlife has now become a source of income that they need to protect (Wijngaarden 2012).

Ecotourism, despite its very broad and good intentions, has its limitations and cannot always satisfy everybody, either human *or* animal. For example, for international nongovernmental conservation organizations, ecotourism can be a means to save and protect natural habitats and animal species; for ecotourists it can provide an interesting travel destination where they can encounter wild animals in nature in a sustainable way; for tour operators it can increase their green, eco-friendly image through marketing; for countries it can be a welcome addition for their national economies; and for local inhabitants ecotourism can be a provider of jobs. But for other local people it can also mean that they lose land because their traditional homes have been converted to parks where visitors do not expect to see people, thereby creating poverty. When indigenous people lose grazing or gathering lands, this too changes their relations with the animals that live in their environment. In the Masai Mara and Serengeti, local Maasai will explain to tourists that they do not hunt for wild meat, but off the record it turns out that this still happens because they must provide for their families—the fact is that, while couched in ecotourist ideals of supporting local peoples, the revenue from ecotourism bypasses most Maasai; only a small number are able to profit from the wildlife financially, while the majority experience various restrictions due to ecotourism regulations (Wijngaarden 2012).

With respect to animal species, while they are often protected through ecotourism, they can also be disadvantaged. For example, studies of boat tours to watch whales and dolphins have increasingly been shown to affect the behavior and stress levels of these large sea mammals, sometimes even causing deaths. With an enormous expansion of tourists joining such trips (from 4 million in 1991 in 31 countries to 13 million in 2008 in 119 countries), this type of tourism, often considered ecotourism, has become a troubling activity (Cressey 2014). Along these same lines, it was also found that wild dolphins in an Australian resort that were fed every day for tourists became dependent on the food from humans. This created lower birth rates and a shorter life expectancy (Brockington, Duffy, and Igoe 2008). The financial value of animals increases even more in the case of trophy hunting. In countries like Botswana and Canada, wealthy tourists shoot large mammals (such as elephants, antelopes, or polar bears) for sport so they can have the animals mounted on the walls of their homes. This is often considered ecotourism by hunting operators and tourists because the revenues that it creates are partly returned to local communities and conservation activities (Dowsley 2009; Gressier 2014).

Altogether, ecotourism is an instigator of change and its value depends on a person's viewpoints, values, and socioeconomic position. All these different interests are important in their own way with regard to the changes that ecotourism can bring to human-animal relations.

Stasja Koot

See also: Indigenous Rights; Taxidermy; Trophy Hunting; Wildlife

Further Reading

Brockington, D., Duffy, R., and Igoe, J. 2008. *Nature Unbound: Conservation, Capitalism and the Future of Protected Areas*. London: Earthscan.

Cater, E. 1994. "Introduction." In Cater, E., and Lowman, G., eds., *Ecotourism: A Sustainable Option?*, 3–17. New York: John Wiley & Sons.

Cressey, D. 2014. "Ecotourism Rise Hits Whales." *Nature* 512(7512): 358.

Dowsley, M. 2009. "Inuit-Organised Polar Bear Sport Hunting in Nunavut Territory, Canada." *Journal of Ecotourism* 8(2): 161–175.

Fennell, D.A. 2003. *Ecotourism,* 2nd ed. New York: Routledge.

Gressier, C. 2014. "An Elephant in the Room: Okavango Safari Hunting as Ecotourism?" *Ethnos: Journal of Anthropology* 79(2): 193–214.

The International Ecotourism Society. Accessed February 18, 2015. http://www.ecotourism.org

Wijngaarden, V. 2012. " 'The Lion Has Become a Cow': The Maasai Hunting Paradox." In Van Beek, W. E. A., and Schmidt, A., eds., *African Hosts & Their Guests: Cultural Dynamics of Tourism*, 176–200. Woodbridge, Suffolk: James Currey.

Elephants

There are two families of elephants: African (consisting of the Savanna elephant and Forest elephant) and Asian (composed of the Borneo Pygmy elephant, the Sri Lankan elephant, the Sumatran elephant, and the Indian elephant). Elephants display a range of behaviors that indicate intelligence, emotionality, and complex sociality. Both African and Asian elephants are threatened with extinction: African elephants because of the trade in their ivory tusks, and Asian elephants because of habitat loss, human-elephant conflict, and the illegal trade in live elephants. The Asian elephant is revered as sacred in Hinduism and Buddhism. Elephants are used for tourism and religious ceremonies in Asia and are held in zoos in North America and Europe, but it is debatable whether elephants' welfare needs can be met in captivity.

Elephants evolved from Proboscidae, trunk-snouted mammals that emerged in northern Africa 40 million years ago. Two million years ago, three species gained prominence: *Elephas* (which became the Asian elephant), *Mammuthus* (the now-extinct mammoth), and *Loxodonta* (precursor of the African elephant). African elephants weigh 12,000 pounds and are larger than Asian elephants at 11,000 pounds. Wild elephants feed for approximately 12 hours daily and travel far in search of food. An elephant's trunk has several hypersensitive nerve bundles (called Ayer's nerve endings) and 100,000 muscles but no bone, which allows for dexterity. The social structure of elephant herds is matriarchal: Female relatives remain together for life. An elephant calf has a strong bond with his or her mother and allomothers (related female elephants). Bulls (male elephants) leave the matriarchal herd from 12 to 13 years old. Adult bulls periodically enter musth, a state characterized by increased testosterone and aggression.

Elephants communicate through scent, touch, and infrasonic vocalizations (low-frequency sounds outside the range of human hearing). They cooperate with one another, use tools, display empathy, solve problems, recognize related elephants over long periods of time, and appear to grieve deceased elephants. Elephants are one of the few species to pass the mirror self-recognition test, which indicates that elephants have self-awareness (a sign of advanced intelligence). The matriarchs (older female elephants) in a herd teach younger generations information vital for social functioning.

Elephants who experience a disruption in this social structure due to poaching (illegal killing), culling (legal killing), translocation (movement away from their home range), or capture for captivity often display impaired social and emotional functioning. These impairments resemble the symptoms of human survivors of trauma, leading some scholars to speak of elephant PTSD (post-traumatic stress disorder) (Bradshaw 2009).

Although populations of African elephants have increased in some countries, overall the species is in decline. From 2008 to 2012 alone, tens of thousands of African elephants were killed for their tusks, which are sent to Asia to be carved into trinkets and jewelry (Wittemyer et al. 2014). The demand for ivory has reached epic proportions, mainly due to an expansion in population and wealth in Asia, the primary source of demand. Elephants are being killed faster than they can reproduce. Asian elephants are also endangered; there are only 25,600 to 32,750 wild individuals remaining (WWF 2015). The human population in Asia is expanding; 20 percent of the world's population now lives in or near the Asian elephant's habitat. This combination of habitat loss and human population growth has increased human-elephant conflict. All African elephants have tusks, but only male Asian elephants have tusks. Therefore, although poaching for ivory still occurs in Asia, a larger issue is the illegal trade in wild-caught live elephants for the tourism trade, particularly at the border of Myanmar and Thailand.

African elephants were once used in battle and to transport military supplies, most notably in the army of Alexander the Great (356–323 BCE). The ivory tusks of these animals are symbols of wealth and status in Asia. Asian elephants are central in Hinduism and Buddhism. In Hinduism, two elephants, Mahapadma and Saumanasa, support the world. Ganesh, the elephant-headed god known as the Remover of Obstacles, is worshiped across India. In Buddhist mythology, Queen Sirimahamaya dreamed of a white elephant with a lotus in his trunk and awakened impregnated with the Buddha. The mythological white elephant is associated with Buddha and is worshiped in many Asian countries, particularly Thailand and Burma. Elephants are a common sight at temples and sacred ceremonies in Asia.

In Asia, many captive elephants endure a training procedure (known as the *phajaan* or "crush method" in Thailand and the *ketti-azhikkal* in Kerala, India) in which the elephant is constrained and beaten with nails or a bullhook (sharpened tool) for several days. The "broken" elephants are then used to entertain tourists. Captive elephants in Asia commonly experience premature weaning (separation from the mother), chronic physical exhaustion, social isolation, the inability to engage in natural behaviors, and forced breeding. In North America and Europe, captive elephants in zoos often exhibit foot problems, arthritis, reproductive health issues, obesity, aggression, and stereotypic behavior (repetitive movements indicative of psychological suffering) (Forthman et al. 2009). Several zoos have decided to no longer exhibit elephants. Others have attempted to improve elephant welfare through environmental enrichment and the use of protected contact (which does not require a bullhook). The question of whether captivity can ever meet the needs of elephants, who in the wild roam long distances and form complex social attachments, is still debated.

Jessica Bell Rizzolo

See also: Black Market Animal Trade; Ecotourism; Endangered Species; Ivory Trade; Wildlife Management

Further Reading

Bradshaw, G. A. 2009. *Elephants on the Edge: What Animals Teach Us About Humanity*. New Haven, CT: Yale University Press.

Forthman, D. L., Kane, L. F., Hancocks, D., and Waldua, P. F., eds. 2009. *An Elephant in the Room: The Science and Well-Being of Elephants in Captivity*. North Grafton, MA: Tufts Center for Animals and Public Policy.

McComb, K., Moss, C., Durant, S. M., Baker, L., and Sayialel, S. 2001. "Matriarchs as Repositories of Social Knowledge in African Elephants." *Science* 292(5516): 491–494.

Wittemyer, G., Northrup, J. M., Blanc, J., Douglas-Hamilton, I., Omondi, P., and Burnham, K. P. 2014. "Illegal Killing for Ivory Drives Global Decline in African Elephants." *Proceedings of the National Academy of Sciences* 111(36): 13117–13121.

WWF. 2015. "Elephants." Accessed March 26, 2015. http://www.worldwildlife.org/species/elephant

Wylie, D. 2008. *Elephant*. London: Reaktion Books.

Emotions, Animal

In 1872, Charles Darwin (1809–1882) was writing about animal consciousness and sentience (i.e., the capacity to feel, perceive, or experience subjectively). However, in the last century there has been little interest in studying animals' emotions in

Western science until the recent rise of public concern for animals and the fields of animal welfare science and ethology.

An emotion can be defined as an intense but short-lived mental response to an event associated with specific body changes (Veissier et al. 2009). Emotions are different from sensations that are physical consequences of exposure to particular stimuli (e.g., cold, heat, pressure), and from feelings, which are more subjective labels for emotions associated with behavior, not body changes (e.g., two people can feel the same emotion but give it different names). Moreover, emotions, from a scientific perspective, are interpreted as the outcome of an individual's assessment of several characteristics (abruptness, newness, predictability, and/or appeal) of a particular event or situation. Emotions are also linked to the individual's evaluation of the consequences of a certain situation and her or his ability to control it. From an evolutionary perspective, emotions are basic mechanisms that developed over many generations and enabled animals to avoid dangerous situations, to hunt for desirable resources and rewards, and to otherwise adapt to their environment.

The psychologist Magda Arnold (1903–2002) pioneered work in "appraisal theories," which outline the response steps of an emotional situation wherein a human evaluates how he or she will be affected by an event, interprets the various aspects of the event, and arrives at a response based on that interpretation. These theories are now being used to study animal emotions. For example, studies of food deprivation have shown that it is often an animal's representation of an event, rather than the event itself, that determines its reaction because it is not so much the lack of food as the perception of deprivation that induces stress. Thus, cognitive processes (i.e., mental information processing and representation of the environment) have to be taken into account in order to better assess emotional experiences of an animal.

In recent decades, ethologists (scientists studying animal behavior) and animal welfare scientists have focused their attention on developing measures for emotions such as pain and suffering. This focus is partly because negative emotions have been considered more important measurements for animal welfare laws and public concern (especially in the West) but also because they are easier to measure. Fear, defined as a reaction to the perception of danger, generally produces observable behaviors such as active defense (e.g., attack, threat posture) or avoidance (e.g., flight, hiding, escape) and passive avoidance (immobility). Additionally, stress can be measured through the increased presence of the hormone cortisol in the blood. Joy and other positive emotions, however, are currently considered less observable and more difficult to measure. Ethologists such as Marc Bekoff (1945–) have done work on joy and play in dogs, but the limited number of assessment methods are not considered sufficient to meet scientific standards for observable evidence.

Despite the increasing number of studies, the exact nature of emotional experiences in animals remains poorly understood, and the existence of subjective states common to both humans and other animals is not readily accepted in the scientific community. Most studies about animal emotions have been conducted on mice or dogs, and it seems easier to demonstrate that an animal is physically experiencing gratification or enjoyment than happiness, if by happiness we mean the emotion that occurs after the animal *evaluates* a pleasurable bodily experience. Early studies showed that rats may be able to laugh when tickled because they make ultrasonic chirps we know to be linked to finding food or other pleasurable experiences. However, there is no consensus about rats' ability to *evaluate* these positive experiences.

More recently, one area in animal science that has seen advances is the emotions of farm animals. This research is important for addressing increased public concern about their quality of life and for growing consumer demand for products obtained from "happy animals." For example, significant progress has been achieved in recent studies of sheep emotions. Basic research models rely essentially on facial expression and to a lesser extent on body posture and physiological responses. Recent studies have indicated that in order to evaluate their environment and events, sheep use the same factors as used by humans (e.g., abruptness, newness, predictability, and possibility of control). This assessment influences emotional responses and behavior that is consistent with the responses and behavior observed in humans, such as changes in heart rate or facial expressions. For example, fear is recognizable by the movement of certain facial muscles in humans and in sheep, only in the latter this is shown by the position of the ears and in the former by the position of the eyebrows. Based on their observations and measurements, recent studies have shown that sheep are able to experience a range of emotions from fear to happiness and that the triggers of these emotions are very similar to the ones identified in humans. So for example, in humans, unpredictable and uncontrollable situations are likely to lead to despair, while a very monotonous and predictable environment is likely to elicit boredom. Sheep responses to these situations are similar to human responses. These results have significant implications for the design of animal housing and for the management of animals on farms and during transport. In order to increase the welfare of farm animals, some control should be offered to them over their environment, and attention should be paid to avoid sudden and unpredictable events (Veissier et al. 2009).

Mara Miele

See also: Advocacy; Anthropomorphism; Dogs; Ethology; Evolution; Humane Farming; Intelligence; Livestock; Research and Experimentation; Sentience; Welfare

Further Reading

Arnold, M. B. 1970. "Perennial Problems in the Field of Emotion." In *Feelings and Emotions: The Loyola Symposium,* 169–186. New York: Academic Press.

Bekoff, M. 2000. "Animal Emotions: Exploring Passionate Natures." *BioScience* 50(10): 861–870.

Dantzer, R. 1986. "Behavioral, Physiological, and Functional Aspects of Stereotyped Behavior: A Review and a Re-interpretation," *J. Anim. Sci.* 62: 1776–1786.

Darwin, C. 1872. *The Origin of Species by Means of Natural Selection*. London: John Murray.

Désiré, L., Veissier I., Després G., Delval E., Toporenko, G., and Boissy, A. 2006. "Appraisal Process in Sheep: Synergic Effect of Suddenness and Novelty on Cardiac and Behavioural Responses." *Journal of Comparative Psychology* 120: 280–287.

Duncan, I. J. 2006. "The Changing Concept of Animal Sentience." *Applied Animal Behaviour Science* 100(1): 11–19.

Ekman, P., Levenson, R. W., and Friesen, W. V. 1983. "Autonomic Nervous System Activity Distinguishes among Emotions." *Science* 221: 1208–1210.

Miele, M. 2011. "The Taste of Happiness: Free Range Chicken." *Environment and Planning A* 43(9): 2070–2090.

Veissier, I., Boissy, A., Désiré, L., and Greiveldinger, L. 2009. "Animals' Emotions: Studies in Sheep Using Appraisal Theories." *Animal Welfare* 18(4): 347–354.

Empathy

The role of empathy in human-animal relationships has been an important one. Without an easily accessible common language with which to communicate, the empathetic connection offers humans a means to imagine the perspectives of other animal species. When manifested through sympathy and altruistic behavior like caregiving, empathy also enables humans to help other animal species and enables nonhuman species to help each other. An absence of empathy may enable hurting others. However, empathy may be intentionally cultivated through education about and time spent with others, and in this way humans may learn to cultivate relationships with other species with more awareness of their preferences and needs.

Empathy is an ability to understand another's situation through experiencing emotions as if one *were* the other. This ability may have evolved to enable social animals to more quickly and accurately navigate relationships. Some researchers believe that the experience of empathy is created by "mirror neurons," nerve cells in the brain that link the body parts of another person with those of one's own body (Gallese et al. 1996). The experience can coincide with body synchronization, such as "infectious" yawns, laughter that transfers from one being to another, or mimicry. Frans De Waal (1948–), a primate researcher who has studied empathy in and between different animal species, has found yawns, laughter, and mimicry to occur across many species. He and other researchers have found examples of empathy in cetaceans (e.g., dolphins), nonhuman primates, dogs, cats, mice, and possibly in other animal species.

When experienced by a caring being, empathy enables sympathy, or the experience of *concern* for another's situation. If translated into action, sympathy leads to targeted helping of another being, even sometimes against the instinct for self-preservation.

By this means, empathy also prevents hurting. It is difficult to watch another's misery when one both identifies with and has intimate knowledge of another's situation. Their pain can be painful to the individual watching it, even though it does not directly cause that individual harm. Some recent studies with rodents and primates have shown possible evidence of empathy when they observe negative things happening to others of their kind. For example, rats helped release previous cage mates from traps even when doing so did not directly benefit them (Bartal et al. 2014). Human caregivers of animals may utilize empathy in this way to gain knowledge about their patients in order to appropriately help their situations. In addition, having an understanding of suffering and abuse on the part of animals can draw attention to creating more ethical standards of care for them.

The idea that empathy might be present in nonhuman animals was first suggested by the American biologist Edward O. Wilson (1929–), who explored altruism in humans and other animals. Wilson especially highlighted the altruistic tendencies of African wild dogs, who regularly brought food to other dogs that had not participated in their hunt. The idea of nonhuman animal empathy, however, continues to be contested. Some researchers argue that what seems to be empathy can be explained by other reasons, such as the simple desire for social contact. In addition, the perception of animal empathy by researchers has been criticized as inaccurately ascribing anthropomorphic (human-like) characteristics to them.

Whether or not it exists for nonhuman animals, human researchers have gained novel insights into nonhuman animal lives through the utilization of their own capacities for empathy. Jane Goodall (1934–), for example, used her ability for empathy to help her discover tool use by chimpanzees. Her empathy for the chimpanzees allowed her to discover that they were using stems as tools to fish insects out of tiny holes. Though the inclusion of empathy in research remains nontraditional because its personal nature makes it difficult to quantify and monitor through objective means, Goodall continues to advocate for it as an important tool.

Empathy may be induced through identification with another being, and it may also be blocked through the lack of that identification. For example, if a human does not believe that a nonhuman animal individual shares a similarity, that human will have difficulty empathizing with that animal. Education at any age, but especially in childhood, about other species' abilities to experience empathy, sympathy, and altruism, and also to feel things such as pain and fear similar to ourselves, may help humans who lack empathy learn to empathize with them. Stories that demonstrate these abilities—such as those of an orphaned grizzly bear cub who cared for her injured sibling, and a group of rhesus monkeys who stopped traffic to protect a baby monkey who was hit by a car (Bekoff 2010)—increase the human ability to feel connections to members of these species.

Humans are especially affected by nonhuman animals who appear to act empathetically toward members of our own species. One such story, told by Goodall, describes a situation in which she handed a nut to a chimpanzee whom she had named David Greybeard. Though he took the nut, he dropped it, and instead held her hand—appearing to understand her intent to communicate friendliness. Other popular stories tell of cetaceans who protect humans from sharks, or companion animals who save their human guardians' lives.

Heather Pospisil

See also: Anthropomorphism; Emotions, Animal; Research and Experimentation

Further Reading

Bartal, I. B., Rodgers, D. A., Sarria, M. S. B., Decety, J., and Mason, P. 2014. "Pro-Social Behavior in Rats Is Modulated by Social Experience." *eLife* 3.

Bekoff, M. 2010. *The Emotional Lives of Animals: A Leading Scientist Explores Animal Joy, Sorrow, and Empathy—and Why They Matter*. Novato, CA: New World Library.

De Waal, F. 2009. *The Age of Empathy: Nature's Lessons for a Kinder Society.* New York: Harmony Books.

Gallese, V., Luciano, F., Fogassi, L., and Rizzolatti, G. 1996. "Action Recognition in the Premotor Cortex." *Brain: A Journal of Neurology* 119: 593–609.

Goodall, J. 1999. *Reason for Hope: A Spiritual Journey*. New York: Warner Books.

Silberberg, A., Allouch, C., Sandfort, S., Kearns, D., Karpel, H., and Slotnick, B. 2014. "Desire for Social Contact, Not Empathy, May Explain 'Rescue' Behavior in Rats." *Animal Cognition* 17: 609–618.

Wilson, E. O. 1975. *Sociobiology: The New Synthesis*. Cambridge, MA: Belknap Press.

Endangered Species

Endangered species are wild animals (and plants) classified by biologists and ecologists as being at risk of extinction (no longer in existence) in the near future. Causes of endangerment include the destruction of habitats, overhunting, pollution, climate change, barriers to migration, reduced food sources, and the introduction of non-native species as competitors—all of which are exacerbated by an expanding human population. Endangered animals and their habitats may become subjects of legal protection and conservation action until the species reaches an agreed-upon target level of population stability. Whether or not an animal is classified as endangered or threatened has many social and environmental consequences; as a result, the listing of endangered species for protection under the law is often a long and contentious political process.

The overwhelming majority of species that have ever lived on Earth are extinct. Extinction is natural and estimated to occur at a "background rate" of a small number of species naturally dying off each year, offset by the emergence of new species. The rise of humans, however, has spelled disaster for many animals. With dozens

The California condor, the largest North American land bird, went extinct in the wild in 1987. Captive breeding programs began releasing the birds back into the wild in 1991, and the small but stable population remains one of the major success stories for saving a species. (Fischer0182/Dreamstime.com)

estimated to be going extinct daily, the world may be losing species at 1,000 to 10,000 times the normal background rate of extinction. For example, when the United States underwent industrialization in the 19th century, many animals and habitats were eradicated; losses include the passenger pigeon (*Ectopistes migratorius*) and the Carolina parakeet (*Conuropsis carolinensis*), with populations of the American buffalo (*Bison bison*) and the gray wolf (*Canis lupis*) nearly hunted to extinction. By the 1900s, such negative impacts inspired concerned citizens and politicians to pass wildlife conservation laws, including the Migratory Bird Treaty Act (1929), the Fish and Wildlife Act (1956), the National Environmental Policy Act (1970), and the Endangered Species Act (1973).

Globally, the International Union for the Conservation of Nature (IUCN), a network of scientists and partner organizations, works in almost every country to maintain their Red List of Threatened Species. Classifying animals from least concern to critically endangered, according to total population size and range, area/quality of habitat, and rate of decline, the Red List prioritizes conservation activities for governments, nongovernmental organizations, and scientific institutions. Additionally, to protect endangered species from being bought and sold around the world, the 1973 Convention on International Trade in Endangered Species of Wild Flora and Fauna (CITES) regulates the trade of over 5,000 animals between 180 countries.

Through such mechanisms, negative human impacts at the global scale are quantified and reduced.

Many strategies organized by citizens, community groups, park rangers, enforcement officers, biologists, and wildlife managers protect endangered animals. Solutions vary by species and location, from short-term to long-term actions, and through implementation at local, regional, national, or global scales. The most popular model of protecting species is the establishment of parks and/or protected areas, which today cover approximately 12 percent of Earth's land and 2 percent of the ocean. By force, fines, or fences, protected areas (often designed around the needs of flagship species such as the rhino, whale, or elephant) exclude local people from the use of areas important to endangered wildlife. This is a problem because many times these residents lose the ability to use the preserved land for subsistence. Other problems with this model include lack of enforcement, poaching, inadequate size, little to no buffer zone between animals and people, isolation from other habitat patches, park locations near or in human conflict zones, negative impacts from tourism, porous and poorly defined borders, and restrictive or insufficient boundaries that inhibit species migration and genetic mixing.

Examples of notable large-scale projects to save species include the World Wildlife Fund's panda efforts in China, their 2011 *Year of the Tiger* campaign, or their "Global 200" campaign, which aims to preserve representative areas of every identifiable ecosystem by focusing on "hotspots"—places rich in unique wildlife, such as Madagascar or Indonesia. More ambitious projects by other groups include the Yellowstone to Yukon rewilding project, which aims to restore a wildlife corridor across the Rocky Mountains (in progress), and the proposed translocation scheme to place free-ranging, large African mammals such as endangered elephants on U.S. grasslands (Donlan et al. 2006). Costing an estimated $35 million, perhaps the most expensive and extreme effort to save an endangered animal species in North American history has been the campaign to save its largest land bird, the California condor (*Gymnogyps californianus*), still classified as critically endangered. Condors have been negatively impacted by humans in many ways, including overharvesting and eradication by hunters and ranchers, ingesting the poisonous insecticide DDT as well as lead ammunition used by hunters, loss of habitat due to urban land and infrastructure development, power-line collisions from use of power-poles as perches, and ingestion of trash. By 1982, only 23 condors survived, and in 1987, all remaining wild condors were taken from the wild and managed under the U.S. Fish and Wildlife Service's California Condor Recovery Program. Though wild condors remain dependent on intensive conservation management efforts, the total population now stands at around 410 birds.

While protection may help endangered species recover, people may suffer when conservation is enacted. Some examples include the loss or restricted use of one's land, economic hardship resulting from regulatory compliance, and changes in lifestyle imposed by restrictions on hunting, harvesting, or fishing. Battles between defenders of wildlife and those injured by the law, such as big businesses, homeowners, or small-scale farmers, often make protection of species a long and costly legal process. Once conservation actions are authorized, controversies persist regarding enforcement techniques, surveillance concerns about habitat and species monitoring, and the powers of unelected, transnational conservation organizations.

Jenny R. Isaacs

See also: Biodiversity; Bushmeat; Flagship Species; Northern Spotted Owl; Poaching; Wildlife Management

Further Reading

Convention on International Trade in Endangered Species of Wild Fauna and Flora (CITES). 2015. "What Is CITES?" Accessed June 29, 2015. http://www.cites.org/eng

Donlan, C. J., Berger, J., Bock, C. E., Bock, J. H., Burney, D. A., Estes, J. A., Foreman, D., Martin, P. S., Roemer, G. W., Smith, F. A., Soule, M. E., and Greene, H. W. 2006. "Pleistocene Rewilding: An Optimistic Agenda for Twenty-First Century Conservation." *The American Naturalist* 168(5): 660–681.

International Union for the Conservation of Nature (IUCN). Red List of Threatened Species. 2015. "Overview of the IUCN Red List." Accessed June 29, 2015. http://www.iucnredlist.org

U.S. Fish and Wildlife Service. 2015. "Endangered Species." Accessed June 29, 2015. http://www.fws.gov/endangered

U.S. Fish and Wildlife Service. 2015. "California Condor." Accessed June 29, 2015. http://www.fws.gov/cno/es/CalCondor/Condor.cfm

Wilson, E. O. 2002. *The Future of Life*. New York: Alfred A. Knopf.

Endangered Species Act. *See* Human-Wildlife Conflict; Wildlife Forensics; Wildlife Management; Primary Documents

Ethics

Ethics is a phenomenon spanning all cultures and places. Its manifestation through social institutions and theories varies greatly, but moral norms are inescapable in human groups and societies; "honor among thieves" is one wry expression of this. Some explain ethics as deriving from divine commands, others through rules of logic, and still others through humanity's evolution as a social species. Whatever its origins, ethics is central to the human experience—so much so that human beings are quintessentially *moral primates* living in *mixed communities* of people, animals, and nature. So too, our basic ethical orientations—like love, friendship, mutual aid, and fairness—are shared on a continuum with at least some other animals.

For a formal definition of ethics, we are well served by Socrates (470–399 BCE). He conceptualized ethics as the study and deliberation over *how we ought to live.* Ethics is thus a conversation about the moral values that inform (or fail to inform) our way of life. It is a concern for what is good, right, and just in our individual and collective lives. This involves a process of critique and vision. We criticize what detracts from the well-being of ourselves and others (human or nonhuman), while at the same time we envision how we might improve that well-being. It is for this reason that politics and public policy always have a moral dimension, and all social movements for justice, animal protection, or conservation are motivated by ethical concerns. Hence ethics and politics are two sides of the same coin, with politics being *ethics writ large.*

The single most important concept in ethics of every sort is that of *moral value.* Whether a being or thing has *intrinsic* or *extrinsic value*—that is, value in and of themselves versus value in terms of their use to others—is the point of departure for all ethics. This distinction helps us decide who or what matters ethically and to whom or what we have moral responsibilities. When someone or something matters from an ethical point of view, they are considered part of a *moral community.*

This is not simply an either/or choice, however, as intrinsic and extrinsic values are often intermixed, resulting in *co-values* that require nuanced ethical interpretation. For instance, human beings have intrinsic value, rooted (at least in part) in the fact that we are a thinking (*sapient*) and feeling (*sentient*) species. This means we are self-aware and capable of feeling emotions and making decisions. The evidence is clear that some other animals, such as wolves and deer, are also sapient and sentient. For this reason (at least in part) they have intrinsic value, too. Moreover, because of our mutual sentience and sapience, the well-being of people, wolves, and deer can be harmed in similar ways. Each of us can feel pain, experience suffering, and/or be frustrated when we are not able to achieve goals appropriate to our individuality and species.

Yet deer, wolves, and people all have extrinsic value too (sometimes called *instrumental value*). For instance, deer are an important source of food for wolves in North America and were an important part of subsistence hunting (hunting for food, not sport or trophies) for First Nations. So wolves also were extrinsically important to these cultures as teachers of hunting skills, and even partners in the hunt itself. At the same time, they were trapped and their bodies were made into tools, clothing, ceremonial dress, and the like.

In the cultural geography of North America, wolves are also instrumental symbols for larger worldviews of nature and society. For some, wolves have intrinsic value and are a flagship (iconic) species for the protection of wild spaces and biodiversity everywhere. For others, wolves have no intrinsic value. They have been and are villains, varmints, and vermin, predators of innocent domestic animals and wildlife, and creatures that should be wiped off the landscape.

And still, people are extrinsically valuable to deer and wolves when we help them to thrive by preserving their habitat or protecting them from unnecessary harm. Domestic and international policies for the protection of the environment and wildlife, and the educational, legal, and political activities of the animal and environmental movements, are examples of how humans prove valuable to deer, wolves, other animals, and their habitat.

Overall, scholars of human-animal relations frequently consider people to be members of a *more-than-human* moral community. This is true for many animal advocates and members of the general public as well, even if nonscholars do not explicitly use ethical terms and arguments to frame their points of view. Yet in the vast majority of academic fields (including geography), anthropocentrism has been the dominant ideology. *Anthropocentrism* (human-centered) is the belief that only human beings have intrinsic value; animals stand outside the moral community, and we need not trouble ourselves about their well-being.

The dominance of anthropocentrism is slowly changing, however, with the emergence of subdisciplines like animal geography and the ongoing concern for animals in interdisciplinary fields like environmental studies. In these venues a variety of alternative ideologies has arisen, collectively known as *nonanthropocentrism*. These alternatives take many forms, inspired by diverse moral commitments to individual animals (*biocentrism*), ecosystems (*ecocentrism*), or the whole community of life (*geocentrism*). Yet despite the differences that exist between forms of nonanthropocentrism, they all share a belief that the moral community extends beyond *Homo sapiens*, and that other animals (sometimes ecosystems too) deserve ethical consideration.

William S. Lynn

See also: Advocacy; Animal Law; Human-Animal Studies; Wolves

Further Reading

Lynn, W. S. 1998. "Contested Moralities: Animals and Moral Value in the Dear/Symanski Debate." *Ethics, Place and Environment* 1(2): 223–242.

Lynn, W. S. 2010. "Discourse and Wolves: Science, Society and Ethics." *Society & Animals* 18(1): 75–92.

Lynn, W. S. 2010. "Geography and Ethics." In Warf, B., ed., *Encyclopedia of Human Geography*, 1013–1016. Newbury Park, CA: Sage.

Midgley, M. 1998. *Animals and Why They Matter*. Athens: University of Georgia Press.

Peterson, A. L. 2013. *Being Animal: Beasts and Boundaries in Nature Ethics*. New York: Columbia University Press.

Sheppard, E., and Lynn, W. S. 2004. "Cities: Imagining Cosmopolis." In Harrison, S., Pile, S., and Thrift, N., eds., *Patterned Ground: Entanglements of Nature and Culture*. London: Reaktion Press.

Ethology

Ethology can best be defined as "the science whereby we study animal behavior, its causation, and its biological function" (Jensen 2002, 3). Ethology is performed using objective, observable definitions of performed behaviors. These measures can then be used to draw conclusions regarding animal behavioral responses within and between animal species. Research in animal behavior is not only concerned with what behavior the animal is performing (doing or engaged in), but also why it is being performed and how the behavior helps the survival of the individual. Behaviors are regarded as being concerned with both the individual's survival and the survival of future offspring.

Animals' behaviors are observed and defined using an ethogram, which is a list of mutually exclusive behaviors accompanied by an objective definition of the behavior. For example, an observable behavior that a dog performs may be "standing alone" and defined as "all legs are erect with all four feet in contact with the ground; animal is at least five feet away from any contact with other animals." Objective definitions are critical, as ultimately we do not know how an animal feels in relationship to how they behave. These definitions can be used to define empirical (e.g., simply *observing* a dog exposing its teeth) and functional (e.g., seeking to *understand why* a dog exposes its teeth) research, both of which are important to our understanding of animal behavior today.

Nikolaas Tinbergen (1907–1988) and Konrad Lorenz (1903–1989) are frequently regarded as the fathers of modern ethology. Tinbergen and Lorenz were awarded a Nobel Prize in Physiology or Medicine (1973), along with Karl von Frisch (1886–1982), for their work focused on behavioral stress in animals, which was regarded as applicable to human suffering. This was a remarkable achievement because much of their work with behavior was initially disregarded. Other scientists referred to them as "mere animal watchers" prior to their award (Tinbergen 1974, 20). Each man specialized by asking key questions about animal behavior and therefore shaped the way ethology is practiced today.

Tinbergen's four questions for determining the causation of an animal's particular behavior are regarded as the core principles of ethology. The four questions cover causation of a behavior (What causes the behavior?), development (How does this behavior develop during the animal's lifetime?), function (How does this behavior contribute to the survival of the animal?), and evolution (Was this behavior performed by this animal's ancestors?) (Tinbergen 1963). Though each area can be researched simultaneously, scientists frequently focus on one question for their entire careers.

Tinbergen's first two questions are considered to relate to behaviors that can occur within an animal's lifetime. Causation of a behavior can be illustrated by male horses that will exhibit a lip-curling behavior when smelling pheromones (chemicals secreted by the body) produced by a female during mating periods (Stahlbaum and Houpt 1989). Development of a behavior is associated with learning and changes in a behavior during an animal's life. For example, what is known as "imprinting" occurs in multiple species. Through imprinting, the young animal will associate a specific adult as a caregiver (usually the mother). This can be observed by watching ducklings walk in a line behind a female duck (Lorenz 1937).

Tinbergen's last two questions, concerning survival and evolution of a species, are considered to occur throughout multiple generations and are therefore observed across generations. Behaviors associated with survival of a species can be found in black-headed seagulls. Seagulls have been observed removing fragments of a broken shell after their young hatch because the sight of white shell fragments serve as a signal to predators that newborn chicks are near (Tinbergen et al. 1961). In evolutionary terms, behaviors that are helpful to a species' survival will continue, while those that are harmful to a species will likely result in death to those individuals (and therefore will not continue). Flies can be observed mating at different speeds with either fast or slow mating occurring. In some situations, fast mating may be beneficial by allowing a single male to mate with more females in a given time period. This behavior will continue across generations as individuals who mate quickly produce more offspring, and those offspring produce more offspring. In situations with low food availability, mating slowly may require less energy and therefore allow for a longer life. In this situation, mating slowly will occur throughout multiple generations, and those who mate slowly will have more copulations during a longer life (Manning 1961).

Other significant contributors to the field of ethology include Dmitry Belyaev (1917–1985), who worked with silver foxes and demonstrated that an animal can be domesticated by selective breeding for the animal's willingness to approach humans, which changed how evolution and animal domestication are considered today (Belyaev 1978). Gilbert Gottlieb (1929–2006) investigated how birds identify their parents through calls and found the influence of both environment and genetics in duckling communication (1961). Perhaps the most well-known ethologist is Jane Goodall (1934–), who is famous for her work with chimpanzees, living with them and observing their social patterns (Goodall 1971). These scientists have demonstrated the complexity of animal behavior and elaborate biological processes associated with behavior.

Currently, ethology is frequently applied to answer other questions in biology. Using early concepts developed by the founders of ethology, today's research is applied to animal welfare, conservation, and behavioral manipulation (such as animal training). Today, the field of ethology is a rapidly growing, dynamic research area.

Nichole Chapel

See also: Bees; Domestication; Evolution; Extinction; Husbandry; Research and Experimentation; Welfare; Wildlife Rehabilitation and Rescue; Working Animals

Further Reading

Belyaev, Dmitry. 1978. "Destabilizing Selection as a Factor in Domestication." *The Journal of Heredity* 70: 301–308.

Goodall, J. 1971. *In the Shadow of Man.* New York: Houghton Mifflin Company.

Gottlieb, G. 1961. "The Following-Response and Imprinting in Wild and Domestic Ducklings of the Same Species (*Anas platyrhynchos*)." *Behaviour* 18(3): 205–228.

Jensen, P., ed. 2002. *The Ethology of Domestic Animals: An Introductory Text.* Oxon: CABI.

Lorenz, K. 1937. "The Companion in the Bird's World." *The Auk* 54: 245–273.

Manning, A. 1961. "The Effects of Artificial Selection for Mating Speed in *Drosophila melanogaster.*" *Animal Behavior* 9(1–2): 82–92.

Stahlbaum, C. C., and Houpt, K. A. 1989. "The Role of the Flehmen Response in the Behavioral Repertoire of the Stallion." *Physiology and Behavior* 45(6): 1207–1214.

Tinbergen, N. 1963. "On Aims and Methods of Ethology." *Ethology* 20(4): 410–433.

Tinbergen, N. 1974. "Ethology and Stress Diseases." *Science* 185(4145): 20–27.

Tinbergen, N., Broekhuysen, G., Feekes, F., Houghton, J., Kruuk, H., and Szulc, E. 1961. "Egg Shell Removal by the Black-Headed Gull, *Larus ridibundus L.*; a Behaviour Component of Camouflage." *Behavior* 19(1): 74–116.

Euthanasia. *See* Shelters and Sanctuaries; Spay and Neuter

Evolution

In the 1700s, the concept of evolution emerged when French naturalists Georges Buffon (1707–1788) and Jean-Baptiste Lamarck (1744–1829) suggested organisms change over generations due to environmental influences. It wasn't until the mid-1800s, however, when the idea of evolution became widely known with the publication of *On the Origin of Species* by Charles Darwin (1809–1882). Although British biologist Alfred Wallace (1823–1913) also proposed the same ideas during the same time, it was Darwin's book that gained popularity. Today we define evolution as change in heritable traits passed on from generation to generation. Evolutionary biologists study and compare history and development of organisms to better understand relationships between different species and organisms and how evolutionary change occurs. Understanding these relationships gives us a better understanding of the history of Earth, animals, and humankind.

Evolution is typically divided into two categories: microevolution and macroevolution. Microevolution (small-scale evolution) is observed as a change in the occurrence of a gene or allele within a single population of individual organisms.

Genes are the basis for the physical characteristics of an individual (known as the individual's phenotype), and alleles are the possible variants of a gene. Each gene has two alleles, one inherited from an individual's father and one from the mother. Theoretically, say there is a population of frogs with a gene responsible for skin color. For this gene suppose there are two alleles, "G" (green skin) and "g" (yellow skin). In hypothetical initial data collection, 20 percent of the frog population had green skin and 80 percent had yellow skin. If one year later data reported 60 percent of frogs in the population had green skin and 40 percent had yellow skin, this would indicate the frequency of the alleles had changed and therefore changed the frequency of the phenotypes.

Macroevolution (large-scale evolution) is observed as character change (gain or loss of a functional characteristic over time), speciation (formation of new species), and extinction (species elimination). Common examples of character changes are gradual gain or loss of arms, legs, fins, and/or wings. Speciation occurs in multiple ways. A new species may arise from an existing species through multiple micro-evolutions or from mating between two distinct existing species, producing a unique offspring.

Evolution does not randomly happen on its own. Although evolution occurs in many ways, there are two primary mechanisms: natural selection and variation/mutations. Natural selection was Darwin's primary explanation of evolution. It occurs when individuals with one or more inheritable characteristics are more likely to survive, reproduce, and pass on these characteristics to offspring, resulting in an increase in the frequency of the characteristic(s) within the population. One of Darwin's most popular findings, which illustrates this process, was his research on the variation in beak size of finches that inhabit the Galapagos Islands. Darwin noted two important observations: Each of the Galapagos Islands consists of varying environments, and there was a large variety of finch species that occupied the islands. These observations led him to conclude that specific species of finches varied on each island due to the resources available. For example, on an island with large seeds as a food resource, the frequency of finches with larger beaks was higher because they were better able to crush and consume the seeds and therefore were more likely to survive and reproduce than finches with smaller beaks, which were unable to consume the seeds.

Variation was also proposed by Darwin, although he was unaware of its complexity. Darwin understood that each individual has its own combination of characteristics, which can be passed down through generations. What we understand today, that Darwin did not, is that this is due to genetic variation. Each individual has a unique genetic code which determines every physical aspect that will be produced. While members of a species often share segments of this code, no two genetic codes are the same, which results in the variation we see in individuals. This variation naturally fluctuates within species and populations, but can be influenced by mutations (unusual, sporadic changes in the genetic code). If two green frogs of the previously mentioned population mated, an offspring with green skin would be expected. However, a mutation could cause the offspring to have yellow skin. Evolution occurs when mutations are inherited from generation to generation.

In the 1700s, scientists were afraid to boldly state and publish their ideas. Their fears were not unfounded as the public initially rejected the theory of evolution due to its conflicts with the Bible and church teachings. While the Catholic Church has since accepted evolution, the controversy remains as religious groups such as Young Earth Creationists, who believe Earth was created by God in 4000 BCE as opposed to a natural phenomenon 4.6 billion years ago, continue to reject evolution. Although Young Earth Creationists are found worldwide, the controversy appears most prevalent in the United States, where many lawsuits have been filed over which version of Earth's creation should be taught in American schools. In 1925, this controversy fronted the page of every newspaper in the United States with the publicity of *The State of Tennessee vs. John Thomas Scopes*, a court case in which high school teacher John Scopes was accused of teaching evolution in his biology class, a violation of state law at the time. Although Scopes was ultimately found guilty, "The Scopes Trial," as it is known, shed new light on the creation versus evolution debate, and evolution would be reintroduced into educational curriculum years later.

Breanna Ten Eyck

See also: Animal Culture; Animals; Extinction; Humans; Species; Speciesism

Further Reading

BBC. 2015. "Charles Darwin." Accessed March 23, 2015. http://www.bbc.co.uk/history/historic_figures/darwin_charles.shtml

Brooker, R., Widmaier, E., Graham, L., and Stiling, P. 2010. *Biology,* 2nd ed. New York: McGraw-Hill Publishing Company.

Darwin, C. 1859. *On the Origin of Species by Means of Natural Selection, or the Preservation of Favoured Races in the Struggle for Life.* London: John Murray.

Linder, D. "State v. John Scopes ('The Monkey Trial')." University of Missouri-Kansas City. Accessed March 23, 2015. http://law2.umkc.edu/faculty/projects/ftrials/scopes/evolut.htm

University of California Berkeley. 2008. "Understanding Evolution." Accessed March 23, 2015. http://evolution.berkeley.edu/

Exotic Pets

Against skyrocketing extinction rates, the number of wild animals kept and traded as exotic pets is booming. Over eight million exotic birds are kept as pets in the United States alone. More tigers are estimated to live in captivity in the United States than live in the wild worldwide. Recent research suggests that the exotic pet trade

Christie Carr gets a lick from her pet kangaroo, Irwin, at her home in Broken Arrow, Oklahoma. The exotic pet trade is booming globally, although exact figures are hard to come by. Despite most countries having signed the Convention on International Trade in Endangered Species of Flora and Fauna, exotic pet ownership is still largely unregulated, frequently creating risks for both the animals and the humans with which they interact. (AP Photo/Sue Ogrock)

drives not only species loss at an aggregate level but also a loss of quality of life for individual animals in captivity (Baker et al. 2014). Experts also raise concerns about the risks exotic pets pose for human health and safety (Smith et al. 2009).

Exotic pets are curious beings. Whether birds or alligators, turtles or tigers, their defining characteristic is that they are out of place. Exotic pets were either born elsewhere—in another country and vastly different ecosystem—or very recently their ancestors were. They are also considered undomesticated. After two generations of captivity, an animal is considered "captive bred," so this designation can be used for the offspring of any captive-born parents. Domestication, on the other hand, is a much longer process of intervening in animal bodies, genetics, and behaviors through selective breeding. Unlike such domesticated animals as cats and dogs, then, exotic pets have not been bred to live in close proximity with humans, although some humans have been capturing and displaying wild animals from all reaches of the globe for thousands of years.

The first exotic pets appeared over 2,000 years ago in ancient empires such as Greece, Egypt, and Rome, when imported animals were gifted in royal courts.

With the onset of European imperial expansion and the so-called "Age of Exploration" in the late 15th century, exotic animal trade escalated (Belozerskaya 2006). From that point on, animals were imported from colonies into royal menageries and zoological gardens in Europe, where they served as symbols of the conquest of distant lands. The Victorian era ushered in a flood of exotic animals in England, and private menageries and pets became common outside royal courts. The 20th century marked a dramatic increase in private exotic animal ownership worldwide. Today, millions of animals are in circulation as exotic pets at any given moment, sold in greater and greater numbers at pet stores, online, and at exotic animal auctions.

Data on the global exotic pet trade are scarce. Legal international wildlife trade, of which the exotic pet trade is a part, is estimated to be worth as much as $150 billion, and the multi-billion dollar illegal wildlife trade is widely considered the third-largest black market in the world. Trade predominantly flows from biodiversity-rich, economically poor nations in Southeast Asia, Central America, and Africa to monetarily wealthy countries like the United States, Japan, the United Kingdom, and China, with the Middle East recently emerging as a significant importing zone (Bush et al. 2014). The United States is the world's leading importer of exotic pets (Smith et al. 2009). Between 2000 and 2006, the United States imported 1.48 billion live animals. Over half of the individual specimens were aquarium fish, and 90 percent of the shipments were designated as exotic pets. Eighty percent were wild-caught. The highest volume of trade is in fish, followed by amphibians, reptiles, birds, and mammals. Trade volumes and distributions can shift quickly, however, in response to changing demand, as the exotic pet trade is a demand-driven economy. For example, the film *Finding Nemo* caused a spike in demand for clownfish (Nemo's species) that has led to their near extinction in the wild.

Attempts to regulate the trade occur at all levels of government, from municipal to international. Most countries have signed the Convention on International Trade in Endangered Species of Flora and Fauna (CITES), an international agreement that came into effect in 1975 "to ensure that international trade in specimens of wild animals and plants does not threaten their survival" (CITES 2013). CITES decides which species can be legally traded worldwide and in what amounts. In North America, however, laws concerning exotic pet ownership largely fall to states and provinces, many of which do not regulate exotic pets at all.

For many experts, this lack of regulation is worrying. Exotic pets pose a risk to human health and safety, particularly because some infectious diseases they carry are transmittable to humans (Smith et al. 2009). Ecological risks are also significant. Species loss due to the exotic pet trade can be so dramatic that experts have coined the term "empty forest syndrome" to describe some of these exporting zones. In importing regions, too, exotic pets can escape or be illegally released into non-native environments, where they may become invasive. The most famous case of this is the breeding population of Burmese pythons now established in the Florida Everglades. But the traded animals themselves arguably bear the risks of the exotic pet trade most profoundly. Prepurchase mortality rates within the trade are as high as 70 percent for reptiles and some birds, or 80 percent for wild-caught marine fish, with similar mortality rates persisting within the first year after purchase (Ashley et al. 2014). Experts argue it is difficult if not impossible to provide adequate care for exotic pets (Baker et al. 2014). If an animal does survive, negative effects often plague its captive life, including disease, post-traumatic stress disorder, anxiety, lethargy, and a generally diminished quality of life. These drastic effects are leading many researchers and public officials to call for an end to the exotic pet trade.

Rosemary-Claire Collard

See also: Black Market Animal Trade; Domestication; Extinction; Invasive Species; Pets; Welfare; Wildlife; Zoonotic Diseases; Zoos

Further Reading

Ashley, S., Brown, S., Ledford, J., Martin, J., Nash, A-E., Terry, A., Tristan, T., and Warwick, C. 2014. "Morbidity and Mortality of Invertebrates, Amphibians, Reptiles, and Mammals at a Major Exotic Companion Animal Wholesaler." *Journal of Applied Animal Welfare Science* 17(4): 308–321.

Baker, S., Cain, R., van Kesteren, F., Zommers, Z., D'Cruze, N., and Macdonald, D. 2013. "Rough Trade: Animal Welfare in the Global Wildlife Trade." *BioScience* 63(12): 928–938.

Belozerskaya, M. 2006. *The Medici Giraffe: And Other Tales of Exotic Animals and Power.* New York: Little, Brown and Company.

Bush, E., Baker, S., and MacDonald, D. 2014. "Global Trade in Exotic Pets 2006–2012." *Conservation Biology* 28(3): 663–676.

CITES. 2013. "What Is CITES?" Accessed March 21, 2015. http://www.cites.org/eng/disc/what.php

Smith, K., Behrens, M., Schloegel, L., Marano, N., Burgiel, S., and Daszak, P. 2009. "Reducing the Risks of the Wildlife Trade." *Science* 324: 594–595.

Extinction

Extinction is the permanent disappearance of a species or a group of species. Although the moment of extinction is usually considered to be the death of the last individual of the species, the capacity to reproduce and recover may have been lost before this point. Extinctions have enabled other species to invade new areas and, in doing so, have significantly affected the distribution of animals across Earth.

The earliest writings about nature, such as Aristotle's (384–322 BCE) 10-volume *History of Animals*, seldom considered that life on Earth had a history. By the end of the Enlightenment (1650s–1790s), however, some intellectuals—including American politician and fossil-collector Thomas Jefferson (1743–1826)—rejected extinction, claiming instead that unidentifiable fossils were the remains of creatures still in existence but living in unexplored parts of the continent. When, as U.S. president, Jefferson sent Meriwether Lewis (1774–1809) and William Clark (1770–1838) on their famous expedition to the Northwest, he hoped they would find these creatures.

Extinction was not scientifically documented as a real event until 1796, when French naturalist and aristocrat Georges Cuvier (1769–1832) used fossils to prove that extinctions had occurred. Cuvier attributed extinctions to catastrophic events (e.g., earthquakes, floods); after each catastrophe, organisms from other areas repopulated Earth. Later, Charles Darwin (1809–1882) believed that extinction was a natural consequence of evolution by natural selection, in which differential reproductive success results in biological traits becoming more or less common in a population. To Darwin, extinction and living species were like dead and living branches of his Tree of Life.

Darwin was right: Extinction is a common event. Indeed, more than 99 percent of all species that have lived are extinct. Almost 2 million species have been described by science, and scientists estimate that there are as many as 11 million species on Earth. However, millions of species are also extinct, and most of these extinctions occurred before humans appeared. Although some species have lasted hundreds of millions of years, most species have become extinct within 10 million years after they appear. For example, the fossil record shows that species of marine invertebrates have an average lifespan of 5 million years, whereas species of mammals have an average lifespan of 1 million years.

During at least five periods in Earth's history, extinction rates have peaked; these peak times of extinction, which each destroyed more than half of all species, are called mass extinctions. The largest mass extinction occurred at the end of the Permian Period (250 million years ago), and the most famous mass extinction occurred at the end of the Cretaceous Period (66 million years ago). The end-Cretaceous extinction, which is linked with a meteor hitting Earth, is famous because it annihilated dinosaurs, the arch symbols of extinction. Mass extinctions have been correlated with changes in Earth systems, such as volcanic eruptions and changes in oceans' circulations, which made it impossible for organisms to find suitable habitats.

Extinctions are vital for evolution, for they open new habitats; this is why the evolution of new species increased after each mass extinction. For example, just as the end-Permian extinction offered opportunities for reptiles, the end-Cretaceous extinction offered opportunities for mammals to diversify into a multitude of forms. As Elisabeth Vrba (1942–) has shown, the same factor—namely, the altered nature and distribution of habitats—that causes extinction promotes the evolution of new species.

Before humans, the average extinction rate was approximately 1 extinction per every 10 million species per year. Today, however, that rate has increased by at least 1,000-fold because of overharvesting (for example, of fish such as tuna and cod) and altered habitats caused by urban sprawl, mining, logging, disease, pollution, and the introduction of invasive species. One of the earliest species driven to extinction by humans was the dodo (*Raphus cucullatus*), a flightless bird last seen in 1662 on the island of Mauritius in the Indian Ocean. Since then, humans have driven several other species to extinction.

In the early 1800s, billions of passenger pigeons (*Ectopistes migratorius*) flew across North America. However, the last passenger pigeon on Earth—a female named Martha, who lived her entire life in a cage—died in 1914 in the Cincinnati Zoo. The last pair of great auks (*Pinguinus impennis*), a flightless coastal bird of the North Atlantic, was killed by a collector in 1844. California's Tecopa pupfish (*Cyprinodon nevadensis calidae*) lived only in outflows of North and South Tecopa Hot Springs, just southeast of California's Death Valley National Park. Development at the springs led to the pupfish being the first animal to be officially declared extinct according to provisions of the U.S. Endangered Species Act. The Tasmanian tiger (*Thylacinus cynocephalus*), the largest known carnivorous marsupial of modern times, was declared a protected species in 1936—the same year that it became extinct. The last Tasmanian tiger died when it was locked out of its shelter and froze to death in a Tasmanian zoo.

Unlike mass extinctions of the past, today's ongoing mass-extinction is being driven not by galactic or planetary processes, but instead by humans. Indeed, extinction has accompanied the colonization of an area by humans. Today, 12 percent of mammals, 31 percent of reptiles, 12 percent of birds, 30 percent of amphibians, and 37 percent of fish are threatened with extinction, the most serious and irreversible effect of humans on Earth's other forms of life. Many organizations, such as the International Union for Conservation of Nature, promote sustainability, which presumes that humans will allow for all species to live.

Randy Moore

See also: Animals; Biodiversity; Climate Change; Coral Reefs; Deforestation; Endangered Species; Evolution; Flagship Species; Habitat Loss; Human-Wildlife Conflict; Hybrid; Indicator Species; Invasive Species; Keystone Species; Species; Wildlife; Zoos

Further Reading

Eldredge, N. 2014. *Evolution and Extinction: What Fossils Reveal about the History of Life*. Buffalo, NY: Firefly Books.

Kolbert, E. 2014. *The Sixth Extinction: An Unnatural History*. New York: Henry Hold and Co.

MacLeod, N. 2013. *The Great Extinctions: What Causes Them and How They Shape Life*. Buffalo, NY: Firefly Books.

F

Factory Farming

"Factory farming" is a term that refers to the rearing of animals for meat, milk, or eggs using practices geared toward maximum output per animal. Although not synonymous, the term is often used interchangeably with CAFO, a U.S. Environmental Protection Agency (EPA) acronym for "concentrated animal feeding operation." Although begun and still most prominent in North America and Western Europe, factory farming is rapidly increasing worldwide. Animal advocates, environmentalists, public health officials, and worker health and safety advocates have all challenged factory farming practices.

The term factory farming was coined by British author and animal advocate Ruth Harrison (1920–2000) in 1964 to indicate that industrialized farms were treating animals more as machines than as living beings. At that time, the intensification of animal production and focus on maximum output were new. For example, farmed animals had previously frequently obtained food on their own through grazing or foraging. Under industrial production, animals began to be housed in close quarters and exclusively fed manufactured feed. Primarily, two developments allowed for this change. First, mid-20th-century animal nutrition science showed how to produce feed that met nutritional needs more efficiently and allowed for more and/or faster weight gain than grazing or foraging. Second, antibiotics developed for preventing disease transmission between World War II soldiers in close quarters began to be used for the same purpose on the now-confined animals. Since those early years, developments in genetic science and farm mechanization (e.g., automated milking systems) have further increased animals' productivity.

Factory farming has increased livestock consumption and changed agricultural landscapes. For example, in 2014, over 9 *billion* total cattle, chickens, hogs, ducks, sheep, lambs, and turkeys were slaughtered for U.S. consumption (HSUS 2015), and more than 99 percent of these animals came from factory farms (ASPCA 2015). Additionally, numbers of farms have decreased while animals per farm have increased. For example, in the United States between 1970 and 2006, the number of dairy farms decreased by 88 percent, while the average herd size increased by over 600 percent (MacDonald et al. 2007). Worldwide, almost 60 billion animals are raised and slaughtered in food production annually (Worldwatch Institute 2013). With the global demand for meat increasing, the number of animals slaughtered and the proportion of demand met by factory farms are also expected to increase.

Jack Salzsieder, manager of The Odor Control Company, checks an odor meter in a hog confinement barn on the Tom Uthe farm near Slater, Iowa, in 2001. Such farms raise concerns about not only animal welfare but also human health, in terms of air and water quality in surrounding areas and antibiotic overuse more broadly. (AP Photo/Charlie Neibergall)

This type of farming is growing globally, especially in Asia, Latin America, and Eastern Europe. Although rightfully criticized on a range of issues, factory farming does substantially lower the purchase price for meat and reduces the amount of open land needed for grazing/foraging. For example, as factory farming has grown in Australia over the past several decades, poultry prices have decreased by 40 percent (ACMF 2012).

The close confinement of animals, however, raises important animal advocacy and welfare concerns, especially with chickens, pigs, and veal calves. Chickens raised for meat (called "broilers") are kept in large "houses," with many in the United States containing tens or even hundreds of thousands of birds, frequently living in one square foot (approximately 30 cm) or less of space each. Sows (female pigs) are kept in metal crates while gestating (pregnant) and shortly after farrowing (giving birth). Gestation/farrowing crates are so confining that sows can only lie down in one position and cannot turn around or interact freely with their piglets. Calves that go into veal production are typically males that are not useful to the dairy industry. Almost immediately after birth, these calves are placed into crates in which they are not able to turn around or interact normally with others of their species and are kept there until transported for slaughter.

More light is also being shed on factory farming's role in global climate change and other environmental issues, similar to other polluting entities such as coal-fired power plants. According to the Food and Agriculture Organization of the United Nations (FAO), meat production accounts for more greenhouse gases (GHGs) than transportation or industry. For example, producing a hamburger contributes as much or more GHGs to the atmosphere as driving an average car to work (FAO 2006). With the concentration of large numbers of animals together, waste containment and disposal is an issue, with the gases released by urine and feces decreasing air quality and increasing the risk that this waste will contaminate surrounding water bodies.

In terms of human health and safety, a primary concern is the high use of antibiotics in factory farming. Many of these drugs are the same as, or very similar to, ones used to fight human disease. The extensive use of these drugs is a main contributor to antibiotic resistance in strains of bacteria, creating risks to both human and animal health. In the United States, 70 percent of all antibiotic use is on agricultural animals. According to a U.S. Food and Drug Administration (FDA) 2014 report, antibiotic use for livestock increased by 16 percent from 2009–2012. Another concern for humans relates to injury rates for those who work in the industry, frequently immigrants and minorities.

There are ongoing efforts to challenge, reform, and/or do away with factory farms. For example, many individuals choose to become vegetarian or vegan, or to only purchase animal products from nonindustrial farmers. Institutions such as the Humane Society of the United States have anti-factory-farm campaigns, and the organization Compassion in World Farming is dedicated to ending these practices.

Connie L. Johnston

See also: Advocacy; Aquaculture; Bovine Growth Hormone; Climate Change; Concentrated Animal Feeding Operation (CAFO); Cruelty; Deforestation; Ethics; Humane Farming; Livestock; Meat Eating; Meat Packing; Pastoralism; People for the Ethical Treatment of Animals (PETA); Shelters and Sanctuaries; Slaughter; Veganism; Vegetarianism; World Organisation for Animal Health; Zoonotic Diseases

Further Reading

ACMF. 2012. "The Australian Chicken Meat Industry: An Industry in Profile." Accessed February 15, 2016. http://www.chicken.org.au/industryprofile/page.php?id=4.4_Consumption

ASPCA. 2015. "Farm Animal Cruelty." Accessed September 24, 2015. https://www.aspca.org/fight-cruelty/farm-animal-cruelty

Compassion in World Farming. 2015. Accessed September 30, 2015. http://www.ciwf.org.uk

FAO. 2006. *Livestock's Long Shadow: Environmental Issues and Options.* Rome: United Nations.

FDA. 2014. *2012 Summary Report on Antimicrobials Sold or Distributed for Use in Food-Producing Animals.* Washington, DC: Department of Health and Human Services.

Harrison, R. 1964. *Animal Machines: The New Factory Farming Industry.* London: Vincent Stuart Publishers.

HSUS. 2015. "Farm Animal Statistics: Slaughter Totals." Accessed September 24, 2015. http://www.humanesociety.org/news/resources/research/stats_slaughter_totals.html

MacDonald, J., O'Donoghue, E., McBride, W. D., Nehring, R., Sandretto, C., and Mosheim, R. 2007. *Profits, Costs, and the Changing Structure of Dairy Farming.* Washington, DC: Economic Research Service/USDA.

Nierenberg, D. 2005. *Happier Meals: Rethinking the Global Meat Industry.* Washington, DC: Worldwatch Institute.

Worldwatch Institute. 2013. "Meat Production Continues to Rise." Accessed September 30, 2015. http://www.worldwatch.org/node/5443#notes

Feedlots. *See* Concentrated Animal Feeding Operation (CAFO); Factory Farming

Feral Animals

Feral animals are commonly defined as those from a domesticated species who live and/or reproduce without direct and/or intentional human supervision, control, or support. Based on this definition, a feral animal is distinct from both domestic animals and wild animals, including wild species living in captivity (such as those in zoos). The term is sometimes used interchangeably, or overlaps, with such descriptors as "stray," "free-ranging," "free-roaming," or "wild." This is particularly common with dogs, cats, horses, and pigs. Feral populations have been identified in nearly every region of the world and in a diverse assortment of species, such as domestic dogs (as distinct from "wild" dogs such as the African wild dog or the dingo), domestic cats, camels, goats, pigs and wild boars, equines (horses, donkeys, and burros), rabbits, water buffalo, and domesticated birds (including select doves, parrots, and waterfowl).

Although the aforementioned definition is used widely, perceptions of what constitutes a feral animal can vary according to species, academic discipline, and geography. In particular, discussions of feral companion animal species (cats and dogs) in veterinary literature often refer to specific behaviors of feral animals in relation to humans. Within this context, feral animals have been defined variously as those that are not socialized to humans, continuously avoid direct human contact, do not tolerate handling by humans, and/or are aggressive toward humans when trapped. Notably, these criteria mean that "feral" need not be a permanent or absolute status for an individual animal. A once-socialized animal may be abandoned and become feral; an animal may also be born feral but later socialized, especially at an early age, to accept human contact.

There is one exception to the core definition of the term "feral," which connotes members of a domesticated animal species. The Commonwealth of Australia uses the term to describe its deer, red fox, and cane toad populations. All three species were introduced to the continent for recreational hunting, as game, and for insect population control, respectively. Within this context, "feral" is used to describe a population of wild animals introduced (but not domesticated) by humans, and which in turn had adverse impacts on native plants and animals.

The way in which "feral" is defined in a particular context can influence how populations are managed and what protections are afforded individual animals. Feral animals may be categorized as "invasive" and/or "pest" species due to their capacity to adversely impact native wildlife, native plants, domestic livestock, agriculture, biodiversity, ecological integrity, and regional economies.

Humans have utilized a variety of approaches to try to manage, reduce, or eliminate populations of feral animals. Non-lethal approaches include surgical and non-surgical fertility control, which have been used in species such as dogs, cats, deer, horses, and pigs. Trap-neuter-return, for example, is the process of surgically sterilizing (spaying or neutering) and vaccinating feral cats. Contraceptive vaccines are currently being used on limited scales to suppress fertility in feral horses and pigs, as well as in wildlife species. Diversionary feeding (use of food to attract animals to an alternate location), control of resources, fencing, relocation, and repellents are additional non-lethal strategies used to manage populations of feral animals, reduce their damage to sensitive ecological areas, and/or mitigate conflict with humans and other species.

Management is also often pursued through lethal means (often referred to as "culling"), including shooting, poisoning, trapping, and biological control (introducing another living organism to control a targeted species). Lethal control of feral populations is controversial. Used in isolation, it necessitates ongoing removal of large numbers of individuals to offset births and immigration of new animals. Moreover, some eradication efforts have had unintended consequences—for example, increased numbers of rats following culling of feral cats—prompting a call for careful evaluation of complex and altered ecosystems prior to pursuing eradication of a feral and/or introduced species (Zavaleta 2002).

Most fundamentally, lethal control prompts ethical debate about the intrinsic worth of feral animals relative to native species, native ecosystems, and human interests, as well as the suffering that some culling methods cause to animals. It is important to recognize that stakeholders may have varied and often incompatible views about an animal's or species' intrinsic, cultural, or subsistence value. As one example of the latter, although feral pigs can transmit disease and threaten biodiversity, agriculture, and ecological integrity, they are recognized as serving as an important source of food and economic revenue for some Australian Aboriginal communities (Koichi et al. 2012; Robinson and Wallington 2012). Emotional reactions, attachment, and resistance have historically been particularly strong regarding culling of charismatic, iconic, or companion species such as horses, dogs, and cats. This is arguably to be expected, given that the line dividing individuals of the same species who are valued companions and who are considered problematic or "invasive" is a gray area that can change.

The complex circumstances in which feral animals, domesticated animals, wildlife, and humans coexist mean that decisions for managing feral populations require consideration of more than the animals' biological, ecological, or epidemiological (disease- and health-related) impacts. Scholarship is increasingly recognizing cultural, social, economic, policy, ethical, and moral considerations in developing protocols to manage feral populations.

Valerie Benka

See also: Animals; Biodiversity; Cats; Dogs; Domestication; Ethics; Invasive Species; Welfare

Further Reading

Commonwealth of Australia. 2015. "Feral Animals in Australia." Accessed April 1, 2015. http://www.environment.gov.au/biodiversity/invasive-species/feral-animals-australia

Koichi, K., Sangha, K. K., Cottrell, A., and Gordon, I. J. 2012. "Aboriginal Rangers' Perspectives on Feral Pigs: Are They a Pest or a Resource? A Case Study in the Wet Tropics World Heritage Area of Northern Queensland." *Journal of Australian Indigenous Issues* 15: 2–19.

Levy, J. K., and Crawford, C. P. 2004. "Humane Strategies for Controlling Feral Cat Populations." *Journal of the American Veterinary Medical Association* 225: 1354–1360.

Robinson, C. J., and Wallington, T. J. 2012. "Boundary Work: Engaging Knowledge Systems in Co-management of Feral Animals on Indigenous Lands." *Ecology and Society* 17: 16–27.

Singer, P. 1997. "Neither Human Nor Natural: Ethics and Feral Animals." *Reproduction, Fertility and Development* 9: 157–162.

Zavaleta, E. S. 2002. "It's Often Better to Eradicate, but Can We Eradicate Better?" In Veitch, C. R., and Clout, M. N., eds., *Turning the Tide: The Eradication of Invasive Species,* 2nd ed., 393–403. Gland, Switzerland, and Cambridge, UK: World Conservation Union.

Fighting for Human Entertainment

In nature, individual animals fight others of the same or different species for reasons such as food, territory, protecting offspring, and territorial dominance. Animal fighting as entertainment for humans, also called "blood sports," possibly evolved from Neolithic (ca. 20,000–6000 BCE) observations of combat between male animals for mating privileges. Humans have staged contests between animals such as horses, cocks, camels, boars, dogs, scorpions, and crickets.

In antiquity (before the fourth and fifth centuries CE), several cultures, notably the Roman, regularly staged elaborate fights between animals, as well as against humans.

It is thought that the first animal fighting events in the Roman Colosseum were executions of criminals by means of predatory animals, and that these events evolved to include both gladiatorial events between humans and fights between animals for entertainment. For example, dogs were regularly matched against lions, boars, bulls, aurochs (an extinct wild ox), and monkeys. These animal spectacles were so popular that a supply chain that stretched to Africa and India was used to provide an estimated 10,000 animals annually, including lions, hyenas, rhinoceroses, zebras, crocodiles, hippopotami, tigers, leopards, and giraffes.

Such blood sports continued during the Middle Ages (ca. 5th through 15th centuries CE), and "bull baiting" and "bear baiting" events, in which bulls and bears were pitted against dogs, are well represented in the historical records in England from the reign of King John (1199–1216) onward. Under King James I (1566–1625), the role of "Master of the Games, Beares, Bulles and Dogges" was created to ensure a constant supply of well-run animal fighting events at Court. It was about this time that the "*bulldogge*" was established by crossing a mastiff with a terrier in order to create a compact dog with both strength and tenacity. The large mastiffs (40–60 kg/88–132 lbs.) used in bull baiting attacked from the front and were easily gored. The *bulldogge*, the forerunner of the pit bull, was low and light enough (20–25 kg/44–55 lbs.) to crawl under a bull and lunge upwards at the neck or nose without being gored. In the 18th century, however, all forms of animal fighting were made illegal in the United Kingdom because of evolving concerns over gambling, a perceived association with cruelty between humans, and growing concern for the animals involved.

In the United States, dogfighting and cockfighting have historically been the two dominant blood sports. Dogfighting was a popular tradition carried over by British and Irish immigrants in the 19th and early 20th centuries. The bloodlines for fighting dogs are sometimes elaborately documented in studbooks (registries that detail the ancestry of purebred dogs) dating back to the early 20th century. However, the inbreeding (the mating of closely related individuals) required for maintaining certain bloodlines—notable especially among the "bluenose pit bulls"—has resulted in higher incidence of certain congenital (inherited) diseases. Cockfighting, a contest between two roosters or "gamecocks" who have razors attached to their feet, has historically been popular with other immigrant groups such as Filipinos, descendants of Caribbean slaves, and Mexicans, as well as descendants of European immigrants living in the rural South. Many different breeds are used, but the best gamecocks are thought to be from the Philippines and the Dominican Republic.

Dogfighting and cockfighting in America takes place in barns, warehouses, basements, and parks equipped with "pits," or fighting rings. Dogs are typically trained for fighting first by play fighting with an older fighting dog, then by being released upon a "bait dog" who has had its mouth taped shut, then through practice fights. Fights are usually to the death or until one animal refuses to fight. Cockfighting involves similar rituals.

Dogfighting and cockfighting are illegal in all U.S. states, but gambling and narcotics trafficking are also associated with dogfighting, with spectators frequently violating multiple statutes in organizing and attending dogfights. In the famous 2007 case, American football player Michael Vick was implicated in multiple crimes such as operating a dogfighting enterprise at his farm, engaging in animal torture and killing, gambling, drug possession, racketeering (running an illegal business), and conspiracy, but was never actually charged with dogfighting in federal court. Instead, Vick negotiated a federal plea agreement to a much lesser charge and paid a minimal fine in state court for dogfighting. Dogfighting and/or cockfighting are legal in many other countries, such as China, the Philippines, and the Dominican Republic. Countries such as Pakistan, Serbia, and South Africa have banned dogfighting.

Animal fighting is not limited to birds and mammals. Cricket fighting, indigenous (native) to China, dates back over 1,000 years to the Tang Dynasty. Chancellor Jia Sidao (1213–1275) wrote what is considered the foundational treatise on cricket fighting in the 13th century. The book describes the origins of cricket fighting in Shandong Province and, among other things, describes how to anger a cricket into becoming ready to fight by teasing its tendrils with a small wooden shaving. Briefly banned under Chairman Mao Zedong (1893–1976), cricket fighting is legal today, although gambling on the fight remains illegal.

Animal fighting has never been more popular than it is today when measured by actual numbers of fights and animals bred and used despite legal restrictions. However, the increasing number of laws banning fighting, and displays of public outrage over such cases as Michael Vick's, indicate that there is broad sentiment against these activities. Additionally, animal lawyers have begun to advocate for fighting dogs to be recognized as crime victims and represented by legal guardians.

John T. Maher

See also: Animal Law; Breed Specific Legislation; Bullfighting; Cruelty; Dogs; Pit Bulls

Further Reading

Coley, J. 2010. "Roman Games: Playing with Animals." In *Heilbrunn Timeline of Art History*. New York: The Metropolitan Museum of Art. Accessed February 4, 2016. http://www.metmuseum.org/toah/hd/play/hd_play.htm

Fleig, D. 1996. *The History of Fighting Dogs*. Neptune, NJ: TFH Publications.

Gorant, J. 2008. "What Happened to Michael Vick's Dogs." *Sports Illustrated* 26: 72–77.

Jacobs, A. 2011. "Chirps and Cheers: China's Crickets Clash." *New York Times*, November 5. Accessed February 4, 2016. http://www.nytimes.com/2011/11/06/world/asia/chirps-and-cheers-chinas-crickets-clash-and-bets-are-made.html?_r=0

Jennison, G. 2005. *Animals for Show and Pleasure in Ancient Rome*. Philadelphia: University of Pennsylvania Press.

Lorenz, K. 1966. *On Aggression*. Orlando, FL: Harcourt Brace & Company.

Stratton, R. 1992. *The Book of the American Pit Bull Terrier*. Neptune, NJ: TFH Publications.

Toynbee, J. M. C. 1973. *Animals in Roman Life and Art*. Ithaca, NY: Cornell University Press.

Fisheries

Fisheries are aquatic systems managed for the harvesting or raising, processing, and selling of fish. The majority of fisheries focus on wild capture, although aquacultural production has grown substantially in past decades. Fisheries can be formally or informally defined and managed, but all involve human intervention in the interest of harvesting species of commercial, recreational, or subsistence value. A fishery is typically defined by a commonality in species sought and human extractive activity within a bounded area, with fishers fishing for certain species with similar gear types in a specific region. Fisheries are significant contributors to employment and economy, with 10 to 12 percent of the world's population relying on fisheries for their livelihoods (UN FAO 2014). While fisheries face great challenges in preserving stocks at levels for continued harvest, implementation of sustainable fishing methods and effective management strategies can work to ensure the future of species, ecosystems, and livelihoods.

Fishery environments vary from saltwater to freshwater and include river, coastal, and marine systems. The majority of saltwater fisheries are located in coastal regions, as these shallow-water environments typically contain greater fish species diversity and abundance, although productive fisheries also exist in open ocean environments including seamounts, or underwater mountains that do not reach to the water's surface, where red snapper (*Lutjanus campechanus*), tuna (*Scombridae*), and orange roughy (*Hoplostethus atlanticus*) are commonly harvested. Freshwater wild fisheries include inland lakes and rivers, while fish farming frequently occurs in lakes, ponds, tanks, and man-made aquatic enclosures.

In 2012, wild fisheries contributed more than $274 billion to the global economy, with aquaculture contributing an additional value of $137.7 billion (World Bank 2012). This represented a total global production by fishers of 158 million tons, with a wild capture of 91.3 million tons and aquaculture production at 66.6 million tons (UN FAO 2014). China is the top-ranking fishing country in terms of total quantity produced, followed by Indonesia, the United States, India, and Peru (UN FAO 2014).

Within fisheries, the term "fish" applies to a broad array of species, including molluscs (e.g., clams, scallops, oysters), crustaceans (e.g., crabs, lobster, shrimp), echinoderms (e.g., sea urchins, sea cucumbers), and any other aquatic animal that is harvested by humans for commercial gain, recreation, or subsistence. Species meeting the biological definition of a fish are referred to as "true fish" or "finfish."

A very small number of species support the majority of the world's fisheries. The Peruvian anchoveta (*Engraulis ringens*) is the most widely wild-caught species in the world as measured by total tonnage, followed by Alaska pollock (*Theragra chalcogramma*), skipjack tuna (*Katsuwonus pelamis*), Atlantic herring (*Clupea harengus*), and chub mackerel (*Scomber japonicus*). These six species alone contributed to nearly 20 percent of the total global wild catch in 2012 (UN FAO 2014).

In recent decades, improvements in capture technologies, including changes in net materials, electronic fish detection, and the use of global positioning (GPS) technology, have led to the more rapid location of fish and more efficient harvests, generally increasing yield while also lowering the unit costs. This has vastly expanded the scale on which fishing operations take place, resulting in rapid depletion of stocks in different fisheries, including cod (*Gadus morhua*) in the North Atlantic, hake (*Merluccius merluccius)* in the Mediterranean sea, and blue crab (*Callinectes sapidus)* in the Chesapeake Bay in North America. This process, known as *overfishing*, involves the taking of fish beyond sustainable levels. A stock that has undergone overfishing may be designated as *overfished*, indicating the population is too low to be stable. Frequently, target species populations are not the only species impacted by fisher behavior, as fishing activity typically leads to *bycatch*, or the unintentional capture and mortality of nontarget fish, birds, reptiles, amphibians, and mammals, contributing to fishery decline and impacting the stability of ecosystems.

Many experts argue that overfishing makes evident the need for effective management strategies to aid recovery and ensure the sustainability of fish stocks and fishery-related livelihoods worldwide. Management interventions include catch and bycatch limits, gear restrictions, and restrictions on when and where fishing may occur. Social, political, economic, and logistical challenges present significant barriers for effective management, including the inherent difficulty of dealing with issues of economic and social importance within human communities; the difficulty in identifying biological objectives for stocks; the availability, quality, and quantity of the data required; the difficulties of implementation; and the costs and logistical challenges of enforcement.

In areas where sustainable management methods have been introduced, stocks are starting to recover. For instance, lingcod (*Ophiodon elongates*) stocks are rebounding on the west coast of North America, where they are managed under a "catch share" management plan where managers divide the total allowable catch for the fishery into shares controlled by fishers, allowing them the flexibility to calibrate their fishing activity to seasonal target stock variation, bycatch species variation, weather conditions, and fluctuations in market price. This demonstrates that managing for conservation of target species is not contrary to economic productivity. The World Bank suggests that, with full implementation of sustainable fishing measures, the capacity of global fisheries could be increased by an additional $50 billion annually (2012).

Sharon Wilcox

See also: Biodiversity; Fishing; Wildlife Management

Further Reading

Food and Agriculture Organization of the United Nations (UN FAO), Fisheries and Aquaculture Department. 2014. *The State of World Fisheries and Aquaculture.* Rome.

Greenberg, P. 2010. *Four Fish: The Future of the Last Wild Food.* New York: The Penguin Press.

Kurlansky, M. 1997. *Cod: A Biography of the Fish That Changed the World.* London: Walker and Company.

Schrope, M. 2010. "Fisheries: What's the Catch?" *Nature* 465: 540–542.

United States National Oceanic and Atmospheric Administration: NOAA Fisheries. 2015. Accessed April 1, 2015. http://www.nmfs.noaa.gov

World Bank. 2012. *Hidden Harvest: The Global Contribution of Capture Fisheries.* Washington, DC.

Fishing

Fishing has been a part of human existence since before the dawn of civilization and documented history. Indeed, it is not an exclusively human endeavor: Fishing may be equally observed being practiced by countless animals in their quest to secure aquatic life as a source of food. The animal world avidly competes with humans over access to the best fishing and, as such, fishing is an activity much contested within its various multispecies relationships. Within this space there is not room to explore a more diverse and multispecies account of "fishing." However, this exploration does make clear that the relationships embodied in fishing, between humans, animals, and technologies of capture, have changed little in terms of their mechanics for many thousands of years. Matters of scale and impact, however, have changed dramatically. Thus, fishing may be considered an allegorical lens through which a broader environmental struggle between humans and other coexistent beings may be viewed.

Early evidence of fishing can be found in many regions of the world but can prove difficult to fully understand due to shifts in tidal reaches and loss of evidence due to land erosion (Torben and Erlandson 2000). However, sites in East Timor, South Africa, and California have contributed significant insights and material artifacts to the archaeological record. The oldest known find of a single-piece, bent fishing hook, dated at approximately 16,000 to 23,000 years old, was at a site in East Timor. It was accompanied by bones of "pelagic" fishes (those at home in the open ocean) that were up to 42,000 years old. This implies that deep-sea fishing, which demands

A Nova Scotia fisherman hauls in trawl nets full of cod. For most of human history fish have been an important food source. Smaller-scale systems of pole fishing or netting are still used by many people; however, large-scale industrial fishing by huge trawler boats is rapidly reducing the number of fish and the overall health of the ocean. (Corel)

a significant level of advanced maritime and technical skill, was taking place at this time. Similarly, in the "Blombos Cave" of South Africa, evidence of inshore fishing without bent hooks (of the kind found in East Timor) dates back 140,000 years. The southern California coast has yielded evidence of a diversity of fishing techniques utilized in the area over a period of many millennia. Some 10,000 years in the past, various nets, hooks and lines, harpoons, spears, and canoes were all tools in the human quest for food in the form of aquatic animals. The success of these innovations promoted periods of high human population density, complex cultures, and technical specialization.

Globally, fishing can be divided into three broad categories: subsistence, recreational, and commercial. Each of these encompasses a vast diversity of practices and economic relationships that differ as much within themselves as with each other. Subsistence fishing, as an example, is particularly difficult to define. It is so widespread throughout the globe, and so closely intertwined within local communities, that any attempts to regulate its practices or measure environmental impacts are almost impossible. Many communities in Asia, Africa, Latin America, and the Pacific Islands rely on small-scale fishing as a source of food and as an income supplement through local trading, but this is far from a comprehensive picture. Subsistence fishing for crab and fish (as a food source) remains widespread even in developed regions such as the U.S. East Coast.

Subsistence fishing serves multiple purposes and is closely integrated with local communities through knowledge sharing, tradition, and the exchange value of fish to meet other specialized needs. Some subsistence fishing practices are "artisanal"—that is, the art and skill of the activity itself holds considerable cultural significance, irrespective of any captured food value. An excellent example of this may be found within indigenous communities permitted to continue hunting otherwise protected whale species as a part of their traditional heritage.

Once removed from a requirement to provide food, the practice of fishing becomes a recreational celebration of diverse traditions and methods that specialize in targeting specific species of fish. Most recreational fishing utilizes either an artificial lure or food bait containing one hook or several hooks arranged by the angler on a "longline" to induce a predatory or feeding response from their chosen aquatic quarry. Other techniques do exist (bow and spear fishing, for example); however, these can result in the death of the target animal. Recreational fishing may still result in fish death and eating the catch in countries with ample food supply (Burger 2002). Releasing fish alive after capture often depends on the species, local culture, and the conservation context in which the activity takes place. Species considered as geographically invasive or ecologically damaging are often killed with impunity.

The global demand for protein and convenience foods drove an increasing harvest of pelagic fish species through large commercial fishing boats in the 20th century. This activity reached a peak in the 1950s and 1960s, with the full mobilization of many nations' factory ship fleets. These vessels could haul up to 450 metric tonnes of fish a day and deliver a pre-filleted product directly to retailers and consumers. After a steep decline in the 1990s, capture activity is being slowly replaced by fish farming practices (aquaculture), which have a more consistent capability to meet global demands for protein. As such, the associated problems of intensive meat production have now become a feature of human-fish relationships.

John Clayton

See also: Aquaculture; Fisheries; Hunting; Invasive Species

Further Reading

Bear, C., and Eden, S. 2011. "Thinking Like a Fish? Engaging with Non-Human Difference through Recreational Angling." *Environment and Planning D: Society and Space* 29: 336–352.

Burger, J. 2002. "Consumption Patterns and Why People Fish." *Environmental Research* 90: 125–135.

Corbyn, Z. 2011. "Archaeologists Land World's Oldest Fish Hook." Accessed March 17, 2015. http://www.nature.com/news/archaeologists-land-world-s-oldest-fish-hook-1.9461

Schumann, S., and Macinko, S. 2007. "Subsistence in Coastal Fisheries Policy: What's in a Word?" *Marine Policy* 31: 706–718.

Sowman, M. 2006. "Subsistence and Small-Scale Fisheries in South Africa: A Ten Year Review." *Marine Policy* 30: 60–73.

Spitzer, M. 2007. "Man v. Gar: The Nature of a Relationship." *Ecotone* 3: 34–42.

Torben, R. C., and Erlandson, J. M. 2000. "Early Holocene Fishing Strategies on the California Coast: Evidence from CA-SBA-2057." *Journal of Archaeological Science* 27: 621–633.

Five Freedoms. *See* Factory Farming; Welfare; Primary Documents

Flagship Species

Flagship species are charismatic animals that tend to evoke emotional responses including affection and sympathy, and thus are likely to arouse public support for their conservation. While biodiversity and ecosystem protection can be vague concepts that are not easy to promote for public awareness and support, these flagship species are readily identifiable animals whose lives, plights, and survival resonate with the general public. Unlike ecologically significant classifications like "keystone," and "umbrella," flagship species are not necessarily critical to an ecosystem's function. Instead, their importance is defined by their appeal to human audiences. Because flagship species are not always ecologically vital, some experts believe that funding dedicated to their conservation and protection would be better allocated to species that have potentially more significant impacts on their ecosystems. The utilization of flagship species can also present a challenge to conservation as a whole, particularly if a flagship species becomes extinct or extirpated (extinct in part of the species' range) from part of its range where conservation strategies had been in place. This could lead to damaged attitudes and involvement of local people, conservationists, and the general public that supports species-specific conservation campaigns.

Due to their ability to attract significant financial support from the public, flagship species are selected as icons or symbols for habitats, ecosystems, environmental issues, conservation programs, and causes. For instance, lions (*Panthera leo*) frequently represent African savanna ecosystems, polar bears (*Ursus maritimus*) are an important symbol of climate change, and the Florida panthers' (*Puma concolor)* plight in the swamps and forests of Florida has come to signify challenges facing the entire geographic region. Many conservation organizations utilize flagship species in their fundraising and promotion materials. The giant panda (*Ailuropoda melanoleuca)* logo of the World Wide Fund for Nature (WWF) is one of the most identifiable logos in the world today. The panda is an excellent example of a species strategically selected not only for its role in an ecosystem but also for socio-economic factors. Dependent on a specific habitat of bamboo forests in central China, the panda's habitat has been greatly reduced by forest fragmentation, farming, deforestation, and other development. The WWF utilizes the giant panda as the organization's ambassador based on the species' ability to succinctly represent complex issues and inspire financial donations, allowing their conservation programs to reach across the globe.

Many flagship species are selected for their broad, international appeal. In particular, large carnivores command considerable global fundraising power. The Bengal tiger (*Panthera tigris tigris*) is frequently chosen for conservation promotion materials and advertising based on its appeal as a rare, beautiful, and powerful animal. International donors contributed approximately $41 million to wild-tiger conservation initiatives from 1998–2005 and have attracted notable celebrities like Hollywood actor Leonardo DiCaprio, who donated $3 million to tiger conservation in 2013 (Linkie and Christie 2007). Funding donated for tiger conservation is also important for other species, as tiger habitats boast incredible biodiversity across the many ecosystem types they inhabit, including savannas, evergreen forests, tropical rainforests, grasslands, mountains, and mangrove swamps. As a result, conservation initiatives promoting tiger conservation directly and indirectly protect many other endangered animals and threatened ecosystems.

Large, charismatic mammals (known as megafauna), including elephants and whales, are also frequently selected as flagship species. Asian elephants (*Elephas maximus*) and African elephants (*Loxodonta Africana)* inspire support from both their range countries as well as Western donors. Like the giant panda and Bengal tiger, the Asian and African elephants face significant challenges to survival. Both elephant species remain threatened by further habitat loss, fragmentation, and poaching fueled by the global illegal ivory trade. Elephants have been used as flagship species in southern Africa and India to conserve important habitat and engender public support for protecting biodiversity. Because elephants require expansive habitat, support for them also benefits other species and ecological preservation over large landscapes. In this way, flagship species can be seen as ambassadors that bring in funding to help conserve other, potentially overlooked, species and ecosystems.

While nongovernmental organizations (NGOs) embrace the use of flagship species, scientists and conservation practitioners share concerns that this approach may be shortsighted. Because these species are selected for their appeal and not the services or benefits they provide to an ecosystem, they are not necessarily the most strategic species to focus on for protection, conservation, and restoration. Less charismatic but nonetheless important animals, including snakes, insects, worms, and crustaceans, are thus typically not the focus of public fundraising campaigns. In extreme cases, identifying a conservation priority based on a flagship species may cause damage to a more threatened species. For instance, the Everglades snail kite (*Rostrhamus sociabilis plumbeus*) is an endangered bird of prey with a slender, curved beak that allows it to eat apple snails from their spiraled shells. The snail kite's restricted diet makes it a habitat specialist (an organism with a limited diet or that depends on confined habitat conditions). In the same ecosystem, the wood stork (*Mycteria Americana*) is classified as a threatened species. Both birds are flagship species for the everglades and are under threat of extinction. However, they have received different attention. The possible detriment to one species for the benefit of another was evident when the Everglades National Park proposed adapting the water flow to enhance stork habitat but was opposed by the U.S. Fish and Wildlife Service on the grounds it would be harmful to the snail kite population. Flagship species are therefore a powerful conservation tool but require caution when considering their role relative to financial allocation and wildlife management strategies.

Kalli F. Doubleday

See also: Endangered Species; Keystone Species; Species; Wildlife Management

Further Reading

Caro, T. 2010. *Conservation by Proxy: Indicator, Umbrella, Keystone, Flagship, and Other Surrogate Species*. Washington, DC: Island Press.

Linkie, M., and Christie, S. 2007. "The Value of Wild Tiger Conservation," *Oryx* 41: 415–416.

Veríssimo, D., MacMillan, D. C., and Smith, R. J. 2011. "Toward a Systematic Approach for Identifying Conservation Flagships." *Conservation Letters* 4: 1–8.

WWF.Panda.Org. "Priority Species." Accessed April 4, 2015. http://wwf.panda.org/what_we_do/endangered_species

YouTube. 2008. " 'Flagship Species' Lions and Tigers." Accessed April 7, 2015. https://www.youtube.com/watch?v=LiKSvdtIJiw

Fox Hunting

Fox hunting has been defined as a "sport" or "blood sport" (i.e., a sport involving the hunting, wounding, or killing of an animal) and entails the use of a pack of fox hounds to chase and often, but not always, kill a fox. The pack is normally followed by members of a hunt on horseback as well as members on foot or in cars. An exception to this has been practiced in Cumbria in the north of England, where fell hounds (lighter, more athletic and independent) instead of fox hounds are employed and followed on foot. A "hunt" is effectively a club that consists of all the associated entities and individuals involved in the upkeep of the fox hound pack and the organization of fox hunts and related social events. The oldest hunt in the United Kingdom, the Bilsdale Hunt in North Yorkshire, was formed in 1668. However, the Quorn Hunt in Leicestershire also claims to be the oldest, though it was formed in 1696. In comparison, the Montreal Hunt (now officially known as the Club de

A rider with a pack of beagles departs on a fox hunt. Fox hunting is a longstanding yet controversial sport in Great Britain. In the early 2000s, traditional fox hunting was banned in England, Scotland, and Wales. Hunt supporters continue to lobby for repeal of the ban. (Corel)

Chasse à Courre de Montréal) in Canada is the oldest hunt in North America, having been formed in 1826.

Fox hunting as we know it today has a long tradition in England, with it being possible to trace the sport back to the 17th century. However, it has been practiced in a number of other countries including America, Australia, Ireland, France, Canada, India, and Italy, among others. While it has a significant history, fox hunting has been and continues to be a divisive issue. The debates for and against this sport are focused on issues of animal rights and human cultural traditions and the right to uphold them.

While it may be said to have a role to play in the regulation of fox populations, its primary function is, and has been, as a sport to be enjoyed by humans. The emergence of this sport followed on from earlier forms of sport hunting on horseback when larger animals such as deer were the preferred prey and more widespread than today in the United Kingdom. The fox, by contrast, became the focus of sport hunting when more preferable animals became scarce. In the process, the fox was transformed from an animal hunted as vermin for eradication to a venerated foe of the hunt. As a venerated foe, a fox that provided a "good" chase and escaped the pack was celebrated.

Fox hunting is clearly a rural sport, and thanks to its longevity it has become an ingrained part of the English rural culture. This can most easily be seen in the contemporary era through an analysis of the significant number of pubs in the English countryside that overtly display their links to the fox hunt through their names (e.g., The Fox and Hound) and associated signage. The link between the pub—itself a cultural icon of the British countryside—and the fox hunt is strengthened through the fact that these places often act as the official gathering place and starting point of hunts.

Fox hunting was banned in England and Wales in 2004, having been banned two years earlier in Scotland. Today, hunts still exist in the United Kingdom, but they cannot actively chase and kill a fox. Instead, hunts are restricted to setting fake trails for their hounds to follow. If during this a fox is disturbed, it can be chased, but setting out with the deliberate intention of hunting a fox is illegal. The banning of fox hunting was driven by the animal rights lobby, which hunt supporters attempted to depict, not without at least some justification, as an urban lobby (i.e., a predominantly city-dwelling population), one out of touch with rural society. The ban brought to a culmination years of work, often contentious, by animal rights groups and activists. Yet today in the United Kingdom, fox hunting remains a contested issue, and hunts continue to pressure Parliament to repeal the ban.

What may seem at first sight to have been a victory for the rights and welfare of the fox deserves further investigation. It is clear that the hunts have at least been partially responsible for the survival of the fox and the health of the fox population in the United Kingdom. Hunting undoubtedly was not undertaken with the rights of the animal as the guiding principle, but it was based on the notion that a fox population was a necessity for hunting to be able to continue and that a healthy population would provide for challenging opposition in the sport. Whatever the underlying motive, the role of hunting in the survival and health of the fox population should not be entirely disregarded.

Fox hunting, not unfairly, attracts the attention of the general public and media, yet that attention arguably represents only a small part of what the hunt actually is. This is one reason why hunts continue to exist and even grow in popularity in the face of the fox hunting ban in the United Kingdom. In addition to the hunting of the fox, the breeding of fox hounds and the maintenance of a balanced pack of hounds represents a serious leisure practice in its own right. Furthermore, the social activities associated with hunts do not require the hunting and killing of a fox in order to continue.

Neil Carr

See also: Advocacy; Animal Liberation Front (ALF); Dogs; Ethics; Horses; Hunting; Wildlife; Wildlife Management

Further Reading

Carr, N. 2014. *Dogs in the Leisure Experience.* Wallingford, UK: CABI.

Griffin, E. 2007. *Blood Sport: Hunting in Britain since 1066.* New Haven, CT and London: Yale University Press.

Free Range. *See* Humane Farming

Fur/Fur Farming

In early modern history, fur products became increasingly fashionable in Europe, and high-quality furs began to attain luxury status. During this period, the fur trade played a key role in the development of international trade and the acquisition of territories in Eurasia and North America by Western European and Russian colonizers. Fur farming developed during the mid-19th century as a more economically efficient and reliable alternative to hunting and trapping furbearing animals in the wild. Since then, fur farming has grown to become a multibillion-dollar industry supplying most of the world's fur. Toward the end of the 20th century, growing public concerns regarding the treatment of animals on fur farms had led to the institutionalization of regulations concerning fur farming practices, as well as the decline of fur as a commodity.

Historical records place the first known established fur trade system in the 6th century, in the northwestern region of present-day Russia, primarily supplying European markets. In the 15th century, the Russian empire began to expand eastward across the Ural Mountains to Siberia, where fur traders discovered vast populations of furbearing animals. Indigenous, or native, peoples who inhabited this region were required to pay a fur tax, or *yasak,* in the form of animal skins, or pelts. From the 16th century until the mid-19th century, the Russian empire expanded its frontiers to the Far East and North America, becoming the primary fur supplier to meet the high demand in Europe and China.

In the eastern region of North America at the beginning of the 17th century, French explorers began making inroads in the fur trade by establishing trading relationships with indigenous societies in what is today the Great Lakes region between Canada and the United States. Dutch merchants soon secured their own fur trading posts on the continent, followed by English merchants toward midcentury. By the 18th century, Russia's dominance of the global fur trade had begun to be steadily eclipsed by Western Europeans' North America-based fur trade.

Historical records of raising furbearing species in captivity, or *fur farming,* date the practice back to 1866 in Ontario, Canada, where the first mink fur farms were established. As fur trapping had begun depleting wild populations of furbearing species, fur farming provided an alternative that ensured a more stable supply of pelts. Since the late 19th century, fur farms have been established throughout North America, Europe, China, and Russia. Today, fur farms supply 85 percent of the fur sold as a commodity, while the rest is sourced from animals hunted or trapped in the wild (Peterson 2010).

Mink and foxes make up the largest share of animals bred on fur farms. The European Union is the leading producer of farmed fur, having produced 64 percent, or

30 million pelts, of mink fur and 56 percent, or 2.1 million pelts, of fox fur in 2010 (European Fur Breeders' Association 2010). China follows closely behind the European Union, along with the United States, Canada, and Russia as major suppliers of farmed fur. In 2013, the estimated value of total fur farms operating across the world amounted to $7.5 billion, while global fur retail sales totaled $35.8 billion (International Fur Federation 2014). Additionally, China has become the largest importer of farmed fur and now holds the largest export share of finished fur products.

While fur farming practices vary widely across the world, most fur farm operations are characterized by the intensive confinement of animals, as well as care specifications defined primarily for fur quality, not animal welfare. Common slaughtering methods include carbon monoxide gas poisoning, electrocution, or by breaking the animals' necks (Peterson 2010; North American Fur Industry Communications 2013). Public awareness and concerns regarding the ethical treatment of animals killed for their fur increased during the latter decades of the 20th century. Animal advocacy organizations have also drawn attention to systemic abuse on fur farms, such as the overcrowded and unsanitary confinement of animals and inhumane slaughter practices such as the painful electrocution method.

Many governments have instituted regulations concerning the humane treatment of animals on fur farms; however, the regulation and methods of enforcement vary widely across the world. For example, European Union legislation includes guidelines for the humane treatment and slaughter of agricultural animals raised for food as well as clothing, including fur. While these guidelines are enforced by routine monitoring carried out by state-authorized agencies, observers find variable results among countries within the European Union. In China, a major supplier of farmed fur, minimal regulatory oversight exists regarding fur farms. Total bans on fur farming are in effect in the United Kingdom and Austria; in the Netherlands, Croatia, and Switzerland, heavy restrictions or partial bans have been established.

Outside of government jurisdiction, fur industry organizations have responded to public concerns regarding the treatment of animals by establishing self-regulatory and certification programs based on ethical codes of practices concerning animal care, management, and slaughter (Fur Commission USA 2015). Fur farms are inspected by authorized veterinarians in order to verify compliance and award certification as a humane fur farm. While such self-regulatory measures foster greater accountability, animal advocates question their effectiveness, citing inconsistency in the implementation and enforcement of such measures throughout the world.

In addition to public concerns regarding the ethical implications of the fur trade, the increasing availability and quality of synthetic fur, or *faux fur*, has contributed to the declining popularity and profitability of fur, with the exception of growing markets in China and Russia.

Rosibel Roman

See also: Ethics; Wildlife

Further Reading

Dolin, E. J. 2010. *Fur, Fortune, and Empire: The Epic History of the Fur Trade in America.* New York: W. W. Norton & Company.

European Fur Breeders' Association. 2010. "Fur Farming in Europe." Accessed June 5, 2015. http://www.efba.eu/fact_sheet.html

European Fur Breeders' Association. 2012. "Welfur: The Animal Welfare Project on Fur-Farmed Species." Accessed June 5, 2015. http://www.efba.eu/welfur/

Fisher, R. 1943. *The Russian Fur Trade, 1550–1700.* Berkeley: University of California Press.

Fur Commission USA. 2015. "Humane Care Certification." Accessed June 5, 2015. http://www.furcommission.com/welfare/humane-care-certification/

International Fur Federation. 2014. "New Global Figures Revealed." Accessed May 27, 2015. http://www.wearefur.com/latest/news/new-global-figures-revealed

North American Fur Industry Communications. 2013. "Truth about Fur: Q & A." Accessed June 5, 2013. http://truthaboutfur.com/en/qa

People for the Ethical Treatment of Animals (PETA). 2014. "Inside the Fur Industry: Factory Farms." Accessed June 5, 2013. http://www.peta.org/issues/animals-used-for-clothing/animals-used-clothing-factsheets/inside-fur-industry-factory-farms

Peterson, L. A. 2010. "Fur Production and Fur Laws." Accessed March 21, 2015. https://www.animallaw.info/intro/fur-production-and-fur-laws

G

Game Preserves. *See* Wildlife Management

Geography

The word "geography" was coined by the ancient Greek intellectual Eratosthenes (285–205 BCE) and is a combination of the Greek words for "Earth" (geo) and "writing/description" (graphy). Geography today is an academic discipline that studies where, how, and why things happen with respect to humans, our societies, and the natural world of which we are a part. For geographers, *where* is the most foundational concept. Geography is one of many academic disciplines that study animals and the human-animal relationship. Because of its overarching interest in understanding all Earth's physical and human systems, nonhuman animals are a key topic of research.

Before geography became a formal academic discipline in the latter half of the 19th century, it had largely been the domain of explorers, naturalists, and military/political leaders. As explorers, especially European, encountered different lands and people, they developed the two early forms of geography that have remained key even as geography has become the diverse field it is today. The first is regional description, which describes the combination of natural and human features in a given area. This descriptive form of geography was, and is, essential to understanding how and where the world differs. The second form is the use of maps. Cartography, or map making, is probably what most people think of today when they think of geography. Maps are a visual summation of a particular topic in the world and allow us to simultaneously understand single or multiple phenomena in one view.

Geography today has several areas of focus. Physical geographers study topics such as plate tectonics and volcanoes, the physical processes of Earth's weather and climate, and Earth's flora (plants) and fauna (animals). Human geographers study the mosaic of human cultures: how they developed and where; the rise of human practices such as religion, economics, and politics; and migration. A key part of the human mosaic is how humans interact with the physical environment. Environment and society geographers explore this specific topic both culturally (e.g., how do different religions view other species?) and/or more directly physically (e.g., how does human activity impact biodiversity?). Mapping scientists focus on technologies like space satellites for remote sensing (data-gathering) of Earth;

global positioning systems (GPS) to track locations of events, migrations, or landscape changes; and cartography/map making. Map making today is largely done digitally through geographic information systems (GIS), which enable a faster, more flexible process than hand-drawn maps. Importantly, geographers may focus on one area but use multiple areas for particular projects or to fully understand a topic. In this way, geography is a multifaceted and synthesizing (bringing together) field of study. Geographers are trained in a variety of methods, from studying primary source documents (like diaries or government reports), field work (e.g., interviewing people, collecting environmental data like water samples, or directly observing animals, people, or events), or data collection (gathering census data or GPS coordinates for a section of forest) to develop maps to conduct their research.

Several key, interrelated concepts are essential to a geographic perspective. *Place* means a specific location, such as Missouri in the United States or Kruger National Park in South Africa. Geographers seek to understand specific places in order to develop an understanding of the uniqueness of particular locations and to compare different places. *Space,* for geographers, is not the outer space of astronomers, but is seen as both a relational concept (e.g., where in space is Massachusetts related to Missouri) and also as a way of grouping places. For example, zoos are a general type of *space* where animals are kept in captivity, but the Kansas City Zoo is a *place* with its own specific development. *Scale* can refer to the ratio of a map to the real world (e.g., one inch on a map equals one mile in the real world), a nested set of scales of analysis (e.g., world, state, city, neighborhood, home, body), or the extent of something (the scale of automobile production). Finally, geographers are also interested in exploring how *place*, *space*, and *scale* interact to form particular landscapes, which geographers study as reflecting all human and natural activities and events. Geography is often closely linked with history, because both fields are interested in processes of change. The difference is that history focuses on changes over time, whereas geography focuses on changes across space.

Geography has advanced the study of human-animal relationships in several ways. Biogeographers, who study the distribution of life on the planet and how life relates to environments, have used mapping sciences to help reveal where different species live, patterns of animal migration, and the impact of human behaviors, such as cutting trees or building dams, on specific species. Zoogeomorphologists study the ways in which animals themselves change their landscapes. Beavers, bears, and prairie dogs are all examples of species who actively alter their environments by building dams and digging holes that, in turn, impact local ecosystems. Environment and society geographers have developed a specialized subfield called animal geography that specifically studies the cultural, political, economic, and landscape aspects of human-animal relations. This includes such topics as the role of ecotourism in both raising awareness about other species and having an impact on the animals themselves, human-wildlife conflict issues such as those involving people living near wild elephants who might destroy their gardens, and how different types of livestock farming have changed landscapes.

Julie Urbanik

See also: Animal Geography; Biodiversity; Biogeography; Ethics; Zoogeomorphology

Further Reading

De Blij, H. 2009. *The Power of Place: Geography, Destiny, and Globalization's Rough Landscapes.* Oxford and New York: Oxford University Press.

Gregory, D., Johnson, R., Pratt, G., Watts, M., and Whatmore, S., eds. 2009. *Dictionary of Human Geography.* Malden, MA: Wiley-Blackwell.

Holt-Jensen, A. 2009. *Geography: History and Concepts, A Student's Guide.* London and Thousand Oaks, CA: Sage Publications.

Thomas, D., and Goudie, A., eds. 2000. *Dictionary of Physical Geography.* Malden, MA: Wiley-Blackwell.

Urbanik, J. 2012. *Placing Animals: An Introduction to the Geography of Human-Animal Relations.* Lanham, MD: Rowman and Littlefield Publishers.

Gestation Crate. *See* Factory Farming

Great Apes

Some 15 million years of evolution separate modern humans (*Homo sapiens*) from the other six surviving species of great apes, or Hominids (chimpanzees [*Pan troglodytes*], bonobos [*Pan paniscus*], gorillas [eastern lowland: *Gorilla berengi* and western lowland: *Gorilla gorilla*], and orangutans [Sumatran: *Pongo ableii* and Borneian: *Pongo pygmaeus*]). DNA evidence leaves no doubt that humans belong to the same family of primates as do these other tailless beings who, like us, walk upright (bipedalism) and have forward-facing eyes and eye sockets; limber, gripping thumbs and fingerprints; large brains; and extended childhoods. According to the UN Environment Program all species of ape but the human are, in the ever-shrinking wild, on the brink of extinction. Likely, "by 2032 less than ten percent of the ape habitat in Africa—and less than one percent in Asia—will remain untouched by human development" (Anthes 2015). Even now, in the human-dominated world, great apes other than humans live on preserves or in captivity, relocated for human purposes—research, conservation, exhibition and/or education, entertainment, and as companion animals purchased from the illegal wildlife trade.

When apes emerged from other primates during the Miocene Era (26 to 25 million years ago), there were likely many more varieties than have survived. Fossil evidence is sparse, and we know only a few of our shared ancestors. A fossil of an ancestor of the African great apes and humans found in Africa dates from 7 million years ago, and fossils of gorilla-like apes found in Africa date to 10 million

Activist and primatologist Jane Goodall with Tess, a female chimpanzee at the Sweetwaters Chimpanzee Sanctuary. There are six surviving species of great apes: chimpanzees, bonobos, eastern and western lowland gorillas, Sumatran and Bornean orangutans, and humans. All of these highly intelligent species, except for humans, are facing extinction. (AP Photo/Jean-Marc Bouju)

years ago (Wadams 2007). Orangutans split from the lineage about 15 million years ago; gorillas followed around 6.5 million years ago (Sorenson 2009). The bonobo, recognized as a separate species from the chimpanzee only in 1933, remains the least studied of the nonhuman apes largely because they were not included in the field studies launched by Louis Leaky (1903–1972) in the mid-20th century. He had Jane Goodall (1934–) live with chimpanzees, Dian Fossey (1932–1985) cohabit with gorillas in Africa, and Birute Galdikas (1946–) study orangutans in Borneo. Extended field studies allowed these remarkable women to earn "trust . . . on the animals' terms" in their natural environments (Montgomery 1998, xvi). Their reports on ape behavior, appearing worldwide in newspapers and books, in articles in both scientific and popular magazines, and in films and documentaries, allowed readers to share the other apes' unique *umwelts* (habitats, worlds).

Goodall observed that, like humans, chimpanzees not only used tools but also scavenged, hunted, and waged war. Recent field studies of a chimpanzee community in the hills surrounding Bessau, a small town in southeastern Guinea, reveals how stealing human crops seems to be altering the behavior of wild chimps. Typically boisterous, congregating in large numbers when eating wild fruits and vegetation, the Bessau chimps form small, cohesive groups and exhibit quiet stealth to steal crops. Rather than eating as they gather, they carry off the crops, assuming bipedal posture in order to carry off larger loads. Additionally, researchers have noted that these chimpanzees have learned to cross roads safely, destroy snares, and share their bounty with others in their group, a form of altruism thought to be infrequently practiced in the wild (Anthes 2015).

On the basis of such studies, Catherine Hill and other anthropologists have suggested that it is essential to study how ape behavior evolves in places where humans and other apes interact, because such studies would provide the basis for effective conservation as apes continue to evolve in the future (Anthes 2015). The primatologist Franz De Waal, whose studies have been based on bonobos only in captivity, stresses that humans inherit what we view as "human abilities," including intelligence, empathy, altruism, and compassion, from our primate ancestors and share those qualities not only with contemporary primates, particularly the other great apes, but also with other mammals and sentient (consciously aware of experience) species (De Waal 2013).

Although both Swedish zoologist Carl Linneaus (1707–1778), in *Systema Naturae*, first published in 1735, and Charles Darwin (1809–1882), in *On the Origin of Species* (1850), recognized that humans were a species of great ape, the kinship was not widely accepted until the public became aware of the pioneering field work of Goodall, Fossey, and Galdikas and of research such as De Waal's, which followed in the later 20th and 21st centuries. To reflect our shared ancestry and genetics, consideration is now being given to altering the scientific names of the subfamily (*Homininae*) that contains humans, bonobos, and chimpanzees to either *Homo-sapiens, Homo-troglodyte,* and *Homo-paniscus,* or *Pan-sapiens, Pan-troglodite,* and *Pan-paniscus,* recognizing all three as either human or chimpanzee (Barnes 2014).

Additionally, courts of law around the world are considering cases that claim that the nonhuman great apes, wild and captive, deserve recognition in our laws equal to that granted humans (Wise 2000). In June 2015, the U.S. Fish and Wildlife Service listed all chimpanzees, wild and captive, under the Endangered Species Act.

Marion W. Copeland

See also: Agency; Animal Cultures; Animal Law; Animals; Bushmeat; Chimpanzees; Deforestation; Emotions, Animal; Endangered Species; Ethics; Evolution; Exotic Pets; Habitat Loss; Multispecies Ethnography; Personhood; Rights; Species; Taxonomy

Further Reading

Anthes, E. 2015. "Apes in a Human World." *The New Yorker.* Accessed April 22, 2015. http://www.newyorker.com/tech/elements/apes-chimpanzees-human-world

Barnes, S. 2014. *Ten Million Aliens: A Journey through the Entire Animal Kingdom.* New York: Marble Arch Press.

Copeland, M. 2015. "Apes of the Imagination: A Bibliography." Accessed November 30, 2015. http://www.animalsandsociety.org/human-animal-studies/apes-of-the-imagination-a-bibliography/

De Waal, F. 2013. *The Bonobo and the Atheist: In Search of Humanism among the Primates*. New York and London: W. W. Norton.

Galdikas, B., and Galdikas, M. F. 1995. *Reflections of Eden: My Years with the Orangutans of Borneo*. Boston, New York, Toronto and London: Little, Brown.

Goodall, J. 1990. *Through a Window: My Thirty Years with the Chimpanzees of Gombe*. Boston: Houghton Mifflin.

Montgomery, S. 1991. *Walking with the Great Apes: Jane Goodall, Dian Fossey, Birute Gildikas*. Boston, New York and London: Houghton Mifflin.

Sorenson, J. 2009. *Ape*. London: Reaktion Books.

Wadhams, N. 2007. "New Fossil Ape May Shake Human Family Tree." *National Geographic News*. news.nationalgeographic.com/news/2007/08/070822-fossil-ape.html

Wise, S. 2000. *Rattling the Cage: Toward Legal Rights for Animals*. Cambridge, MA: Perseus Books.

Greyhound Racing

Greyhound racing first took root in the United States in the 1920s and reached its peak in the 1980s, when the sport was legal in more than 15 states. Legalized greyhound racing presently takes place in five states (Alabama, Arkansas, Florida, Iowa, and West Virginia), with fewer than 25 tracks currently in operation nationwide. A form of legalized gambling known as pari-mutuel wagering (i.e., those betting on competitors finishing in the top places share the total amount wagered, minus a management fee) is sanctioned at these tracks with the oversight of state (or local) racing commissions. In recent decades, the sport has come under heavy criticism by animal advocates and has declined in popularity significantly.

Greyhound racing—often referred to as the "Sport of Queens" by promoters—developed from a sport known as "coursing," which was popularized by Queen Elizabeth I of England and other British aristocrats in the 16th century. Coursing is a competition between two greyhounds as they chase a rabbit. During this period these greyhounds were specifically bred for the sport of coursing. Traditionally, the demonstration of the dogs' athletic abilities (rather than catching the rabbit) was the primary aim of the competition. The sport, which briefly gained a small following in the United States around the turn of the 19th century, was largely subsumed by organized greyhound racing in the early 20th century.

In the decade before World War I, Owen Patrick Smith, a small-town promoter from South Dakota, designed a mechanical lure to replace live quarry. The lure featured an artificial "rabbit" mounted on a moveable, electric-powered device positioned on the exterior rail of the race track. The device has been modified to some degree over the past century, but its creation was the critical step in the development and evolution of greyhound racing.

Greyhound racing began to gain popularity in parts of the United States in the 1920s and 1930s. The standard race featured eight greyhounds simultaneously chasing a mechanical lure around an oval track. The sport slowly gained a grassroots fan base and became popular largely because it was appealing to gamblers. Some people disliked greyhound racing because they did not want gambling in their communities. However, pari-mutuel gambling on greyhound racing was soon legalized in a number of states, with the first being Florida (1931), where the sport grew in popularity largely because of tourism. After World War II, the sport expanded significantly and was legalized in various other states.

American racing greyhounds are registered with the National Greyhound Association (formerly the National Coursing Association), headquartered in the U.S. state of Kansas, and are bred specifically for racing competitions. The greyhound has a short, smooth coat, and appears in a variety of colors and combinations, including white, fawn, brindle, and black. They are naturally lean and muscular, weighing approximately 55 to 80 pounds (25 to 36 kilograms), and are about 27 to 30 inches (68.58 to 76.2 centimeters) tall. Their flexible spines, large hearts, and powerful lungs allow them to achieve tremendous speeds (up to 45 miles per hour/72 kilometers per hour) rapidly. The racing distance is usually 5/16, 3/8, or 7/16 of a mile.

In contrast to thoroughbred horse racing, racing greyhounds are not usually assigned racing appointments as individual competitors. Rather, entire racing kennels acquire a "booking" (racing contract), with a racetrack. Some greyhound kennels specialize in breeding, others focus on training, whereas others are exclusively racing kennels. Breeding kennels today are mostly based in Texas, Oklahoma, and Kansas, but for a period of time, Florida was a top breeding state. The individuals who work with the greyhounds were traditionally known as "dogmen." In the early years in particular, when racing was seasonal, dogmen would travel the "racing circuit," securing, for instance, a booking at a Florida track in the winter and one at a New England track in the summer. A number of such racing circuits existed, but all of them required seasonal travel from track to track.

The causes of greyhound racing's decline in the United States are multifaceted and complicated. The sport's popularity among the American public gradually began to wane in the last decades of the 20th century. Animal protectionists, who had long believed that greyhound racing was cruel, began in the late 1970s to launch aggressive campaigns against it. While early criticism focused on the cruelty of chasing live quarry during training, the objections over the sport gradually changed, and shifted to the welfare and treatment of the dogs. For many years, there were no large-scale adoption programs available for ex-racing greyhounds, and antiracing activists honed in on the frequent euthanasia of greyhounds who had "retired" from the sport but had nowhere to go (with the exception of animals kept for breeding). They also charged that the sport itself was cruel and dangerous, and led to the premature deaths of many racing dogs.

Some of the reasons for the sport's decline are unrelated to the efforts of anti-racing activists. A more competitive and diversified entertainment market rendered it increasingly difficult for greyhound promoters to attract fans. State lotteries, Native American gaming venues, riverboat casinos, and slot machine parlors all emerged as competitors for the gambling dollar. More recent efforts at ending legalizing greyhound racing have focused on eliminating state subsidies designed to buttress the weakening industry. Antiracing groups such as GREY2K USA are also working on "decoupling" legislation—eliminating state laws that require casinos to also feature live greyhound racing on site.

Gwyneth Anne Thayer

See also: Dogs; Horse Racing; Working Animals

Further Reading

Cook, A. 2015. *High Stakes: Greyhound Racing in the United States.* Arlington, MA: GREY2K USA Worldwide.

Greyhound Racing Association of America. Accessed September 26, 2015. http://www.gra-america.org

GREY2K USA. Accessed September 26, 2015. http://www.grey2kusa.org/index.php

Hartwell, P. C. 1980. *The Road from Emeryville: A History of Greyhound Racing.* San Diego: California Research Publishing.

Humane Society of the United States. 2015. "Greyhound Racing." Accessed September 26, 2015. Accessed September 26, 2015. http://www.humanesociety.org/issues/greyhound_racing

National Greyhound Association. Accessed September 26, 2015. http://www.ngagreyhounds.com

Thayer, G. A. 2013. *Going to the Dogs: Greyhound Racing, Animal Activism, and American Popular Culture.* Lawrence, KA: University of Kansas Press.

H

Habitat Loss

Human population growth and consumption of natural resources, like trees and soil, put tremendous strain on ecosystems and are the dominant drivers of habitat loss. Loss of habitat poses the greatest threat to global species decline and to the continued availability of resources on which animal life depends. Extinction rates of 10,000 times the rate found in the fossil record lead scientists to speculate human activities are creating the conditions for the sixth mass extinction event of Earth's species.

A habitat is an area that provides the specific resources a particular species needs to survive. The relationship between a species and its habitat develops through adaptations over potentially millions of years. Within habitats, interspecies dependencies such as mutualism (in which both species benefit) and predator-prey relationships further influence species' survival. Habitats with high biodiversity (variety of organisms) are important because they offer a greater range of ecosystem services (human benefits from ecosystem processes), such as wetlands for clean water, plants for clean air, and trees for carbon storage and lumber.

Today, scientists agree that the rate and magnitude of habitat loss is driven by the increasing consumption of a growing human population. The expansion of agriculture for crops and grazing animals has been the dominant force in the destruction of terrestrial (land-based) habitats since humans became farmers about 10,000 years ago. Agriculture is also the main threat to freshwater and coastal habitats as significant increases in soil erosion cause sedimentation of waterways. After agriculture, deforestation and urban-suburban development are also major contributors to habitat loss. Other major impacts include surface mining, pollution, fire suppression (preventing fire in ecosystems that need it), overgrazing, stream channelization, water diversion, and dam building.

Habitat loss occurs when an ecosystem is damaged to the degree that it cannot support all the species that normally live there. The three basic types of habitat loss are destruction, degradation, and fragmentation, each with varying effects on species, biodiversity, and ecosystem services. Habitat loss by destruction is the elimination of habitat by converting it to another use or as a consequence of resource exploitation. Habitat destruction results in the loss of species, biodiversity, and the ecosystem services the habitat provided. Habitat loss through degradation may not be as easy to identify because habitats can be damaged and incapable of supporting native species, but they might still look like healthy ecosystems. For instance, the Big Island of Hawaii is lush with an abundance of species, yet many are invasive. Native to other areas and introduced by humans, invasive species pushed out endemic (local) species and changed habitats of the island until they could not support native species. Habitat loss by fragmentation is the incremental breakup of an intact landscape into unconnected patches separated by other land uses, like roads or suburban housing. The remaining fragments may be too small or too isolated to support a viable community of species that lived in the intact ecosystem. For example, the critically endangered Sumatran rhino suffers from habitat fragmentation because the remaining pieces of habitat are so distant that the last 100 individuals have difficulty finding mates with whom to reproduce.

Scientists involved with projects like *Living Planet Report*, *State of the World's Forests,* and *Millennium Ecosystem Assessment* collect data on long-term changes in the world's ecosystems. Results indicate a trend of accelerating habitat loss even as data show an already extensive loss of habitats across Earth's major ecosystems. For example, global wetlands dropped by 50 percent in the 20th century, while both mangroves and coral reefs plummeted 20 percent and 38 percent, respectively, since only the 1980s. Additionally, up to 50 percent of the planet's original forest ecosystems have disappeared, and as much as 70 percent of remaining forests are so heavily fragmented that they are no more than a mile (1,600 meters) deep (Haddad et al. 2015). Temperate forests in the Northern Hemisphere were almost wiped out by the middle of the 20th century, and the pressure is now on tropical forests, which have the highest biodiversity and support half of the world's species. The World Wildlife Fund for Nature (WWF) estimates tropical deforestation rates are 10 times the rate of forest regrowth, producing the most rapid biodiversity loss. At this rate, the Food and Agriculture Organization of the United Nations (FAO) predicts all Central American tropical forests could disappear by 2220. Likewise, FAO estimates current rates of global habitat losses will cause the degradation or destruction of more than 70 percent of the planet's terrestrial ecosystems by 2032.

Habitat loss is the primary threat for more than 80 percent of species on the International Union for the Conservation of Nature and Natural Resources' (IUCN) Red List of threatened species. Current rates of habitat loss are leading to extinctions 100 to 10,000 times the background extinction rate (average rate of extinction from natural causes found in the fossil record). Since 1970 alone, both terrestrial and marine species declined up to 39 percent, while freshwater species declined 76 percent (WWF 2014). Furthermore, a third or more of amphibians, highly vulnerable to habitat changes, are in danger of extinction. Scientists believe the severity and extent of current conditions are the result of human consumption exceeding Earth's capacity to replace resources, thereby putting human lives at risk and triggering a global extinction of species not seen since the dinosaurs. Habitat loss, driven by human activity, is contributing to the sixth mass extinction of species.

Michelle L. Shuey

See also: Biodiversity; Climate Change; Deforestation; Extinction

Further Reading

Food and Agriculture Organization of the United Nations. 2012. "State of the World's Forests 2012." Rome. Accessed January 12, 2015. http://www.fao.org/docrep/016/i3010e/i3010e.pdf

Haddad, N. M., Brudvig, L. A., Clobert, J., Davies, K. F., Gonzalez, A., Holt, R. D., Lovejoy, T. E., Sexton, J. O., Austin, M. P., Collins, C. D., Cook, W. M., Damschen, E. I., Ewers, R. M., Foster, B. L., Jenkins, C. N., King, A. J., Laurance, W. F., Levey, D. J., Margules, C. R., Melbourne, B. A., Nicholls, A. O., Orrock, J. L., Song, D-X., and Townshend, J. R. 2015. "Habitat Fragmentation and Its Lasting Impact on Earth's Ecosystems." *Science Advances* 1(2). Accessed April 15, 2015.

International Union for the Conservation of Nature and Natural Resources. 2015. "The IUCN Red List of Threatened Species. Version 2015.2." Accessed March 3, 2015. http://www.iucnredlist.org

International Union for the Conservation of Nature and Natural Resources. 2015. "The IUCN Red List of Ecosystems." Accessed June 21, 2015. http://www.iucnredlistofecosystems.org/

Millennium Ecosystem Assessment. 2005. *Ecosystems and Human Well-Being: Synthesis.* Washington, DC: Island Press.

World Wildlife Fund for Nature. 2014. "Living Planet Report 2014: Species and Spaces, People and Places." Gland, Switzerland: WWF International. Accessed October 3, 2014. http://wwf.panda.org/about_our_earth/all_publications/living_planet_report/

Hoarding, Animal

Animal hoarding involves individuals acquiring and attempting to care for numbers of domesticated animals beyond their means and capabilities such that the animals are severely neglected, often to the point of death by starvation or even cannibalism. In a typical hoarding situation, the median number of animals is 39, but often over 100 animals are involved (Patronek and Nathanson 2009). While neglect due to hoarding is psychologically distinguishable from direct, violent acts of cruelty, from the animal victims' point of view it can be an extreme form of suffering.

The many forms of contemporary human-animal relationships vary in the degree to which they are beneficial or detrimental to the two parties of that relationship. Although during the course of its development hoarding may be beneficial to either or both parties, in its full-blown form hoarding is grossly detrimental to the animals involved and to the well-being of the hoarder as well. Hoarding is a major source of animal abuse, with approximately 3,000 cases seen annually in the United States, and one for which preventative and effective remedial remedies are not yet developed or in place.

Comparison to other forms of contemporary animal abuse illustrates the uniqueness of hoarding and the challenges it presents to public policy. Although public opinion is divided and largely turns a blind eye to most forms of institutional animal abuse (e.g., intensive production of animal-based foods), most forms of individual animal abuse (e.g., unnecessary violence toward a companion animal) are socially unacceptable. However, particularly in its milder forms, hoarding often is viewed sympathetically by media and the public. Indeed, at least initially and in terms of self-presentation, hoarding typically is motivated by real concern for the well-being of animals, with hoarders using phrases such as "my 'babies' need me." Hence, the apparent paradox—hoarders are "caring abusers." This view is buttressed by the fact that, unlike perpetrators of violent individual animal abuse, hoarders are predominantly (76–83 percent) women (Patronek and Nathanson 2009), and women are conventionally seen as more caring and empathic than men—findings with strong empirical support.

Although there are other instances of mixed motives in the many forms and contexts of our treatment of animals (e.g., the companion animal treated as a member of the family but then relinquished at an animal shelter or abandoned on a country road), understanding how hoarders are "caring abusers" is a major challenge for investigators. A number of theories and concepts have been proposed and are actively being investigated. Maintaining a view of themselves as caring in the face of gross neglect of so many animals requires considerable distortion of the situation. Hoarders' failure to see this contradiction is part of a larger deficit in thinking clearly, problem-solving, and managing their lives, which presents a challenge in efforts to work with them in therapy.

Hoarders also have some compulsive features likening them to people who hoard inanimate objects. Like these collectors of papers, books, and various other memorabilia, hoarders create and live in a space that is, from most people's points of view, uninhabitable—with poor sanitation, personal hygiene, diet, and access to living space.

A number of researchers have suggested that hoarders are addicted to the activity of collecting animals and that their treatment might be modeled after the treatment of substance abusers. Yet others theorize that the underlying problem in hoarding is the failure to establish stable emotional attachments. Such secure attachments are the basis of forming and maintaining mature interpersonal relationships. The destructive, distorted, and rigid attachments to the hoarded animals substitute for these more mutually beneficial attachments.

Yet another theory asserts that hoarders do not relate to others, including animals, as fully distinct individuals (Brown 2011). In this view, hoarders are so preoccupied with themselves that they relate to others as if they were extensions or mirrors of themselves—"narcissism." Treatment for disorders of attachment and narcissism are available and might be tested with this population.

None of these theories have yet been validated or led to an effective intervention. Complicating the situation is that there are different types of hoarders: overwhelmed caretakers who begin with good intentions but eventually have inadequate resources to maintain care, rescuers who believe that only they can or are willing to care for the animals, and exploiters who only superficially care and really seek financial gain. Often these types are different stages in the career of an individual hoarder.

Current interventions typically include both criminal justice and human service agencies, and so require effective networking among these disciplines and some degree of cross-reporting and cross-training. Neighbors and family members often alert animal control or police to the problem, based on smells or noise. Dealing with the assessment, treatment, placement, and, most often, euthanasia of a large number of animals strains local resources and takes a long time. As indicated, although borrowing from approaches to the treatment of compulsive behavior, addiction, attachment disorder, and narcissistic personality have been tried, no clear, effective, and generalizable therapeutic treatment of the perpetrator is available. Yet without some form of therapeutic intervention, recidivism (recurrence of the problem) rates are extremely high—approaching 100 percent (Patronek and Nathanson 2009). One interesting approach under study (although a large drain on human services) is to provide the hoarder with regular home visits and support from a human services person to facilitate release of animals and prevent further acquisitions. Another is the adaptation of an approach to the treatment of violent animal abuse that focuses on accountability, empathy, and interpersonal skills training.

Kenneth Shapiro

See also: Bestiality; Cruelty; Domestic Violence and Animal Cruelty; Human-Animal Bond

Further Reading

Brown, S. 2011. "Theoretical Concepts from Self-Psychology Applied to Animal Hoarding." *Society & Animals* 19: 175–193.

Hoarding of Animals Research Consortium. 2010. Accessed March 24, 2015. http://vet.tufts.edu/hoarding/index.html

Patronek, G., and Nathanson, J. 2009. "A Theoretical Perspective to Inform Assessment and Treatment Strategies for Animal Hoarders." *Clinical Psychology Review* 29: 274–281.

Horse Racing

Horse racing is a major sporting activity in many countries. It includes thoroughbred and standardbred/harness racing, plus races with ponies and other equine breeds. Thoroughbred racing includes events on flat courses (known as "the flat"), plus longer "jumps" races over hurdles (brush fences of standard size and configuration)

Seabiscuit crosses the finish line to beat War Admiral in a race at Baltimore's Pimlico Race Course in 1938. Thoroughbreds are the breed that race on these flat courses. The cruelty and mistreatment involved in horse racing in general, and thoroughbred racing in particular, is increasingly coming under fire from animal advocacy groups. (Library of Congress)

and steeplechasing (larger obstacles of varying size and configuration). Harness racing involves two gaits: Pacing horses move both legs on the same side in unison; trotting horses move their legs in diagonal pairs. Horse racing is often supported by gambling, but not everywhere. The racing industry expands by bringing in "new money," either through wealthy individuals (increasingly in developing countries) or through syndication, where people pool resources and purchase a horse. Horse racing has different cultures and characteristics, depending on where it occurs.

Over time, different forms of racing have resulted in horses being bred selectively for desired traits. The thoroughbred breed originated in Great Britain from the cross of local mares with Arabian stallions. The most influential stallions imported into England were the Byerley Turk (in the 1680s), the Darley Arabian (1704), and the Godolphin Arabian (1729), with the pedigree in 95 percent of modern racehorses traced to the Darley Arabian. The resulting bloodlines were codified with the establishment of the General Stud Book in 1791 (McManus, Albrecht, and Graham 2013). The Stud Book is the official register of thoroughbreds. A horse can only be registered as a thoroughbred and be eligible to compete in events such as the Kentucky Derby (United States) or the St. Ledger (United Kingdom) if both parents are registered thoroughbreds and the horse is conceived and born according to the regulations in the Stud Book. In 2013, there were 69 recognized national stud books in the International Stud Book list (International Federation of Horseracing Authorities 2013).

The richest thoroughbred race in the world is the $10 million Dubai World Cup, whereas the richest jumps race is the Nakayama Grand Jump in Japan. Throughout the world, 92,000 thoroughbred foals were born in 2013, down from over 100,000 foals born annually before 2012. The reduction is due to the state of various national economies following the 2008 global financial crisis and concerns about quality. Despite reductions in the number of thoroughbred births, horse racing is still a major activity in many parts of the world. In 2013, there were 148,473 thoroughbred races on the flat held in 50 racing jurisdictions, with most races held in the United States, Australia, and Japan. There were also 8,408 jumps races in 17 countries, with most races held in Great Britain, France, and Ireland (International Federation of Horseracing Authorities 2013). Standardbred racing (also known as harness racing) is popular in parts of the United States, Australia, New Zealand, Canada, and Europe (particularly Sweden, France, Italy, and Finland). It has declined in popularity since the mid-20th century. The standardbred horse is longer in body, generally calmer in temperament, and with shorter legs than a thoroughbred. In this type of racing, horses are harnessed to a sulky—a light two-wheeled vehicle carrying one driver. Trotting is the only standardbred gait used in Europe, while "pacing" horses are more popular in Australia, the United States, Canada, and New Zealand. Pacing races are generally faster than trotting, as pacing is a faster gait and there is more use of a mobile starting barrier (a "rolling" start where horses are moving in the correct gait behind a car that speeds ahead when the race begins). The traditional standing start is difficult for horses to begin to pace/trot correctly.

Horse racing is now controversial for its human-animal relations. While some people reject the use of animals for human entertainment, specific animal welfare concerns in horse racing include deaths and injuries in jumps racing, the whipping of horses, and the "wastage" of racehorses. Wastage occurs because the global thoroughbred breeding and racing industry is prone to "overproduction," or too many horses that are not financially viable. Overproduction will always occur because, structurally, a limited number of horses can succeed. Concerns about unwanted horses being killed is a major welfare challenge for horse racing industries.

While thoroughbred racing, in particular, has become more visible in recent years, the breeding component of the industry is still relatively hidden, despite being the most lucrative part of the industry. The high-value end of the racing industry provides raw materials for the breeding industry, namely young stallions with excellent racing records (particularly as two- and three-year-olds) and broodmares with outstanding pedigree, based either on their own racing achievements or those of their close relatives.

According to its regulations and proponents, thoroughbred breeding is "traditional" and "natural" in that the stallion and mare have to physically mate and the foal has to be born from that same mare. The conception and birth are aided by human intervention, however, including through planned mating, a "teaser" stallion to arouse the mare, and special footpads and neck protectors to prevent injuries to the stallion and/or mare. In contrast, standardbred and other breeds of horses are generally conceived using artificial insemination.

Phil McManus

See also: Advocacy; Horses; Livestock; Welfare

Further Reading

Cassidy, R., ed. 2013. *The Cambridge Companion to Horseracing.* New York: Cambridge University Press.

International Federation of Horseracing Authorities. 2013. "International Federation of Horseracing Authorities Annual Report, 2013." Accessed June 15, 2015. http://www.horseracingintfed.com/resources/Annual_Report_2013.pdf

McManus, P., Albrecht, G., and Graham, R. 2013. *The Global Horseracing Industry: Social, Economic, Environmental and Ethical Perspectives.* New York: Routledge.

Horses

The modern horse's association with humans has changed fundamentally since the relationship began 400,000 years ago. Horses are hoofed mammals (*Ungulata)* classified within the biological family *Equidae*. With one exception, most horses in the wild today are feral (untamed horses descended from domesticated stock). From farming, food, and transportation, to therapy, entertainment, and companionship, the human-horse relationship has been an important part of human life and development.

The original ancestor of *Equidae*, known as the "dawn horse" (*Hyracotherium)*, can be found in the fossil record around 55 million years ago. No larger than a hare of today, they roamed Europe, North America, and Asia, and during the Eocene epoch (55 to 33 million years ago) used land bridges to cross into North America. As *Hyracotherium* subsequently evolved into equids in North America during the Miocene (23 to 5 million years ago), they used the Bering land bridge to cross back into Asia and Europe. During the Late Miocene and into the Pliocene (5 to 2.6 million years ago), these modern horse ancestors migrated to other continents. About 12,000 years ago, the remaining North American equids became extinct, likely from diseases and alterations in vegetation due to a climate change event. During the 16th century, Spanish explorers and soldiers brought horses back to North America with them on their ships. Horses were introduced to Australia with the first ship of immigrants from Great Britain in 1788. Australia and Antarctica are the only two continents that have no fossil records of the horse's ancestors.

Archaeologists believe that early humans used spears to hunt horses almost 400,000 years ago, and that horse meat was a significant food source during the Upper Paleolithic (ca. 40,000–10,000 BCE). Approximately 5,000–6,000 years ago, horses were domesticated, which is currently thought to have occurred in Eurasia in an area that is today part of Ukraine, Russia, and Kazakhstan (Warmuth et al. 2012). Following domestication, the horse was used by humans not only for meat and milk but also in hunting and warfare. After introduction of the wheel for transportation around 2100 BCE, horses not only were used to pull chariots but also became valuable workers on farms, assisting humans with land cultivation by pulling plows.

Over the last few centuries, horses have pulled vehicles, such as trams, for passenger transportation in urban areas like Toronto and San Francisco. The horse tramway was replaced by the cable car during the Industrial Revolution in the 19th century. Today, horse-drawn carriages in cities are controversial. Activists argue that cities are unhealthy and unsafe for horses. In New York City, horses have collapsed in the street due to fear ("spooking"), tripping, extreme heat, and illnesses, and have been involved in accidents with vehicles and pedestrians.

With automobiles and tractors, horses became less useful as workers and modes of transportation, and more popular within entertainment and sport, although horses have been used for riding competitions since the Middle Ages (ca. 5–15th centuries) in Europe. Today these competitions take place worldwide and commonly include horse racing, show jumping, and eventing (an equestrian triathlon that includes dressage [performance of special rider-guided movements in a series of tests], cross-country riding, and show jumping). Horses have been competing with humans in the Olympics for over 100 years and are the only nonhuman animals in the international sporting event. Steeplechasing, a race in which horse and rider must navigate ditches and hedges on a racecourse, was modeled on the obstacles typically encountered in fox hunting. Steeplechasing (also known as jumps racing) is controversial due to the high number of horse deaths and injuries. Horses have been used in rodeos in North America since the 18th century. Animal advocacy groups argue that human rodeo competitors utilize fear, stress, and pain to make horses perform in activities such as bronc/bronco riding (in which a rider attempts to stay on a bucking horse) and wagon racing. Many horses are injured and die in these rodeo events.

Another contemporary dimension of horse-human relationships is therapeutic horseback riding, which is practiced primarily in North America, Europe, and Australia. This form of animal assisted therapy helps children and adults with disabilities develop mobility and improve physical strength and concentration and provides a sense of independence and achievement. The relationship between rider and horse is an important aspect to this therapeutic healing.

Horses continue to be consumed as meat, although controversy surrounds this practice today. While horse meat is commonly consumed in France, Belgium, Central Asia, and South America, the idea of eating horses is unacceptable in the United Kingdom and most of North America. Although, as of 2015, there is no federal law prohibiting horse slaughter in the United States, no slaughterhouses currently do so; however, U.S. horses are exported to Canada and Mexico for slaughter as food for humans and companion and zoo animals. Furthermore, slaughter byproducts are widely used: hides for leather, intestines for sausage casings, tails for paint brushes, and hooves for glue.

Today the only true wild breed of horse is the Mongolian Przewalski horse. Other free-roaming horses found throughout the world are feral. Human treatment of wild horses differs by country. In Australia, they are considered pests and a risk to native ecosystems, while in Namibia, wild horses are widely accepted and promoted in tourism. Proponents of wild horses in Alberta, Canada, support them as native species, effective seed distributors, and as an important part of the province's community and its history.

Angela Dawn Parker

See also: Animal Assisted Therapy; Cruelty; Evolution; Feral Animals; Fox Hunting; Horse Racing; Human-Animal Bond; Meat Eating; Military Use of Animals; Non-Food Animal Products; Slaughter; Working Animals

Further Reading

Franzen, J. L. 2010. *The Rise of Horses: 55 Million Years of Evolution*. Translated by K. M. Brown. Baltimore: The Johns Hopkins University Press.

Iannella, A. 2012. "Protests after Horse Put Down Following Fall at Oakbank Racing Carnival." Accessed September 20, 2015. http://www.adelaidenow.com.au/sport/superracing/horse-dies-in-oakbank-steeplechase/story-fn67mcwv-1226321095790

Kesner, A., and Pritzker, S. R. 2008. "Therapeutic Horseback Riding with Children Placed in the Foster Care System." *ReVision* 30(1/2): 77–87.

Notzke, C. 2013. "An Exploration into Political Ecology and Nonhuman Agency: The Case of the Wild Horse in Western Canada." *The Canadian Geographer* 57(4): 389–412.

Steeplechase Museum. 2015. "History of Steeplechasing." Accessed September 20, 2015. http://www.steeplechasemuseum.org/site/page/history

Vancouver Humane Society. "Rodeos." Accessed September 20, 2015. http://www.vancouverhumanesociety.bc.ca/campaigns/rodeos

Warmuth, V., Eriksson, A., Bower, M. A., Barker, G., Barrett, E., Hanks, B. K., Li, S., Lomitashvili, D., Ochir-Goryaevaf, M., Sizonovg, G. V., Soyonovh, V., and Manica, A. 2012. "Reconstructing the Origin and Spread of Horse Domestication in the Eurasian Steppe." *PNAS* 109(21): 8202–8206.

Human-Animal Bond

The human-animal bond is a positive connection ("bonding") between human and nonhuman animals—mainly pets. Bonding is a hormonal chemical process, occurring neurobiologically (in the brain), that integrates psychological, social, and physical impulses between beings (animals). Bonding emanates from attachments that humans and nonhumans make with one another. English psychologist John Bowlby (1907–1990) developed "attachment theory," based on psychological and biological ideas that there is a universal tendency for animals to seek closeness—or bond—to one another (Bowlby 1969). This bond is characterized by emotions ranging from sympathy and affection to trust, love, and empathy (psychological identification). Understanding the human-animal bond is a key part of learning about why humans care about other species.

The human-animal bond is advanced hormonally by the presence of oxytocin. Research shows that this hormone generates empathic bonding between parent and child, especially during childbirth and breastfeeding. Supported by eye and body contact, oxytocin reduces stress, encourages trust, generates mutual assurance, and bolsters immune response. It also encourages biophilia (which literally means "the love of life")—a feeling of deep connection with other animals.

According to biologist E. O. Wilson (1929–), newborns and young children feel a connection to the nonhuman animal world from birth. Scientists doing research in nurseries containing newborns observe that empathic bonding occurs quite early after delivery. To see if this was true for animals as well, ethologists (animal behaviorists) Konrad Lorenz (1903–1989) and Nikolaas Tinbergen (1907–1988) conducted research into the instinctive behavior of greylag geese after hatching. They concluded that physical and emotional attachment in these animals is elicited by "imprinting," the process by which some avian newborns bond to the first moving object they see, usually their mothers, thus reinforcing the concept of bonding as it occurs in humans (Lorenz 1979).

The human-animal bond is often used to facilitate human therapy through what is called animal assisted therapy (AAT) because the presence of animals can promote calm and relaxed humans. AAT got its start when psychologist Boris M. Levinson (1907–1984) brought Jingles, his dog and co-therapist, to a session in 1953 with a troubled young client and found that the client was able to relax and get more out of the therapy sessions. AAT now relies not only on domesticated pets such as dogs but also farm animals and certain marine animals, like dolphins, who are highly intelligent and generally people-friendly. A variety of psychological and physiological therapeutic goals can be met by using AAT precisely because it creates a human-animal bond. Interventions using AAT improve social and cognitive functioning. For example, humans with attention-deficit disorder (ADD) may undergo behavioral improvement when they interact with trained therapy dogs.

Fifth grader Martha Leonzo reads a book to trained therapy dog Ross during his visit to her elementary school. The human-animal bond is an experience that not only makes us feel good but also reduces anxiety and can actually be measured in our body through lowered heart rates. Understanding the human-animal bond helps us learn that animals are not just objects but beings who experience and respond and can enter into relationships. (AP Photo/Manuel Balce Ceneta)

The human-animal bond is also what allows animals to be trained to perform certain services for "their" humans. Service dogs, because of their keen sense of smell, can be trained to detect physiological changes in humans such as heightened blood sugar levels or chemical changes indicating a possible seizure to help people with diabetes and epilepsy. Service dogs may also be used to calm people when they are having severe panic attacks, thereby helping to lower their blood pressure and stabilize their brain chemistry. Service dogs also provide assistance to people in wheelchairs or who are blind. They can also be taught to read sign language when working with deaf persons. These relationships are successful because of a human-canine bond based on reliance and trust.

Animal companions offer unconditional love and loyalty and are frequently considered to be family members, offering emotional sensitivity and solace. Animal companions can even assist people with feelings of loneliness. They may also help with childhood and teenage development. Companion animals have been used to help male adolescents increase their self-esteem in certain youth programs, thus building social and emotional rapport between human and nonhuman animals. Nonhuman animal partners regularly are included in library reading programs for children because assistance dogs may help to put at ease those readers who feel more secure in the company of nonjudgmental listeners, especially dogs.

Conversely, nonhuman animals benefit from the formation of emotional bonds with humans. Dr. Leo K. Bustad, DVM (1920–1998), hypothesized that in our interactions with other species, a relationship parallel to the parent-child bond occurs, creating a human parent-animal child bond (Bustad 1990). Humans impart love and affection to animals and serve as reliable and constant factors in their lives. They are nurturers and may be able to forestall neglect, abandonment, and other risky situations. Some programs designed to improve the lives of nonhuman animals—for example, those in which prisoners tend to the developmental needs of homeless or companion animals and enrich their lives—afford greater opportunities for adoption and the growth of salutary (positive) emotional bonds between species.

Human animals also care for sentient beings who are homeless, wild, or simply live outdoors, protecting them from danger and other perils. Human animals look after nonhuman animals, healing those who are sick, injured, or incapacitated. They may take overall responsibility for the general welfare of nonhuman animals and sustain them by providing material and emotional necessities of life. Human oversight instills trust, ensures physical safety, and offers emotional security to nonhuman animals.

Barbara Hardy Beierl

See also: Animal Assisted Activities; Animal Assisted Therapy; Pets; Service Animals

Further Reading

Bowlby, J. 1969. *Attachment and Loss.* New York: Basic Books.

Burkhardt, R. W. 2005. *Patterns of Behavior: Konrad Lorenz, Niko Tinbergen, and the Founding of Ethology.* Chicago: University of Chicago Press.

Bustad, L. K. 1990. *Compassion: Our Last Great Hope.* Renton, WA: Delta Society.

Levinson, B. 1969. *Pet-Oriented Child Psychology.* Springfield, IL: Charles C. Thomas.

Lorenz, K. 1979. *The Year of the Greylag Goose.* London: Eyre Methuen.

Olmert, S. D. 2009. *Made for Each Other: The Biology of the Human-Animal Bond.* Cambridge, MA: Da Capo Press.

Wilson, E. O. 1984. *Biophilia.* Cambridge, MA: Harvard University Press.

Human-Animal Studies

Human-Animal Studies (HAS), or Animal Studies in the Humanities, is the umbrella category for all scholarly work that examines the spectrum of relations between humans and other species from social science (history, geography, anthropology) and humanities (literature, art, philosophy) perspectives. Understanding this broad category is key to understanding how different scholarly disciplines not only approach, but contribute to, our understanding of how human-animal relations have developed, how they change, and the impacts for both humans and other animals.

HAS differs in focus from other animal-related or animal-based scholarly fields. Animal studies in the physical sciences (biology, chemistry) generally refers to the study of the physical properties of animals themselves and/or their use as models for studies in human or veterinary health. Anthrozoology is the scientific study of human-animal interactions. Researchers in this field may use quantitative methods (gathering large-scale data or measuring bodily functions like heartbeats or blood pressure) to study the responses of children to different animals or whether or not animals may help people heal faster. Ethology is a science field that focuses on the study of animal behavior in the wild and under controlled conditions. Ethologists might devise tests to see how fast a crow could problem-solve for a treat or study the different physical displays of dogs (ears up, tail down) to learn how they might be communicating with each other.

In contrast, HAS is focused on 1) specific intersections and relationships between humans and animals and 2) how human society is not exclusively human. In this way the field is both multidisciplinary (involving more than one scholarly discipline) and interdisciplinary (one research project may combine methods from multiple disciplines). HAS scholars are interested in such topics as where and how animals have been categorized (food versus pet animals), how animals have been used (pets or research objects), how animals connect to culture (animals as mascots or movie stars), the economies of animal-related industries, and the politics of animals (laws, protests).

The consensus is that the development of HAS is rooted in two books (DeMello 2012): the Australian philosopher Peter Singer's (1946–) *Animal Liberation* (1975) and the American philosopher Tom Regan's (1938–) *The Case for Animal Rights* (1983). Singer's book exposed the often violent and inhumane treatment of animals in industries such as farming, science, and entertainment, shocking many scholars and much of the public because so many people didn't know what was happening with animals in these locations. Regan's work argued that animals, like humans, are "subjects" of a life, with value in and of themselves, instead of objects to be used by humans. Together, these two books helped launch the modern animal rights and advocacy movements, as well as HAS. The field has grown dramatically over the past 25 years, and the Animals and Society Institute (ASI) tracks classes and degree programs (bachelor's, master's, doctoral, and law schools) in North America, Australia, New Zealand, Latin America, Europe, and Israel. In addition, there are now 24 scholarly journals exploring the full spectrum of human-animal relations (ASI 2015).

HAS argues that animals have, in large part, been "erased" from the study of society by a human history written from religious, philosophical, and science perspectives that see humans as the pinnacle of creation—the only ones who make history and the world. This consistent privileging of humans over all other species in society is often referred to as "speciesism," and HAS scholars argue that it can be understood as parallel to racism or sexism. The parallel exists in, for example, the way male-dominated and white-dominated social and political structures have controlled the lives of, and histories about, non-whites and women. It is not that women or non-whites haven't contributed to the history and development of society but that they have not been sufficiently included in the historical record. So in the same way that scholars of such fields as women's and gender studies and African American studies are recovering the invisible histories of these groups, HAS scholars see their role as making animals visible to the history and processes of human society. HAS scholars, like others of specific groups, emphasize that it is not only about recovering and understanding histories and roles but actively changing present-day structures to be more just and humane. HAS scholars do not argue that animals should be treated exactly the same as humans but that human societies must challenge ways in which *all* groups—human or non—can be mistreated, exploited, or made invisible by cultural, economic, ethical, political, and religious processes.

Two of the major challenges for continued expansion of HAS have to do with acceptability and studying animals themselves outside a physical science context. In terms of acceptability, it has often been difficult to get traditional academic disciplines (e.g., history, literature) to grasp the relevance of studying animals, and institutions have often been reluctant to allow courses, open departments, and grant degrees in HAS because it is often confused with what many have seen as solely emotion-based activism instead of evidence-based research. Studying the animals themselves also presents a huge challenge to HAS scholars, as we cannot ask animals questions as we do humans. For example, a dog cannot say if s/he enjoys being in a research lab, nor have dogs left diaries of their past experiences. Therefore, scholars have to use creative and legitimate research methods that often combine ethological work with interpretations of human comments about animals.

Julie Urbanik

See also: Advocacy; Animal Geography; Animal Law; Ethics; Ethology; Human-Animal Bond; Humane Farming; Multispecies Ethnography; Rights; Social Construction; Speciesism

Further Reading

Animals and Society Institute (ASI). 2015. "Human-Animal Studies." Accessed January 14, 2016. http://www.animalsandsociety.org/human-animal-studies/

DeMello, M. 2012. *Animals and Society: An Introduction to Human-Animal Studies.* New York: Columbia University Press.

Regan, T. 1983. *The Case for Animal Rights.* Berkeley: The University of California Press.

Singer, P. 1975. *Animal Liberation: A New Ethics for Our Treatment of Animals.* New York: Random House.

Urbanik, J. 2012. *Placing Animals: An Introduction to the Geography of Human-Animal Relations.* Lanham, MD: Rowman and Littlefield.

Wilke, R., and Inglis, D., eds. 2007. *Animals and Society: Critical Concepts in the Social Sciences.* New York: Routledge.

Humane Education

Humane education is an educational approach that seeks to cultivate students' awareness of, and commitment to, human rights, animal protection, and environmental sustainability while also developing their knowledge about, and understanding of, the traditional curriculum of reading, history, mathematics, and other subjects. It is comparable to other types of education about people, animals, and nature, like social studies, animal studies, and environmental studies. Indeed, humane education often overlaps with these other types of education, as when social studies teachers discuss human rights with their students, animal studies professors address animal protection in their classes, and parents talk about environmental sustainability with their children. However, it is distinct from these other types of education in that it affirms the intrinsic value of people, animals, and nature. That is, humane education is rooted in the beliefs that people, animals, and nature have value in and of themselves, and that education is an opportunity to introduce students to these values. In this sense, it is education that promotes coexistence between humans and animals. According to its proponents, moreover, humane education is the most effective means of achieving such human-animal coexistence on a personal and societal level.

The history of organized humane education begins in Great Britain and the United States in the early 19th century, and it follows two main trajectories. Humane education's first historical trajectory is the broadening of its focus from animal protection to human rights, animal protection, and environmental sustainability. Beginning in the late 18th century, the British and American public became increasingly interested in teaching children to be kind to animals (Unti and DeRosa 2003). As anticruelty societies like the Royal Society for the Prevention of Cruelty to Animals (RSPCA) and the Massachusetts Society for the Prevention of Cruelty to Animals (MSPCA) formed in the mid-19th century, cultivating children's kindness to animals became an important part of their work. Over a century later, it still is. Since the late

20th century, though, humane education's focus has broadened from kindness to animals to compassion for all beings. Many humane educators still teach students about animal protection; however, a growing number of them have begun to recognize the interconnectedness of people, animals, and nature and to cultivate students' understandings of human rights, animal protection, and environmental sustainability.

Humane education's second historical trajectory is the variation of its role in public education. During the mid-19th century, teaching it in American public schools was popular but optional. In 1886, humane education became mandatory in Massachusetts public schools, and by 1920, it was mandatory in 20 other states as well. Nonetheless, it never became compulsory at the national level. By the mid-20th century, humane education's role in public schools had diminished significantly. The recent efforts to standardize curricula may reverse this trend. In the 1990s, humane educators began aligning their programs with public schools' new curricular standards. During the early 21st century, these programs may become more popular in public schools.

Several organizations exemplify the historical trajectories and current state of humane education. The American Humane Association (AHA), founded in 1877, and the National Humane Education Society (NHES), founded in 1948, represent humane education's traditional, narrower focus on animal protection and its role in animal protection organizations. According to NHES, for example, humane education "teaches people how to accept and fulfill their responsibilities to companion animals, such as cats and dogs, and all forms of animal life" (NHES 2010). Such a focus would be appropriate for an education program at an animal shelter but not for one in a school.

The Institute for Humane Education (IHE), founded in 1996, and Humane Education Advocates Reaching Teachers (HEART), founded in 2001, represent humane education's contemporary, broader focus on human rights, animal protection, and environmental sustainability and its role in public schools. HEART's mission, for instance, is "[t]o foster compassion and respect for all living beings and the environment by educating youth and teachers in humane education" (HEART 2015). Clearly, it has developed its education programs for schools, not animal shelters. As their founding dates indicate, IHE and HEART represent not only humane education's current state but also its next steps.

The most critical issue facing humane education is the relative lack of research on its efficacy. To be sure, some research on it does exist (Unti and DeRosa 2003). In 1985, for example, Ascione, Latham, and Worthen published the *Humane Education Evaluation Project*, and in 2001, O'Hare and Montminy-Danna released an evaluation of a humane education program by the Potter League for Animals. Although the findings of Ascione, Latham, and Worthen were ambiguous, O'Hare and Montminy-Danna found that humane education can teach children to be kind and compassionate. Still, they left unanswered the question of how it can most effectively cultivate these qualities. If humane education is to play a meaningful role in public schools in the future, researchers must start answering this question.

Stephen Vrla

See also: Advocacy; Ethics; Human-Animal Studies

Further Reading

American Humane Association. "Humane Education Resources." Accessed July 5, 2015. http://www.americanhumane.org/interaction/programs/humane-education/

Humane Education Advocates Reaching Teachers (HEART). "About HEART." Accessed July 5, 2015. http://teachhumane.org/about/

Institute for Humane Education. "About IHE." Accessed July 5, 2015. http://humane education.org/about-the-institute/vision-mission/

Kean, H. "Humane Education and Children in Britain." Accessed July 5, 2015. http:// bekindexhibit.org/about/humane-education-and-children-in-britain/

National Humane Education Society. "So You Want to Be a Humane Educator . . ." Accessed July 5, 2015. http://www.nhes.org/sections/view/63

National Museum of Animals and Society. "Be Kind: A Visual History of Humane Education, 1880–1945." Accessed July 5, 2015. http://bekindexhibit.org/

Unti, B., and DeRosa, B. 2003. "Humane Education: Past, Present, and Future." In Salem, D. J., and Rowan, A. N., eds., *The State of the Animals II: 2003,* 27–50. Washington, DC: Humane Society Press.

Humane Farming

In recent decades, the suffering of animals raised for food in intensive confinement agriculture systems (also known as "factory farms") has gained prominence on the worldwide animal protection agenda and sparked worldwide campaigns against the confinement of multiple laying hens in small "battery" cages, the extreme confinement of pregnant sows (female pigs) in gestation crates, the keeping of veal calves in narrow stalls, and other practices. Ensuing debates have focused not only on the reform of production systems but on the role of individual food choice in promoting animal welfare. This has linked organized animal protection to the Slow Food, local food, organic, small farmer, and sustainable agriculture movements, which are explicitly focused on food, nutrition, and national food policies around the world.

Humane farming typically refers to the raising of animals with access to fresh air, food, sunlight, and the natural environment, and who are grass-fed, pasture-raised, free of antibiotics, able to move and exercise, and slaughtered so as to minimize pain, distress, and suffering; it has been embraced by many advocates, consumers, public health officials, farmers, ranchers, certifiers, retailers, journalists, and scientists. A number of producers and suppliers have embraced approaches like The Five Freedoms, a rubric for humane animal treatment that first emerged in the United

Kingdom in the 1960s: freedom from hunger and thirst; freedom from discomfort; freedom from pain, injury, or disease; freedom to express (most) normal behavior; and freedom from fear and distress.

Some assert that no animal-based farming can be humane, from a conviction that such husbandry, even at its best, cannot be genuinely cruelty-free. They believe that only plant-based farming can be humane and that veganism (i.e., adopting a completely plant-based diet and, for many, using no animal products whatsoever) represents the best way to prevent the suffering and death of farm animals. It is sometimes stated by those who hold this view that the promotion of humane, non-factory-farmed meat will further entrench meat consumption, retarding progress toward a truly humane world in which no meat is consumed by human beings.

Those who accept the raising of animals under higher welfare standards as humane respond that consumption of the resulting meat, dairy, and eggs does not necessarily lead to complacency or indifference. To the contrary, they suggest, people who consume such products as an alternative to factory-farmed commodities may reflect more deeply about the welfare of animals, and eventually shift their consumption toward plant-based foods. Their choices result in less animal suffering overall and in less rather than more animals consumed because fewer animals will be raised, and generally under better conditions, given the less intensive nature of traditional, small-scale husbandry systems. In this way, the humane farming sector, comprising farmers and ranchers who are good stewards of animals and the environment, provides an alternative to industrial systems of animal agriculture and undermines those systems.

In most nations, legal and regulatory protection for animals raised for food is weak, making consumer choice a promising driver of positive animal welfare reforms. In some economically well-off nations, such as the United States, welfare-focused labeling of animal products, while not well regulated by government, has gained ground. In those nations where the question of farm animal welfare has been explored, polls indicate that a large percentage of consumers say that the humane treatment of animals raised for food is important to them and that they would support meaningful labeling practices. In the United States, the Animal Welfare Approved, Humane Farm Animal Care, and the Global Animal Partnership labels demonstrate the value of strong animal welfare standards with clear, detailed, and meaningful requirements, overseen by third-party certifiers. But the fact that weaker labels can still be found on the products of conventional factory farming suggests that stronger government support for meaningful standards and oversight is needed. Those labels that do not require access to fresh air and the outdoors, indoor enrichment to relieve boredom, and freedom to engage in natural behaviors, and also fail to include prohibitions on physical alterations such as teeth filing and tail docking of piglets or beak trimming of chickens, as well as restriction or prohibition of nontherapeutic antibiotics use, fall outside of the goals of most advocates of welfare labels.

The future of humane farming, whether or not it involves animal husbandry, is inextricably connected to reports that global demand for meat may double by 2050 with anticipated population increase and rising consumption. The production and consumption of animals raised under high welfare standards, and the encouragement of plant-based food alternatives, are both counterweights to the systems of intensive animal agriculture that have taken hold worldwide since the mid-20th century. Fewer animals raised and consumed, under higher standards of welfare, and with less overall suffering, are likely to comprise the common ground for discussions of humane farming in the future.

Bernard Unti

See also: Bovine Growth Hormone; Climate Change; Concentrated Animal Feeding Operation (CAFO); Factory Farming; Husbandry; Livestock; Meat Eating; Slaughter; Veganism; Vegetarianism

Further Reading

Animal Welfare Approved. www.animalwelfareapproved.org

Farm Animal Welfare Council. 2009. "Five Freedoms." http://webarchive.nationalarchives.gov.uk/20121007104210/http://www.fawc.org.uk/freedoms.htm

Global Animal Partnership. www.globalanimalpartnership.org

Hayes, D., and Hayes, G. B. 2015. *Cowed: The Hidden Impact of 93 Million Cows on America's Health, Economy, Politics, Culture, and Environment.* New York: W.W. Norton.

McWilliams, J. 2014. *The Modern Savage: Our Unthinking Decision to Eat Animals.* New York: Thomas Dunne Books.

Niman, N. H. 2014. *Defending Beef: The Case for Sustainable Meat Production.* White River Junction, VT: Chelsea Green Publishing.

Humans

Perhaps the most well-known animal species is *Homo sapiens*, also known as human beings. This mammalian member of the Animalia kingdom is remarkable because of advanced physical and social capabilities, highly developed brain function, and complex interaction with other animal and environmental systems on Earth. The first anatomically modern *Homo sapiens* appeared approximately 200,000 years ago. The evolution of the human species has resulted in both productive and destructive relationships with the living world and other species.

In terms of biological classifications, humans are in the taxonomic class *Mammalia,* the order *Primates* (which also includes apes and monkeys), the family *Hominidae* (great apes), and the genus *Homo*. The species' beginnings remain open to debate, but many scientists, such as evolutionary biologists and anthropologists, who study the development of the human species currently believe that the following three groups of species were branches on the early human family tree. The earliest, 4–5 million years ago, is the genus *Ardipithecus*. They were chimpanzee-like and are thought to be our closest evolutionary link to other primates. Next is the *Australopithecus* genus, dating back 2–4 million years, and fossils show evidence of both tree-climbing and regular upright walking on two feet (bipedal). The *Paranthropus* genus, appearing 1–3 million years ago, had a distinctly larger tooth and jaw structure, signifying an evolving ability for early humans to eat a variety of foods.

The modern *Homo* genus appeared 1–2 million years ago in Africa and once included several other species (e.g., *Homo habilis* and *Homo neaderthalensis*). The only species today is *Homo sapiens*. *Homo sapiens* began migrating into the lower latitudes of East Asia approximately 70,000 years ago and then into Europe. *Homo sapiens* are thought to have arrived in the Americas a minimum of 15,000 years ago. In 2015, the world population of humans surpassed 7 billion.

The human gestation period is 40 weeks, and babies are born less physically developed than many other mammalian species, although they show remarkable physical and brain growth and development in the first year of life. Human beings are altricial, meaning that they require a growth and development phase after birth to be able to move independently, while some other mammal species, such as horses and elephants, are precocial, or able to walk soon or immediately after birth.

The modern human has a distinctly larger brain than our forebears, which has allowed us to adapt, respond, and innovate within our environments. In essence, instead of horns, size, or speed, our brains evolved to be our protection. In addition, human hands co-evolved with the brain to include a wide range of motor skills through the use of opposable thumbs and four fingers, allowing for nimble manipulation of tools.

Neuroscientists who study brains are still in the process of discovering how they work and what makes human brains different. The answer seems to lie in the evolution of a very advanced prefrontal cortex—the part of the brain involved in language and problem solving. The human prefrontal cortex is the largest of any species (Stix 2014).

Without question, however, humans are not the only animal with complex mental lives. Intelligence, culture, communication, tool use, and even warfare have all been documented in other species. It is the extent to which the human brain has developed that sets us apart and has allowed the species to have a dramatic impact on itself and on the planet. For example, the process of domestication of other species and the development of agriculture (both thought to have begun 10,000–12,000 years ago) exemplify one major impact. Archeologists and anthropologists believe that becoming sedentary farmers allowed human society to become more and more complex. For example, because only some people had to farm, other social roles like business, government, military, and even permanent architecture came into being. Unlike other animals, humans today have a range of food-gathering methods—hunting and gathering, nomadic herding, small-scale farming, and industrial farming.

While human language was probably around longer than domestication, our ability to turn vocal communication into written form appears to have occurred between 5,000 and 6,000 years ago in Mesopotamia (in the present-day Middle East) with the Sumerian people (Yu-Chen n.d.). Written language contributes not only to more complex thinking, but also to a more complex society, allowing information to be passed across space and through time. As human societies became more intricate, human groups began consolidating control and spreading across the globe. Human struggles for power have always been linked to negative experiences for certain people (e.g., women, slaves, indigenous [native] people, or ethnic or national groups). However, human imagination has brought us religion, philosophy, science, poetry, music, and everything from fire to iPhones and space travel. For such a young species, we have made incredible accomplishments and had an equally incredible impact.

Our modern day industrial and consumer-oriented global society, in combination with our increasing population and destructive environmental impact, has led to the coining of the term "Anthropocene"—the Age of Humans (Ellis 2013). We simultaneously live in a world where less than half of the population consumes the vast majority of the planet's resources while the other half of the population struggles for daily survival. In terms of our environmental impact, some scientists argue that we are in the middle of the sixth mass extinction event of the planet (Kolbert 2014). Consensus about climate change, biodiversity loss, consumption patterns, and inequalities among humans are making cohesive global action more urgent than ever.

Philip Tedeschi, Erica Elvove, Brooke Harland

See also: Animals; Biodiversity; Climate Change; Communication and Language; Domestication; Evolution; Extinction; Great Apes; Intelligence; Social Construction; Species; Speciesism

Further Reading

Bekoff, M. 2007. *The Emotional Lives of Animals*. Novato, CA: New World Library.

Ellis, E. 2013. "Anthropocene." Accessed January 27, 2016. http://www.eoearth.org/view/article/150125

Harari, Y. 2015. *Sapiens: A Brief History of Humankind.* New York: Harper Collins Publishers.

Kolbert, E. 2014. *The Sixth Extinction: An Unnatural History.* New York: Henry Holt and Company.

Smithsonian. 2016. "What Does It Mean to Be Human?" Accessed January 20, 2016. http://humanorigins.si.edu

Stix, G. 2014. "What's Special or Not About Human Brain Anatomy." Accessed January 22, 2016. http://blogs.scientificamerican.com/talking-back/what-s-special-or-not-about-human-brain-anatomy/

Yu-Chen, K. n.d. "The Five Original Writing Systems." Accessed January 22, 2016. http://rutchem.rutgers.edu/~kyc/Five%20Original%20Writing%20Systems.html

Human-Wildlife Conflict

Wildlife existed millions of years before humans, and conflict between humans and wildlife is as old as our coexistence. Human-wildlife conflict is any interaction that results in negative impacts to humans and/or to wildlife. Early conflict was directly between humans and wildlife and centered on predator-prey relationships. Over time, conflict between humans and animals has shifted as relationships and shared geographies have changed. Although direct human-wildlife conflict persists, contemporary forms are multifaceted and are often related to rising intersocial human conflict over resources and space.

Historically, humans addressed conflict with wild animals by eradicating troublesome wildlife or by establishing controlled reserves to keep terrestrial (land-based) wildlife separate from humans. Wildlife reserves still exist worldwide, though contemporary management also strives to create ways for humans and animals to successfully share space. In the 20th century, recognition of the ecological importance of wildlife prompted efforts to protect many species globally, including species difficult for humans to coexist with. The 1973 establishment of the UN Convention on International Trade in Endangered Species of Wild Fauna and Flora (CITES) prompted protective wildlife legislation in member countries around the world.

Protective legislation is designed to safeguard wildlife and habitat from human activity, and therefore it can interfere with human development, natural resource extraction, and other forms of land use, such as farming, that are economically important or bound to cultural tradition. Changing social and environmental shifts that increase competition for resources and space can increase human-wildlife tension. For example, rapid development in the U.S. Southwest has displaced wildlife living on the edges of urban areas and disrupted traditional migration routes and ranges of wildlife, including coyotes and javelina (a wild mammal that is related to and resembles pigs). Wildlife must now cross highways and travel through suburban yards, where they interface directly with humans in undesirable ways, often getting hit by cars, fighting with pets, digging through trash, or feeding in gardens.

Wild animals may adapt to survive in changing environments in ways that are not favorable to humans. For example, Bengal tigers in the mangrove swamps of India and Bangladesh respond to a lack of native prey and a corresponding abundance of humans by preying on people who venture into the swamps to fish and collect wood. In India and Africa, elephants, rhinoceros, monkeys and apes, and other animals raid crops (graze on or trample human gardens) and destroy human property. Large predators worldwide have learned to prey on easy-to-hunt livestock. Protection of endangered wildlife is a global concern, but impacts from wildlife and conservation legislation are felt strongly at the local level. When wildlife poses a physical or economic threat to people, critical local support for wildlife protection can be constrained.

Another source of human-wildlife conflict is trafficking (illegal trade) of live animals and animal parts, which is largely economically driven. In developing countries poverty motivates local involvement, but globally trafficking is supported by a lucrative black (illegal) market that serves major profit-oriented businesses in medicine, food, fashion, and exotic pet industries. Trafficking involves a global network of activity that includes local poaching (illegal hunting, killing, or trapping of wild animals), transfer to international traders, and sales to wealthy consumers, often in developed countries. The high risk and high profit in illegal wildlife trade parallels that of the illicit drug trade, and wildlife and drugs are often traded by the same individuals and groups operating in the same black markets. In the Democratic Republic of Congo (DRC), a long civil war has prevented a stable society. Human displacement, hunger, and disease have increased economic reliance on wildlife poaching, and poachers will kill park rangers attempting to protect wildlife. Thus, human conflict aggravates human-wildlife conflict.

Competing values and ideologies can contribute to human-wildlife conflict. For example, conflict over spotted owls in the U.S. Pacific Northwest erupted in the 1980s between timber industry professionals placing a high value on profits and traditional livelihoods, and environmentalists arguing that the ethical obligation to protect endangered species habitat was paramount. Many nongovernmental organizations (NGOs) or activist groups monitor protected wildlife, habitats, and the enforcement of protective legislation. Organizations such as Greenpeace and the Center for Biological Diversity regularly engage in legal conflict with government agencies and corporations. NGOs with different goals or values may conflict with each other. Likewise, different agencies of the same government may conflict over different goals, methods, or legal mandates. For example, U.S. state game and fish departments have a goal to protect wildlife and a mission to provide hunting, fishing, and recreational opportunities to people. This can conflict with the U.S. Fish and Wildlife Department's legal mandate to protect wildlife even if it restricts human activities.

Human-wildlife conflict is costly, so conflict resolution is a shared stakeholder goal. Every human-wildlife conflict situation is unique and requires solutions with respect to particular species, habitat, cultures, land use, laws, and economic sensitivities. As human activities become more entangled with wildlife, managers are working to modify both animal and human behavior to reduce negative interactions and increase opportunities for coexistence. Education, alternative livelihoods, loss compensation, and sensitivity to cultural practices can engage people in conservation and help mitigate conflict. Wildlife managers now recognize that animals actively negotiate space with humans and that they exercise agency (power and choice). For example, animals do not always stay in areas they are relocated to. Contemporary management is increasingly focused on predicting and responding to wildlife behavior as opposed to controlling it.

Anita Hagy Ferguson

See also: Agency; Elephants; Endangered Species; Ivory Trade; Northern Spotted Owl; Poaching; Sharks; Tigers; Traditional Chinese Medicine (TCM); Trophy Hunting; Wildlife; Wildlife Management; Wolves

Further Reading

Andre, C., and Velasquez, M. 1991. "Ethics and the Spotted Owl Controversy." *Issues in Ethics* 4(1). Accessed June 19, 2015. http://www.scu.edu/ethics/publications/iie/v4n1/

Goldman, M., Roque de Pinho, J. J., and Perry, J. 2013. "Beyond Ritual and Economics: Maasai Lion Hunting and Conservation Politics." *Oryx* 47: 490–500.

Neme, L. 2014. "For Rangers on the Front Lines of Anti-Poaching Wars, Daily Trauma." *National Geographic*. Accessed June 19, 2015. http://news.nationalgeographic.com/news/2014/06/140627-congo-virunga-wildlife-rangers-elephants-rhinos-poaching/

U.S. Fish and Wildlife Service. 2013. "ESA Basics: 40 years of Conserving Endangered Species." Accessed June 19, 2013. http://www.fws.gov/endangered/esa-library/pdf/ESA_basics.pdf

Wikramanayake, E., Dinerstein, E., Seidensticker, J., Lumpkin, S., Pandav, B., Shrestha, M., Mishra, H., Ballou, J., Johnsingh, A. J. T., Chestin, I., Sunarto, S., Thinley, P., Thapa, K., Jiang, G., Elagupillay, S., Kafley, H., Pradhan, N. M. B., Jigme, K., Teak, S., Cutter, P., Aziz, M. A., and Than, U. 2011. "A Landscape-Based Conservation Strategy to Double the Wild Tiger Population." *Conservation Letters* 4: 219–227.

Woodroffe, R., Thirgood, S., and Rabinowitz, A., eds. 2005. *People and Wildlife: Conflict or Coexistence?* London: Cambridge University Press.

Hunting

Hunting wild animals, which has been done since the dawn of humans, is a practice that has become controversial because it is no longer necessary for human survival in many areas today, thanks to animal farming. Hunting is a complex issue because it encompasses many methods and forms, involves tradition and heritage, and also elicits conversations about humans' rights to kill wild animals, and about the rights of the animals—both species and individuals—themselves. In addition, hunting plays both positive and negative roles in species conservation.

Different types of hunting include subsistence (hunting for needed food and/or products), sport (hunting for recreation), canned (hunting that occurs with captive-raised and/or enclosed animals who are easier to kill), and trophy (hunting to take home "trophies" of the skins, horns, antlers, or skulls). Subsistence hunting,

Bob Hammond registers his turkey on the opening day of hunting season. While many people disagree with hunting wild animals, many hunters say they feel a deep connection with these animals because they must learn so much about them, and they find satisfaction in being self-sufficient. Many hunters also believe eating wild animals is better than eating animals from industrial farms in terms of both animals' quality of life and the environment. (AP Photo/Jim Cole)

where still practiced, is far less contentious than sport hunting—in many indigenous cultures, hunting is viewed as an unfortunate necessity. However, in tropical areas on the planet, bushmeat (meat from wild animals including nonhuman primates) is procured for food for locals as well as exotic food for tourists. It has been linked to the spread of certain notable zoonoses (diseases that can be transmitted from animals to humans) such as HIV and the Ebola virus. Many "exotic" wild animals are also hunted for products made from their bodies, such as elephants for ivory. Rhinoceroses are hunted for their horns for use in traditional Chinese medicine, which has also affected populations of tigers and black bears, among other wild species.

While hunting practices differ widely between countries and cultural groups, they occur on every continent on which humans live. Game (hunted animals) includes reptiles, amphibians, birds, fish, and mammals. Hunters use guns, bows and arrows, or harpoons (for whaling). Hunting occurs on horseback, with dogs or birds of prey, and some indigenous (native) hunters chase animals on foot: North American Navajos chase pronghorn antelopes, Australian Aborigines chase kangaroos, and African bushmen chase zebra and wildebeest. Many indigenous hunting practices engage the hunted animal directly. For example, members of the Ojibwe tribe of North American Indians pray with hunted deer both before and after the hunt to ask permission and to give thanks and ask forgiveness. Though the hunter role in many indigenous cultures has been historically ascribed to the male gender due to time demands of motherhood, women hunters of the Agta tribe in the Philippines successfully maintain both roles. In the modern United States, hunters are primarily men, but female hunters are increasing—many in connection with local food movements (U.S. Census Bureau 2014).

Governmental regulation created the modern system of hunting in the United States in the late 1800s and early 1900s in response to the extreme overhunting of many species (some to the point of extinction, such as the passenger pigeon) for food and other products. The resulting system of licenses, permits, stamps, fees, bag limits (government-imposed limitations on the total number of a species that may be killed by each hunter), and taxes on weapons and ammunition has been effective in conserving game species, though poaching, or illegal hunting, still occurs. A similar system exists in the United Kingdom. Recent awareness of the importance of predators, such as wolves, big cats, and bears, within ecosystems has changed attitudes toward their status as game animals. Hunted close to extinction in many parts of the world, present-day protective measures and relocations, including the reintroduction of wolves to Yellowstone National Park in 1995, have helped increase certain local populations of predators in Western countries. Today, some conservation efforts rely upon income gained through hunting and fishing. For example, the U.S. Wildlife and Sport Fish Restoration Program claims a greater contribution to fish and wildlife conservation than any other effort through offering conservation grants to state and regional organizations.

Global debates about hunting have taken place around ethical and moral issues that discuss cruelty, necessity, and effect on local ecosystems. The anti-hunting stance questions the suitability of any form of killing individual nonhuman animals, but especially without need. In addition, anti-hunters cite species populations that have been negatively affected by hunting. Especially contentious topics include the hunting of baby seals by members of the Inuit tribe, the hunting of whales and other marine mammals worldwide, fox hunting that employs the help of dogs (this has been banned in many countries, including Germany and the United Kingdom), and "varmint hunting," which targets animals who are considered pest species, including crows, prairie dogs, and coyotes, among others. This last form has been criticized severely in the United States, as the animals killed are usually discarded and many may be killed at once during large contests.

Those who are pro-hunting argue that hunting benefits conservation practices. Modern hunters often express a desire to protect wildlife, and they devote time and money to building and maintaining habitat for game. Hunting fills the empty niche left by overhunted predators to help keep game animal populations in check. In addition, many hunters address questions of morality by citing the concept of "fair chase," which requires hunters to give wild animals a fair chance to escape by hunting only on foot and only pursuing wild animals who have the freedom to escape. Therefore, some modern hunting practices, such as canned hunting, may be contentious even among hunters when they violate this code.

Heather Pospisil

See also: Bushmeat; Canned Hunting; Dogs; Elephants; Endangered Species; Extinction; Fishing; Fox Hunting; Human-Wildlife Conflict; Ivory Trade; People for the Ethical Treatment of Animals (PETA); Poaching; Tigers; Traditional Chinese Medicine (TCM); Trophy Hunting; Whaling; Wolves; Zoonotic Diseases

Further Reading

Leopold, A. 1949. *A Sand County Almanac and Sketches Here and There.* Oxford: Oxford University Press.

Marvin, G. 2006. "Wild Killing: Contesting the Animal in Hunting." In The Animal Studies Group, ed., *Killing Animals*, 10–29. Urbana, IL and Chicago: University of Illinois Press.

Merskin, D. 2010. "The New Artemis? Women Who Hunt." In Kowalsky, N., and Allhoff, F., eds., *Hunting Philosophy for Everyone: In Search of the Wild Life*, 225–238. Malden, MA: Wiley-Blackwell.

Nelson, R. K. 1996. "Introduction: Finding Common Ground." In Peterson, D., ed., *A Hunter's Heart: Honest Essays on Blood Sport*, 1–10. New York: Henry Holt and Company.

Reo, N. J., and Powys Whyte, K. 2011. "Hunting and Morality as Elements of Traditional Ecological Knowledge." *Human Ecology* 40(1): 15–27.

Stolzenburg, W. 2008. *Where the Wild Things Were: Life, Death, and Ecological Wreckage in a Land of Vanishing Predators.* New York: Bloomsbury USA.

U.S. Census Bureau, U.S. Department of the Interior, U.S. Fish and Wildlife Service, and U.S. Department of Commerce. 2014. *2011 National Survey of Fishing, Hunting, and Wildlife-Associated Recreation.* Accessed December 7, 2014. https://www.census.gov/prod/2012pubs/fhw11-nat.pdf

Husbandry

The term "husbandry" indicates the art and science of breeding, raising, and caring for domesticated animals for agricultural purposes. (The term can also indicate the cultivation of plants.) Animal husbandry takes a variety of forms worldwide and, in many locations, has changed significantly over the past century. This is especially true as the demand for meat increases globally, leading to larger farming operations and efforts to increase efficiency. Controversy has also arisen over what constitutes good husbandry, especially in locations that have seen dramatic changes in practices.

A basic animal domestication process, and a significant aspect of husbandry, is "selective breeding"—controlling reproduction in order to produce offspring that have the most useful and/or appealing qualities for humans. In its most basic form, this involves bringing males and females with desirable traits (e.g., cattle that have more muscle mass, and therefore meat) together to mate. If this process is continued for a number of generations, over time more and more offspring will have the desired characteristics. Although the original form of selective breeding that brought two animals physically together continues in a number of locations (e.g., by nomadic herders in Tanzania), it is increasingly being replaced by artificial insemination and genetic selection practices, which are now used almost exclusively in commercial livestock production in wealthier countries, such as the United Kingdom.

In meat production, animals may not be raised to maturity (as would be the case if used for breeding or producing milk or eggs), but instead only to the desired size/weight for slaughter. If animals are raised for a family's or community's own consumption, the rate and rapidity of growth may be less important than if the animals are raised as commodities for sale in a profit-driven market and, therefore, husbandry practices will likely be different. In the latter case, more resources may be devoted to producing more/faster weight gain, for example by feeding the animals directly, frequently with a high-calorie manufactured food, rather than having them roam and graze or forage, the more traditional way of feeding. Keeping animals confined also limits the amount of energy spent through moving and, therefore, also contributes to weight gain. In terms of feeding and confinement, an example at one end of the spectrum is the case of the nomadic sheep- and goat-herding Yörüks of Turkey, whose animals graze freely on natural vegetation either along the coast or in the mountains, depending on the season. An opposing example is the raising of beef cattle on a U.S. industrial farm. Between 12 and 18 months of age, these cattle will be kept in an enclosure (known as a "feedlot") with no vegetation and fed a grain diet in order to put on weight quickly prior to slaughter (National Cattlemen 2006).

A major component of husbandry is that of care provided to the animals beyond feeding. This includes housing, veterinary care, and management. Housing standards and practices are widely variable and depend on such things as species/variety, type of production, and cultural traditions. For example, in North America, chickens raised commercially for meat (known as "broilers") are frequently kept uncaged by the thousands or tens of thousands in long buildings, and egg-laying chickens (known as "layers") are frequently housed in tiny, cramped "battery" cages in windowless buildings. In contrast, in rural Laos, chickens raised for a family's (egg and meat) consumption may roam freely in an unfenced yard or around the village.

In terms of veterinary care, husbandry practices will address routine activities, such as castration (removal of testicles) of males not designated for breeding, or medicines given to prevent intestinal parasites. Nonroutine care includes treatment of illness or injury, and in less wealthy countries or communities, this may comprise most of the veterinary care, as there may not be sufficient resources for or access to preventive medicine. Regarding routine castration, this practice has been done for centuries throughout the world in order to manage male animals' sexual behavior and aggression. It can be done both surgically and nonsurgically, most often without anesthesia or painkillers, and is often done by someone other than a trained veterinarian. With respect to nonroutine care, many commercial farms will have isolation areas in which a sick or injured animal can be treated away from the other animals. On very large farms, however, it may be difficult to identify animals that need care. Additionally, if the animal is of little economic value as an individual, the cost of treatment may be deemed to outweigh any benefit gained, and no action may be taken other than to dispose of the individual.

Many management practices are directed toward animals' temperament. Nose rings in bulls are a good example. The rings have been used for centuries around the world to help control large and perhaps difficult animals, by allowing a handler to exert pressure on a very sensitive part of the body. Management issues that are more typical on larger commercial farms are aggression within large groups of unfamiliar animals and also boredom. Aggression and/or boredom can cause piglets to bite each other's tails. Management of this behavior has been to cut off ("dock") the pigs' tails, usually without anesthesia.

Many of the husbandry practices, such as those reviewed above, on larger, commercial farms have been challenged as cruel in recent years. Animal advocates have argued that practices considered to be "good husbandry" in industrial farming are chiefly directed toward maximizing profit and disregard animal welfare.

Connie L. Johnston

See also: Concentrated Animal Feeding Operation (CAFO); Cruelty; Domestication; Factory Farming; Humane Farming; Livestock; Meat Eating; Pastoralism; Slaughter; Veterinary Medicine; Welfare

Further Reading

Ekarius, C. 2007. *Storey's Illustrated Guide to Poultry Breeds.* North Adams, MA: Storey Publishing, LLC.

Gadd, J. 2005. *Pig Production: What the Textbooks Don't Tell You.* Nottingham, UK: Nottingham University Press.

Lewis, C. 2014. *The Illustrated Guide to Cows: How to Choose Them—How to Keep Them.* New York: Bloomsbury USA.

National Cattlemen's Beef Association. 2006. "Fact Sheet: Feedlot Finishing Cattle." Centennial, CO.

Hybrid

The term "hybrid," as a concept, requires a sociological and geographical understanding in addition to the biological. Unpacking the various uses of hybrid (and hybridity) entails putting perspectives from the social sciences and humanities into conversations with perspectives of the natural and biological sciences. The conceptual frameworks associated with hybrid-thinking provide tools for understanding complex issues at the intersections of science and politics, nature and culture, and technology and society, realms essential to the study of animal geographies and human-nonhuman relations.

Simply put, a hybrid is some entity that is a mixture of two or more components. Biological and genetics scientists understand a hybrid as an offspring between two animals or plants of different breeds, varieties, species, or genera (the biological classification above species and below family). For example, a wolf-dog hybrid is a cross between two species—domestic dogs (*Canis lupus familiaris*) and gray, red, or other wolves (*Canis lupus or rufus*). Wolf-dogs are purposefully bred by humans to meet a growing demand for domestic ownership, but wolves and dogs also interbreed in the wild and produce fertile offspring. Another example of a hybrid is the liger, a cross between a male lion (*Panthera leo*) and a female tiger (*Panthera tigris*). Gender specificity is an important determining characteristic of hybrid species, as, for example, a cross between a male tiger and female lion is a tigon, exhibiting different traits from the liger. In these examples the hybrid, as wolf-dog or liger or tigon, emerges from interbreeding of different species (interspecific) under the same genus (e.g., *Canis* or *Panthera*). Hybrids can exist at higher classifications, between animals of different genera or families, but these are extremely rare.

For scholars in the social sciences and humanities, the concept of the hybrid is used as a way to understand "mixtures" of social and natural worlds. In modern Western society, it is typical to think of human society and nature (which includes animals) as two separate entities. However, the French philosopher and social scientist Bruno Latour (1947–) suggests that nothing in the world is completely natural or completely social but, rather, a combination. Latour even takes this a step further and argues that the division of nature and society as distinct and separate entities is incorrect, as everyone and everything is, and has always been, inseparably intermingled. It is primarily only the modern, Western worldview that categorizes things as either natural or social. A clear example of hybrid, as the term is used outside biology, are human embryos that are frozen and later thawed for fertilization. The embryos occur naturally and will contribute to the biological process of reproduction, but that process has been aided by a human-created technology.

Scholars therefore have embraced the term hybrid to investigate their areas of interest as both social and natural at once. For example, the term hybrid is used by the philosopher and biologist Donna Haraway (1944–) in her exploration of the OncoMouse™—a genetically modified mouse created by Harvard University researchers to carry an activated oncogene, a gene used to increase the mouse's susceptibility to cancer. Haraway defines the mouse as a "breast cancer research model produced by genetic engineering" (Haraway 1997, 47). She explains that the

OncoMouse™ is an animal produced with genetic engineering (a biological hybrid) for the purpose of research on breast cancer, but s/he is also a piece of private, intellectual property, a mixture of nature, technology, and capital (money) encapsulated in a research "tool" for the production of scientific knowledge (a social hybrid). The mouse is a hybrid not merely because of genetic changes, but also because s/he is simultaneously a "model," a commodity, a patented animal, and a laboratory device.

The scholarship on hybridity reflects a larger movement toward what is called posthumanism in social science research, which challenges the privileged space of humans as subjects (those who act) and all others as objects (those acted upon). Physical or natural sciences such as genetic engineering and ethology (the study of animal behavior) have illuminated the connections between humans and animals, opening up pathways toward better understandings of human-animal relations. Haraway suggests it is important to acknowledge the ways biological sciences have simultaneously and paradoxically "produced" animals as objects of knowledge and, at the same time, illustrated the genetic similarities of humans and animals, complicating our assumptions of animals' roles in social relations. The human-animal relationship is both simplified and complicated by progress in evolutionary biology and genetic science. Social scientists use the concept of hybridity as a valuable theoretical tool to navigate these intertwining relationships.

Anthony M. Levenda

See also: Biotechnology; OncoMouse; Social Construction; Species

Further Reading

Haraway, D. J. 1997. *Modest–Witness@Second–Millennium.FemaleMan–Meets–Onco Mouse: Feminism and Technoscience*. New York and London: Routledge.

Latour, B. 1993. *We Have Never Been Modern*. Boston: Harvard University Press.

Philo, C., and Wilbert, C. 2004. *Animal Spaces, Beastly Places: New Geographies of Human-animal Relations*. New York and London: Routledge.

Urbanik, J. 2012. *Placing Animals: An Introduction to the Geography of Human-Animal Relations*. Lanham, MD: Rowman & Littlefield Publishers.

Vilà, C., and Wayne, R. K. 1999. "Hybridization between Wolves and Dogs." *Conservation Biology* 13(1): 195–98.

Whatmore, S. 2002. *Hybrid Geographies: Natures Cultures Spaces*. London: Sage.

Wolch, J. R., and Emel, J. 1998. *Animal Geographies: Place, Politics, and Identity in the Nature-Culture Borderlands*. London: Verso.

I

Indicator Species

Indicator species are animals or plants that are sensitive to ecological disturbance and whose health forecasts or reflects changing conditions in their native ecosystems. These species are among the first to be influenced by changes or damage to the environment, providing early warning signs that an ecosystem is experiencing adverse effects. The relative occurrence or absence of indicator species, their distribution, population density, reproductive success, and physical condition can all be useful indices for the health of ecosystems. Monitoring indicator species can provide effective means to detect potential environmental problems before permanent damage is done. Indicator species are also often utilized to monitor the success of environmental management strategies or ecosystem restoration initiatives. Due to their ability to reflect the impacts of environmental disturbances and ecosystem health, indicator species are a cost- and time-effective tool for resource managers and governing bodies.

Both animal and plant species are utilized as indicator species in order to evaluate air and water quality by government agencies, including the U.S. Environmental Protection Agency (EPA), as disturbances due to air and water pollution and climate change are reflected in indicator species' physical condition and population dynamics. Indicators are used to verify the compliance of industries to particular antipollution laws. Indicator plants often accumulate large concentrations of pollutants in their tissues, allowing scientists to identify and monitor contaminant levels. Some fish species are useful indicators, as their constant contact with aquatic conditions and their responsiveness to disturbances including chemical, temperature, or habitat changes provide useful insight into water quality and ecosystem viability.

Natural resource managers and conservation biologists also value indicator species. While it is impossible to account for all factors in an ecosystem, indicator species reduce this complexity by providing a clear and specific set of components that indicate environmental quality. Evaluation of indicator species is frequently incorporated into broader habitat assessment programs implemented by U.S. government agencies like the U.S. Fish and Wildlife Service (USFWS) and the U.S. Forest Service (USFS). The Habitat Evaluation Procedures (HEP) of the USFWS identifies the quality and quantity of available habitat for wildlife, utilizing indicator species in environmental assessments and analyses of habitat quality.

Birds are frequently selected as indicators based on several characteristics: Their vocalizations make them easy to detect and identify; many birds can be monitored effectively over large areas; and their abundance, distribution, and reproductive success are influenced by surrounding habitats. Perhaps one of the most famous examples in the United States, the northern spotted owl (*Strix occidentalis*) requires old-growth forests (natural forests with little disturbance, dominated by old trees) for nesting. If spotted owls are surviving at viable populations (strong numbers not threatened with extinction), the old-growth forests are deemed healthy as well.

Starting in the early 1900s, miners brought canaries into coal mines as an alarm system to alert them to toxic gases such as carbon monoxide or methane. The birds showed visible distress if these toxins were in the air, allowing miners to evacuate. The phrase "canary in a coal mine" has come to indicate something that is sensitive to harmful conditions that can be used as a warning system. This phrase is frequently used to characterize the plight of butterflies, as they are sensitive to habitat or climatic changes and respond rapidly to change. Butterflies respond to changes in their environment more quickly than birds or plants, and because of their extreme sensitivity and short life spans, they are a valuable indicator species. In 2015, the USFWS announced that since 1990, around 970 million monarch butterflies have vanished. Monarch population decline is attributed to the loss of habitat over the last two decades, primarily through the removal of milkweed plants, which are the butterflies' food source, home, and nursery for caterpillars. At the same time, the butterflies' overwintering sites in Mexico are being degraded through illegal and legal forest logging, conversion of land for agriculture, and climate change. Monarch butterflies are indicators for the health of native landscapes across North America and speak to the broader viability and success of pollinator species.

The Nile crocodile, *Crocodylus niloticus,* is an indicator species for the Okavango Delta, the largest inland delta in the world and the most critical wildlife sanctuary in southern Africa. The crocodile is the first to be impaired by pollution, poisons, and endocrine disruptors (compounds found in plastics and pesticides) due to their sensitive nervous systems. The Delta's breeding female crocodile population has been reduced by up to 60 percent as a result of water pollution, a frightening indicator of effects that may be experienced by other species over time (Fraser 2009). Furthermore, the crocodile is the top predator; thus its absence affects all levels of the food chain (producers, herbivores, primary carnivores, secondary carnivores) and the nutrient content in the aquatic ecosystem.

While indicator species are incredibly useful for understanding changes in an ecosystem, the value of any one species should be assessed cautiously. Because no two species occupy the same explicit roles in the ecosystem, no single species can be expected to represent a complete ecosystem. Thus, multiple species should be observed to detect the root causes of change and potential landscape scale effects. Secondly, some experts caution that several factors, separate from the decline of a habitat, may disturb an indicator species community. This makes clearly detecting and understanding indicator demographic trends a complicated process. These concerns require careful and thorough analysis of indicator species population changes, such as distribution and density.

Kalli F. Doubleday

See also: Northern Spotted Owl; Species

Further Reading

Carignan, V., and Villard, M. 2002. "Selecting Indicator Species to Monitor Ecological Integrity: A Review." *Environmental Monitoring and Assessment* 78.1: 45–61.

EnviroScience. 2012. "Fish." Accessed April 7, 2015. http://enviroscienceinc.com/fish-2/

United Kingdom Butterfly Monitoring Scheme. 2006. "Butterflies as Indicators." Accessed April 7, 2015. http://www.ukbms.org/indicators.aspx

USFWS. 1980. *Habitat Evaluation Procedures (HEP).* Washington, DC: US Department of the Interior.

USFWS. 2015. "Save the Monarch Butterfly." Accessed June 1, 2015. http://www.fws.gov/savethemonarch

YouTube. 2013. "120 Seconds of Science—What Is an Indicator Species?" Accessed April 7, 2015. https://www.youtube.com/watch?v=SnHLIRbeCNg

Indigenous Religions, Animals in

Animals play significant, diverse, and complex roles in many indigenous religions. They are often characters in myths and origin stories and might act as metaphors or symbols in indigenous ideologies. Animals are sometimes killed for sacrifices, and parts of their bodies may be used in rituals. Many indigenous groups consider animals as sacred and thus carefully attend to their relationships with these creatures in day-to-day life. Some cultures have animistic beliefs wherein animals are thought to be persons with souls similar to humans. There is no one way to characterize animals in indigenous religions, but case studies demonstrate how geographically and culturally distinct peoples integrate animals into their spiritual lives. By considering how indigenous peoples think about animals, we gain a sense of the global diversity of human-animal relations and of the ways that indigenous lifeways are affected when environments change or when access to animals becomes limited.

Indigenous peoples are sovereign cultural groups with distinctive languages and traditions whose ties to certain lands are longstanding and often precede the settler colonialism of about the 15th century onward. For example, the Maori of New Zealand, the Dine (Navajo) of North America, and the Ainu of Japan could all be considered indigenous peoples. There are thousands of culturally unique indigenous groups across the globe, but many share historical and contemporary experiences of forcible removal from land, ecocide (the purposeful destruction of the environment), and genocide at the hands of colonial powers such as the United States or the United Kingdom.

Indigenous religions vary greatly in their form, content, and antiquity and are just as diverse as the peoples who practice them. In some places, indigenous religions exist side-by-side with global religions like Christianity, Islam, or Buddhism, and elements of these are sometimes incorporated into indigenous belief systems. Despite their diverse histories, indigenous religions do have some shared qualities—they are generally central to cultural identities, they tend to integrate spirituality into all aspects of life, and they often value continuity over time, with many celebrating longstanding connections between people and land.

The Maya, a large and diverse indigenous group from Mesoamerica (including parts of Mexico, Guatemala, Belize, and Honduras) consider their Creator to be one and the same with Earth and all its creatures, suggesting the importance of respect for nature and the interconnectedness of human and animal lives. In the Mayan account of creation, Earth was originally a Cipactli, a being with zoomorphic (or animal-like) features including the body of a crocodile or snake, the tail of a fish or crab, the paws of a reptile or jaguar, and the ears of a deer. Later in time, mountain cat, coyote, parrot, and crow helped to create humans by supplying the corn that formed the bodies of the first people. The Maya believe that the Creator gave each animal its own habitat and place in the world, thus ensuring their right to respect and continued existence. Although many animals serve as food for humans in Mayan cosmologies (or systems of belief about the universe), some play different roles—dogs, for instance, can help human souls pass through a river of alligators flowing through the underworld, but only if treated well by their owners in life. The *Popul Vuh*, a sacred Mayan text, emphasizes the importance of ethical treatment of animals in its story of the wooden people, a group of humans who brought about their own demise through abusing animals and plants.

The Yup'ik, an indigenous group from southwestern Alaska, rely on animals for many of their subsistence needs (e.g., basic food, clothing, shelter) and carefully manage their relationships to these creatures through religion and ritual. The Yup'ik believe that animals, like humans, are persons possessing free will, agency, and awareness and can thus choose which hunters are worthy of taking their lives. Hunters demonstrate their worthiness by remaining humble, refraining from teasing animals, and killing prey quickly and humanely to minimize suffering, as well as through ritual acts like providing hunted seals with a drink of water to welcome their spirits. In the past, communities would celebrate annual festivals wherein seal bladders, thought to hold the animals' souls, were released back into the sea to ensure that seal spirits could return home and offer a bountiful harvest the next year. The value of respecting animals is passed on to children through stories, one of which tells of a young hunter sent to live as a seal for a year to observe whether the hunters in his community were acting properly. As this tale demonstrates, human-animal transformation is a reality for Yup'ik people, with animals sometimes appearing as humans, and humans able to travel into animal worlds.

Indigenous religions are often deeply affected when human-animal dynamics shift due to environmental degradation and land development. In Africa, indigenous groups like the Maasai have been forcibly removed from the sacred lands where they once hunted, grazed livestock, and practiced their traditional religions to make way for private and public developments, ranches, and wildlife reserves (e.g., Kenya's Maasai Mara National Reserve) that are meant to protect animals and promote sustainable tourism. It is important to honor indigenous peoples' deep understandings of ecology and their rights to religious sovereignty when considering how to best protect animals in our changing world. As indigenous religions suggest, we cannot separate the destinies of humans and the animals that are so often the subject of our spiritual beliefs.

Anna C. Sloan

See also: Agency; Biodiversity; Indigenous Rights; Personhood

Further Reading

Bernal-Garcia, M. E. 2001. "The Life and Bounty of the Mesoamerican Sacred Mountain." In Grim, J. A., ed., *Indigenous Traditions and Ecology: The Interbeing of Cosmology and Community,* 325–349. Cambridge: Harvard University Press.

Fienup-Riordan, A. 1994. *Boundaries and Passages: Rule and Ritual in Yup'ik Eskimo Oral Tradition*. Norman, OK: University of Oklahoma Press.

Grim, John A., ed. 2001. *Indigenous Traditions and Ecology: The Interbeing of Cosmology and Community*. Cambridge: Harvard University Press.

Kipuri, N. 2006. "Human Rights Violation and Indigenous Peoples of Africa: The Case of the Maasai People." In Kunnie, J. E., and Goduka, N. I., eds., *Indigenous Peoples' Wisdom and Power: Affirming Our Knowledge Through Narratives*, 246–256. Burlington, VT: Ashgate.

LaDuke, W. 1999. *All Our Relations: Native Struggles for Land and Life*. New York: South End Press.

Montejo, V. D. 2001. "The Road to Heaven: Jakaltek Maya Beliefs, Religion, and the Ecology." In Grim, J. A., ed., *Indigenous Traditions and Ecology: The Interbeing of Cosmology and Community,* 175–195. Cambridge: Harvard University Press.

Viveiros de Castro, E. 1998. "Cosmological Deixis and Amerindian Perspectivism." *The Journal of the Royal Anthropological Institute* 3(4): 469–488.

Indigenous Rights

Although there is no universally accepted definition for "indigenous peoples," there are general characteristics commonly associated with them: Indigenous peoples self-identify as the first and native inhabitants; they tend to have smaller populations

For native people around the world who depend on wildlife as not only a food source but also an integral part of their cultural identity, restrictions on access to animals is seen as a direct attack on their survival. Whale hunting, as seen here, has been controversial because many conservationists and animal advocates believe that whales should be protected because of their high intelligence. (AP Photo/Gregory Bull)

compared to the dominant culture of their country; and they tend to retain their own language, distinctive cultural traditions, and territory. Animals are an invaluable resource that allows indigenous communities to survive in difficult environments such as the Arctic, Amazon, Southeast Asia, Central Asia, and Sub-Saharan Africa, and they play a vital role in sustaining indigenous culture. Indigenous rights is a key topic for Human-Animal Studies because it interconnects the well-being of humans and animals with global sociopolitical phenomena.

In recent decades, indigenous peoples worldwide have demanded the right to be involved in the policy-making processes that affect their lives and those of animals with whom they share an environment. As one manifestation, the UN Declaration on the Rights of Indigenous Peoples (UNDRIP) advocates the freedom of indigenous peoples to keep and protect their animals to maintain and develop their traditional cultural and economic practices. It also supports their rights to conserve and protect the environment and the productive capacity of their lands and resources as the basis of their cultural identity, self-rule, and heritage. Responding to rapid social change and threats to the global environment, demands for land claims and self-government have been based on historical and cultural rights to both lands and animals.

Indigenous hunting rights are central to indigenous self-rule and the continuity of their cultures. The fact that many indigenous groups have their own governments that are in relation to the dominant government complicates the configuration of their hunting rights and relationships to animals. Therefore, the goals of animal conservation groups and the goals of indigenous peoples wishing to hunt are not compatible in all cases. Indigenous peoples' beliefs can conflict with the conservation view that animals are not for human consumption. Whales provide a significant range of important resources, including meat, blubber, bone, baleen (a filter-feeder system found in the upper jaws of baleen whales), sinew, and internal organs in addition to architectural materials and spiritual fulfillment for several indigenous groups. For example, Alaska Native tribes such as the Iñupiat and Yup'ik have experienced conflicts and cooperation with the International Whaling Commission to secure the annual whaling quota for subsistence whaling. To them, the retention of whaling rights means cultural survival. Similarly, a variety of indigenous protests have facilitated the formation of self-governing organizations in order to secure access to important animal species. Another example is that tribes such as the Nez Perce, Umatilla, Yakama, and Warm Springs organize the Columbia River Inter-Tribal Fish Commission, which proudly adopts the salmon as a cultural icon and the organization's logo. One of their major ceremonies is called the First Salmon Feast, in which they honor the salmon and intertribal kinship and fellowship. Retaining and building two-way relationships with animals characterize indigenous ways of life, and many believe that animals "give themselves" to humans who are worthy of such gifts.

On an economic front, some indigenous subsistence practices have turned into commercial ventures, blurring the boundary between indigenous rights for cultural survival and local economic fulfillment. For example, the Miskito people on the coast of Nicaragua once embraced green turtles as the center of their cosmology (worldview) as well as a major part of their traditional subsistence economy. This relationship changed dramatically with the increased demand from international markets for turtle meat, calipee (the fatty gelatinous substance found immediately over the lower shell of a turtle), and cartilage, resulting in a change in hunting seasons, methods, and social relations linked to the distribution of turtle meat as foreign companies hired Miskito turtlemen to harvest turtles year-round. This process led to a serious depletion of the green turtle population, and tribal members were confronted with rising social tensions. Despite this historical struggle, the Miskito people see the importance of codifying indigenous rights, and today harvesting turtles is evolving into a symbol of resistance to colonial injustice. This effort was reflected in the unilateral declaration of independence from Nicaragua under the name Community Nation of Moskitia.

Another facet of indigenous rights is manifested in how indigenous populations attempt to adapt to climate change. This global phenomenon challenges the traditional reciprocity between indigenous peoples and animals with its impacts on biodiversity, cultural diversity, and indigenous observations of animals and environment. Indigenous peoples almost universally use local biodiversity as a buffer against variability, change, and catastrophe in their environment to minimize the risk due to harvest or hunting failure. Adoption of many different crops and varieties that have different susceptibility to droughts and floods traditionally made indigenous survival possible. Indigenous peoples are fighting loss of biodiversity and adapting to climate change through migration, irrigation, water conservation techniques, land reclamation, and changes in hunting and subsistence techniques. For example, in northern Finland among the Sámi, reindeer herding is at the heart of their culture and way of life, although it has been threatened by the increasing unpredictability of winter weather patterns. Sámi herders, in order to retain their human-reindeer relations, are now working to solidify indigenous rights to revitalize land-based traditions through the active participation in indigenous-driven international organizations. Similarly, global indigenous populations attempt to advance the role of traditional knowledge in environmental policy and practice, which contributes to the enhancement of indigenous rights based on historical interactions with animals.

Chie Sakakibara

See also: Biodiversity; Indigenous Religions, Animals in; Whaling

Further Reading

Nietschmann, B. 1979. *Caribbean Edge: The Coming of Modern Times to Isolated People and Wildlife*. New York: Bobbs-Merrill.

Rose, D. B. 2011. *Wild Dog Dreaming: Love and Extinction*. Charlottesville, VA: University of Virginia Press.

Sakakibara, C. 2011. "Climate Change and Cultural Survival in the Arctic: Muktuk Politics and the People of the Whales." *Weather, Climate and Society* 3(2): 76–89.

UN General Assembly. 2007. *United Nations Declaration on the Rights of Indigenous Peoples: Resolution/Adopted by the General Assembly*. Accessed February 3, 2015. http://www.refworld.org/docid/471355a82.html

Wenzel, G. 1991. *Animal Rights, Human Rights: Ecology, Economy, and Ideology in the Canadian Arctic*. Toronto: University of Toronto Press.

Institutional Animal Care and Use Committees (IACUCs)

Institutional Animal Care and Use Committee (IACUC) is a term used to describe committees that oversee the use of nonhuman animals in laboratory research in the United States. IACUCs are required at institutions that undertake federally funded laboratory research, like those funded by the National Institutes of Health

(NIH). It is important to understand IACUCs because of their role in governing the lives and deaths of animals in research. For many animals that will spend their entire lives in research labs, IACUC guidelines and practices have real impacts on animals' experience of research protocols and experimentation.

The NIH Office of Laboratory Animal Welfare defines policies for animal welfare in laboratories, which are then instituted by the IACUC (which is overseen by an Institutional Official) at each university. Researchers using animals in their studies must submit a summary of their research protocol to the IACUC for approval. The IACUC also oversees twice-a-year inspections of laboratories where animals are present to ensure adherence to animal welfare policies.

In 1966, the first U.S. federal law was passed protecting animals in laboratory research—the Laboratory Animal Welfare Act, which would later become the Animal Welfare Act (AWA). This passed in a climate of public outcry about the plight of animals in research after a 1966 article, published in *Life* magazine, that described the increasingly commonplace theft of dogs and cats from homes by animal dealers who then sold many of these animals to laboratories. Importantly, the AWA only covers some animal species and excludes rats and mice (the majority of species used in laboratory research), birds, farmed animals, and all cold-blooded animals. Prior to the passage of the AWA, researchers were free to determine on their own what constituted ethical care of animals. Through the second half of the 20th century, regulations and policies related to laboratory animal welfare were repeatedly revised and refined. Passed in 1986, the Public Health Service (PHS) Policy on Humane Care and Use of Laboratory Animals introduced IACUCs as we know them today. This policy is the one to which institutions currently adhere, and it states that all vertebrate animals should be covered under its welfare guidelines.

IACUCs are generally comprised of three to five members (although it is permitted for one person to serve multiple roles on an IACUC, it is not recommended) and require a knowledgeable and senior chair of the committee, a veterinarian with experience in a laboratory setting and with the species being used, a nonaffiliated committee member to offer a noninstitutional point of view, a scientist with experience in animal research, and a nonscientist. Decisions and approvals are passed only if there is a quorum (majority) present and a majority voting in favor of the proposed protocol. IACUC programs involve training and education of committee and program members, researchers, and animal care technicians, in addition to reviewing and approving research protocols and conducting inspections of the institution every six months.

IACUCs ultimately report to the NIH Office of Animal Welfare, but at the institutional level IACUCs are the primary body overseeing research involving animals in university settings, which means that even social science research (and other research outside the laboratory setting) is reviewed by the IACUC. One of the issues with this institutional structure is that IACUCs are generally not trained or knowledgeable about forms of research involving animals beyond the laboratory. In real terms, this means that IACUC-required trainings for researchers involve teaching them, for instance, the acceptable methods of euthanizing rats, mice, dogs, cats, primates, and other commonly used species at the end of a study. But IACUCs are ill-equipped to oversee more qualitative, ethnographic research on animals—like cows on farms, for example.

As federally mandated programs in institutions that receive federal funding, IACUC-generated information about animals in laboratories is accessible to the public through Freedom of Information Act (FOIA) requests, and IACUC meetings are generally open to the public. This makes the number of animals, the nature of the research, and other specific information about the animals accessible to any member of the public concerned about the welfare—or, more fundamentally, the *use*—of animals in laboratories. To give an example, the Beagle Freedom Project (a nonprofit animal advocacy group dedicated to ending the use of all animals—and particularly dogs and cats—for research) launched a program in 2015 called the Identity Campaign, which solicits members of the public to submit requests for information to IACUCs about a singular animal—a beagle, for instance—in a particular lab, to collect as much information about that animal, and to advocate for their release and adoption at the end of the study.

This kind of external pressure from the public and animal advocacy groups highlights the ongoing debate about the role of animals in laboratory research in which IACUCs are enmeshed. IACUCs were formed as a way to implement greater care and ethical practice related to animal use in laboratory settings, but their presence has not eliminated fundamental ethical questions about how and whether animals should be used in research. In fact, IACUCs at many institutions advocate the Three R's approach—Replacement (of animals with nonanimal or *in vitro* models), Refinement (to reduce the pain and improve well-being of animals), and Reduction (to use fewer animals to obtain the same or equivalent results)—as an indication of a need to move toward less invasive practices, as well as toward an overall reduction in animal use for science.

Kathryn Gillespie

See also: Multispecies Ethnography; Research and Experimentation; Vivisection

Further Reading

National Institute of Health Office of Animal Welfare. 2015. *Public Health Service Policy on Humane Care and Use of Laboratory Animals.* Accessed March 24, 2015. http://grants.nih.gov/grants/olaw/references/PHSPolicyLabAnimals.pdf

National Research Council. 2011. *Guide for the Care and Use of Laboratory Animals*, 8th ed. Washington, DC: National Academies Press.

Office of Laboratory Animal Welfare (OLAW). 2002. *Institutional Animal Care and Use Committee Guidebook*, 2nd ed. Accessed March 24, 2015. http://grants.nih.gov/grants/olaw/GuideBook.pdf

Zurlo, J., Rudacille, D., and Goldberg, A. M. 1996. "The Three R's: The Way Forward." *Environmental Health Perspectives* 104(8). Accessed April 9, 2015. http://caat.jhsph.edu/publications/Articles/3r.html

Intelligence

What is intelligence? According to the Merriam-Webster dictionary, intelligence is "the ability to learn or understand things or to deal with new or difficult situations." This general definition captures how the word "intelligence" is normally used. A synonym for intelligence is the word "smart." This definition also focuses on ways in which animals *adapt* to different social and nonsocial environments. Thus, ethologists (scientists who are interested in animal intelligence and who study animal behavior under natural or near-natural conditions) see intelligence as an adaptation that is expressed differently by individual animals, including members of the same species. Applying the renowned ethologist Niko Tinbergen's (1907–1988) ideas about how to further our understanding of intelligence, we need to study the evolution of intelligence, how intelligence allows individuals to adapt to their immediate environments, how individual differences in adapting influence their reproductive success (how many offspring they have who then go on to have offspring of their own), what factors cause various forms of intelligence to evolve, how intelligence develops in individuals, and how and why individual differences emerge.

There are often practical matters associated with the use of the word "intelligence." Some people have argued that less intelligent animals suffer less than more intelligent animals. However, there are no data to support this claim. Also, individuals of supposedly more intelligent species are often claimed to be more valuable and more worthy of protection from harm than individuals of supposedly less intelligent species. Thus, because of these two claims, some conclude that it is more permissible to do things such as conduct physically invasive research on individuals of purportedly less intelligent species.

Recent research has revealed many unexpected results about animal intelligence. For example, it is now known that fish and crocodiles use tools, New Caledonian crows make and use more sophisticated tools than chimpanzees, and birds are able to predict future food resources. Young New Caledonian crows also go to "tool schools" where adults teach them to learn how to make and use tools. It is also known that finches use strict rules of syntax (for humans this refers to how words are arranged to create sentences), great tits (a bird species) learn foraging strategies from other tits and then pass them on to future generations, and fish use what is called "referential" (gestural) communication by nodding their heads in a particular direction to tell other fish where food is located, a capacity once thought only to be found in humans.

Unexpected results also apply to insects. For example, although honeybees have relatively tiny brains compared to birds and mammals, they use abstract thought and symbolic language. They learn how to most efficiently travel between multiple sites (called the "traveling salesman problem"), to mix medications for the hive, and to distinguish between complex landscape scenes including types of flowers, shapes, and patterns. Bees also learn categories and sequences of behavior and adjust them for future rewards. They consider social conditions, locations, time of day, and use multiple senses. They also are masters of mazes and show short-term and long-term memory, ranging from days to entire life spans. The tiny honeybee brain has only around 1 million neurons, and bees contradict the notion that insect behavior is stereotyped and inflexible.

Based on these and other studies, "the cognitive maximization hypothesis" (Bekoff 2013) has been developed and suggests that perhaps small-brained animals maximize the use of the relatively little they have more efficiently than do big-brained animals, such as chimps and dolphins. Future research will be needed to determine if this is so. What is known is that big brains may be useful for some animals, but small-brained animals do very well as long as they can do what is necessary to survive and thrive in their own worlds. Because members of a given species need to do certain things to function appropriately as members of their species, it is not useful to ask, for example, if dogs are smarter than cats, if chimpanzees are smarter than mice, or if birds are smarter than fish. Each does what she or he needs to do to be a dog, cat, chimpanzee, or mouse. The notions that small-brained animals are "less intelligent" than big-brained animals and "suffer less," need to be revisited as they are incorrect and harmful to animals who are known to endure deep and prolonged suffering.

Another major problem with ranking species in terms of intelligence is that it assumes that intelligence is characteristic of a species overall. This view is misguided because intelligence is an *individual* trait, and there are significant individual differences among members of the same species. Therefore, while cross-species comparisons in intelligence can be fraught with errors, comparing members of the same species can be useful in terms of the ways in which individuals learn social skills or the speed with which they learn different tasks. If, for example, one dog learns social skills or a particular task faster than another dog, it would be correct to say that for *that specific task* she was smarter than the other individual. However, just as there are multiple intelligences in humans, there likely are similar traits in nonhumans as well. Thus, some dogs or cats may learn to discriminate among different stimuli better than other dogs or cats, but it is also likely that there will be differences among them in other learning situations. For example, it is said that some dogs are "street smart," and therefore do much better out on their own than do less adaptable or savvy dogs. Future research that focuses on individual differences in learning will surely generate much needed and interesting data.

The rapidly growing fields of comparative cognitive neuroscience and cognitive ethology (the comparative study of animal minds) continue to provide exciting information about the brains, incredibly active minds, and intelligences of the fascinating animals with whom we share our planet. What an exciting future lies ahead in this field of research.

Marc Bekoff

See also: Agency; Animal Cultures; Animals; Anthropomorphism; Bees; Cats; Chimpanzees; Communication and Language; Dogs; Dolphins; Ethics; Ethology; Sentience; Species

Further Reading

Bekoff, M. n.d. "Animal Emotions." *Psychology Today*. Accessed January 2, 2015. http://www.psychologytoday.com/blog/animal-emotions

Bekoff, M. 2013. "The Birds and the Bees and Their Brains: Size Doesn't Matter." Accessed January 2, 2015. http://www.psychologytoday.com/blog/animal-emotions/201304/the-birds-and-the-bees-and-their-brains-size-doesnt-matter

Bekoff, M. 2013. "Do 'Smarter' Dogs Really Suffer More Than 'Dumber' Mice?" Accessed January 2, 2015. http://www.psychologytoday.com/blog/animal-emotions/201304/do-smarter-dogs-really-suffer-more-dumber-mice

Chittka, L., and Niven, J. 2009. "Are Bigger Brains Better?" *Current Biology.* Accessed January 2, 2015. http://www.cell.com/current-biology/abstract/S0960-9822%2809%2901597-8

Invasive Species

Sometimes the coexistence between humans and animals can get quite complicated. One such case is that of invasive species. We can first think of an invasive species as an animal (or plant, or other species) that has found its way out of its established range and into a new location. If that species ends up having a dominant influence in that new location and, as a result, a negative influence on the local ecosystems humans rely upon, we consider it invasive. Today there are over 4,000 invasive species in the United States alone, and many more worldwide. Invasive species are estimated to cause trillions of dollars in damage by harming agriculture and human infrastructure. In addition, and perhaps more importantly, they are a leading driver of biodiversity loss.

The historical context for invasive species begins 300 million years ago. The planet looked very different then, for the continents that we know today were not separated by oceans but existed together as one giant landmass. This "supercontinent" is known as Pangaea. On Pangaea, all land-based species could potentially travel from what is now Africa to what is now North America very easily.

While often it is the animal itself that is marked as invasive and targeted for control or extermination, it is essential to remember that in most cases these animals would not be in their new locations if humans did not bring them there. People often think it is fun to acquire an exotic pet like a python, but when the snake gets too big for the owner to handle, the animal is frequently released and impacts local ecosystems. (AP Photo/J. Pat Carter)

Around 200 million years ago, however, Pangaea began to break apart, and the single landmass slowly separated into the continents and oceans with which we are familiar. As a result, since the breakup of Pangaea, species could only evolve with others in what we now call their "bio-geographical realm," a broad category that describes the large regions in which ecosystems share a relatively similar history. Except under very rare circumstances, species did not move from one realm to another and therefore, importantly, biodiversity in each realm is substantially different.

This isolation began to unravel around the 1500s, the time most scholars point to as the beginning of European colonialism (when they took political control over other parts of the world). As Europeans started to travel and trade beyond Europe, they brought many species with them. These included small rodents unintentionally brought aboard ships, farm plants and animals intentionally introduced in order to replicate the European way of life in a new land, and also diseases (brought both intentionally and unintentionally). While most were not harmful, some of these nonhuman travelers would significantly impact the people, animals, and plants in the places that were new to Europeans, quickly spreading and becoming dominant species in their new homes. In other words, some of these nonhuman travelers became invasive species.

This process has continued ever since—not just between Europeans and the Americas but between other places, too. Over the past 50 years, a period in which global trade and travel have dramatically increased, the rate of species traveling and becoming invasive has also increased. As this history shows, invasive species events and changes in human social processes are tightly connected.

Today, invasive species ecologists examine these events as a series of stages, so that they may recognize a variety of forces that could lead to an invasion event: from nonhuman forces like floods, hurricanes, or predator-prey relationships to human-generated forces like trade, agricultural change, and urbanization. As such, invasive species ecologists see context and a historical perspective to be of primary importance. We can look at two invasive species, nutrias and Burmese pythons, to understand what a problem they can be.

In the early 1900s, fur farmers and government officials introduced nutria (also known as Coypu), a large, semiaquatic rodent that looks like a cross between a beaver and a rat, into North America so ranchers could raise them for their fur. The market for nutria fur did not last long, but that did not mean the nutria went away. Some nutria were released into the wild, where they quickly started to devour tall grasses, rushes, and other plants vital to wetlands as well as destroying irrigation systems and other human-made items. They caused so much damage that by the 1950s places in which they had the biggest impact, the Chesapeake Bay area and farther south, introduced nutria control programs. Since then, millions of dollars have been invested in trying to control the damage nutrias are still causing.

Burmese pythons, on the other hand, likely had a much different route to invasion. They are a major problem in South Florida, devastating many local species and eating their way through the Everglades National Park. While it is impossible to say exactly how they became invasive, most people point to the fact that Miami, the largest city in South Florida, is a hub for the exotic pet trade. It is thus likely that people bought Burmese pythons for pets and then dumped them in the woods when they became too difficult to care for.

Invasive species problems are expected to increase in the future for two reasons. First, global trade is not only continuing to accelerate, but patterns of trade are changing. More countries in the global south (Africa, Asia, and Latin America) are now primarily trading among each other and not predominantly through countries in the global north. This change will likely create new vectors for invasive species. Second, as Earth's poles warm as a result of climate change, many species will migrate toward these poles and thus new areas will be opened to invasion. In addition, warming poles will allow more ships to travel through formerly impenetrable avenues and likely carry novel species with them. In all, barring major structural changes in human social systems, invasive species will become an increasing problem for socio-ecological systems over the next century.

Jordan Fox Besek

See also: Biodiversity; Exotic Pets; Species; Wildlife

Further Reading

Crosby, A. W. 2004. *Ecological Imperialism: The Biological Expansion of Europe, 900–1900*. New York: Cambridge University Press.

Elton, C. S. 1958. *The Ecology of Invasions by Animals and Plants*. London: Methuen.

Lockwood, J. L., Hoopes, M. F., and Marchetti, M. P. 2013. *Invasion Ecology*, 2nd ed. Chichester, UK: Wiley-Blackwell.

Islam. *See* Slaughter; Western Religions, Animals in

Ivory Trade

Ivory is the compositional material of elephant tusks. Because one-third of the tusk resides within an elephant's face, the elephant is usually slaughtered in order to obtain the tusk, which is then carved into trinkets, statues, or jewelry. As a result of the ivory trade, African elephants are being slaughtered faster than they can reproduce, but international regulation of the ivory trade has been fraught with sociopolitical conflict.

Ivory is revered as a sign of wealth and status in many Asian cultures, and most ivory is destined for China, Thailand, and other Asian nations. However, the United States also has a thriving ivory market, particularly in New York and California. Most ivory derives from African elephants. Although some subpopulations of African elephants are increasing, overall the species is being decimated at a rate that exceeds its growth capacity (Wittemyer et al. 2014). In Central, Western, Southern, and Eastern Africa, more than 50 percent of elephant kills are illegal; the illegal ivory trade has doubled from 2007 to 2011 (and tripled from 1988 to 2011) and is currently at its highest level in 16 years (Underwood et al. 2013).

The international trade in ivory is regulated by the Convention on International Trade in Endangered Species of Wild Fauna and Flora, or CITES. In 1989, CITES banned the international trade of ivory. Certain African nations, such as Kenya and Tanzania, have argued that an ivory ban is essential for security, stability, and conservation-related development such as wildlife tourism. Others, such as Zimbabwe and South Africa, contend that a ban on all commercial trade of elephant parts is too restrictive for countries with healthy elephant populations. Since the initial ivory ban in 1989, political pressure has led to the reduction of protections for African elephant populations in Zimbabwe, South Africa, Botswana, and Namibia. The debate over ivory is also nested within larger ecological issues of elephant management. A recurrent argument for loosening ivory regulations is that culling (killing "excess" elephants) could restore ecological balance, prevent habitat degradation, and reduce human-elephant conflict (such as elephants entering agricultural areas for food). However, scientists that view ecological change as natural recommend the creation of wildlife corridors to connect separate elephant populations rather than lethal removal. Furthermore, given elephants' intelligence and the evidence that elephant communities ravaged by culling or killing for ivory display impaired social functioning and violent behavior, many question the acceptability of lethal elephant management (Lavigne 2013).

CITES has twice (in 1997 and 2007) authorized one-time sales of confiscated ivory from selected African countries to Asia. The rationale for these sales was that selling large stockpiles of ivory would lead supply to be higher than demand, thus reducing prices and, therefore, poaching. However, there are several issues with this approach. The 1997 one-time sale yielded a small increase in elephant mortality in Kenya and Zimbabwe; in certain remote areas in these two countries, a dramatic increase in mortality occurred (Bulte et al. 2007). In China the opposite—a ban on ivory auctions—has been most effective in driving down its price (Gao and Clark 2014). One-time sales also create consumer confusion and obstacles to enforcement of ivory regulations. Because there is no cost-effective way to distinguish the origin of ivory, both regulatory officials and consumers are unable to differentiate between "illegal" and "legal" ivory (Lavigne 2013). Furthermore, rampant corruption allows for the transfer of illegal ivory into legal markets, and there is evidence that the ivory trade funds militant and terrorist groups, undermining Africa's political stability. Although these one-time sales were meant to reduce the demand for ivory through monitored sales, the poaching of African elephants has steadily increased (Wittemyer et al. 2014).

In compliance with CITES, the United States implemented a federal ban on imports of ivory produced after 1989. However, in Los Angeles and San Francisco, the number of ivory products from recently killed elephants (in violation of the federal ban) doubled from 2006 to 2014 (Stiles 2015). The United States has proposed shifting the burden of proof (that ivory is legal or illegal) from the government to the seller and banning commercial imports of ivory. However, because elephants killed for sport and pre-ban ivory are not considered "commercial," even these proposed revisions allow for many loopholes. Several states, such as New York and New Jersey, have passed more stringent ivory regulations.

Ivory bans can be effective. Overall, the population of African elephants rebounded after the 1989 global ivory ban (Lemieux and Clarke 2009). However, numerous countries in Central Africa, such as the Democratic Republic of Congo, Central African Republic, Zambia, and Angola, continued to lose substantial numbers of elephants after the ban, a phenomenon largely explained by the proximity of unregulated domestic ivory markets. A patchwork of domestic ivory regulations is unlikely to reduce poaching because the ivory will continue to be transported to neighboring nations where it can easily be sold. Forecasting in ecological economics (a field that looks at the demand and supply of wildlife products) indicates that a trade ban is likely to reduce poaching if a) it produces a stigma effect (a reduction in demand because of social norms), b) it facilitates interception of smuggled goods, c) there is little ivory stockpiled, and d) it does not negatively affect funding for law enforcement (Heltberg 2001).

Jessica Bell Rizzolo

See also: Elephants; Non-Food Animal Products

Further Reading

Bulte, E. H., Damania, R. and Van Kooten, G. C. 2007. "The Effects of One-Off Ivory Sales on Elephant Mortality." *Journal of Wildlife Management* 71(2): 613–618.

Gao, Y., and Clark, S. G. 2014. "Elephant Ivory Trade in China: Trends and Drivers." *Biological Conservation* 180: 23–30.

Heltberg, R. 2001. "Impact of the Ivory Trade Ban on Poaching Incentives: A Numerical Example." *Ecological Economics* 36: 189–195.

Lavigne, D. 2013. *Elephants and Ivory*. Yarmouth Port, MA: IFAW.

Lemieux, A. M., and Clarke, R. V. 2009. "The International Ban on Ivory Sales and Its Effect on Elephant Poaching in Africa." *British Journal of Criminology* 49(4): 451–471.

Stiles, D. 2015. *Elephant Ivory Trafficking in California, USA*. New York: Natural Resources Defense Council.

Underwood, F. M., Burn, R. W., and Milliken, T. 2013. "Dissecting the Illegal Ivory Trade: An Analysis of Ivory Seizures Data." *PloS One* 8(10): e76539.

Wittemyer, G., Northrup, J. M., Blanc, J., Douglas-Hamilton, I., Omondi, P., and Burnham, K. P. 2014. "Illegal Killing for Ivory Drives Global Decline in African Elephants." *Proceedings of the National Academy of Sciences* 111(36): 13117–13121.

J

Jainism. *See* Eastern Religions, Animals in

Judaism. *See* Slaughter; Western Religions, Animals in

K

Keystone Species

Like the keystone that stabilizes an arch, keystone species hold together an ecosystem (a community of organisms and inanimate elements in the environment that together make up a natural system). Keystone species are those that maintain the diversity and functions of an ecosystem and whose impacts are typically disproportionate to their relative population size. Keystone species may be animals or plants, but do not need to be present in large numbers in order to have significant impact on their ecosystems. Without keystone species, an ecosystem can change dramatically, transition to a new state, or collapse entirely. Keystone species vary widely and include large herbivores whose grazing activity shapes the environment and creates habitats for other species, essential plants that support entire food webs starting with insect populations, top carnivores that regulate prey species abundance, and several other kinds of organisms at all levels of the food chain that perform essential ecosystem services. Keystone species are important to environmental conservation because these species provide insight into both the complexity of ecosystems as well as how to better protect and manage them. Like a falling row of dominoes, the loss of a keystone species sets off a cascading effect of species loss. Identifying and conserving keystone species assists in maintaining an ecosystem's functions and helps protect the other species and services that rely on them.

Zoologist Robert T. Paine documented the first evidence of an ecosystem undergoing extensive change in the absence of a keystone species in 1966. For three years, Paine systematically removed a top predator, the purple sea star (*Pisaster ochraceus*), from the tidal plains of Washington state's Tatoosh Island. Once the sea stars were removed, Paine observed that the ecosystem underwent complete transformation. Mussel populations, previously the primary prey of sea stars, grew exponentially and dominated the area, forcing out many other species like algae and aquatic snails. What was once a biodiverse ecosystem became habitat for a single species. This loss initiated a process that Paine later call *trophic cascades*, or the rise and fall of connected species throughout the food web. Landscapes where a keystone species has been eradicated demonstrate the intricate interdependence of species within ecosystems.

Like the sea star, many keystone species are predators that are critical components of their environments despite their relatively small population sizes. Every human-inhabited continent has witnessed the extinction of top predators in at least parts of the predators' range, or the geographic area where the species has historically been present. Resulting changes in these landscapes provide clear evidence that top predators are vital and irreplaceable in their governing roles in ecosystems. The loss of top predators often leads to declines in biodiversity due to the complete loss of species from part of its historic range (referred to as *extirpation*) or even total species extinction. John W. Terborgh and colleagues found a 75 percent decline in the variety of vertebrate species where hydroelectric dams in Venezuela isolated small islands and effectively created predator-free systems (Terborgh et al. 2001). The large body of water produced from the dam restricted predators like the jaguar, puma, and harpy eagle from hunting on the islands. As a result, herbivore populations increased and subsequently overgrazed the land, leading to decreased growth of many canopy tree species. In turn, this change in habitat significantly limited the presence of vertebrate animal species.

The extirpation of predator species often leads to overabundance of herbivorous ungulates (hoofed animals) such as deer, horses, elk, and moose. In Yellowstone National Park, the extirpation of wolves resulted in elk overpopulation. Without the threat of predation, the elk grazed the landscape down to the topsoil in many places. Declines in shrubs and grasses led to weak and unhealthy trees, and rivers slowed as erosion of the topsoil increased with lack of vegetation, brutally impairing the quality of wetlands dependent on higher water flow. Elk browsed dense thickets and along rivers, places they had avoided when wolves were part of the ecosystem. In January 1995, a wolf reintroduction program in Yellowstone revealed the importance of the wolf as a keystone species and the benefits of reintroducing these species once vilified for being predators. Fourteen wolves were released into the park after an absence of more than 70 years. The elk quickly retreated from the valley floor where they had overgrazed for decades, and new plant growth rapidly filled the valleys and lined the waterways in the absence of grazing pressure. Shrubs, grasses, and trees created habitat that facilitated the return of birds and small mammalian species, including mice and rabbits. These species formed the prey base for mammalian, avian, reptilian, and amphibian predators. With the return of willow and aspen trees, beavers also returned, increasing from one colony prior to reintroduction to 112 colonies in 2011 (National Park Service 2015). The reintroduction of wolves introduced cascading effects in the ecosystem, as beavers are also a keystone species whose dam-building activity produces ponds and reservoirs that store water to recharge the water table and provide shaded cold water for fish like trout that are an important food source for many species.

While apex predators significantly impact entire ecosystems, their need for large territories, abundant prey species, and distance from centers of human activity makes them difficult to conserve in a developing world. Examples from environments where large predators are reintroduced validates the importance of protecting keystone species from extinction.

Kalli F. Doubleday

See also: Endangered Species; Extinction; Species; Wildlife Management

Further Reading

Maehr, D. S., Noss, R. F., and Larkin, J. L., eds. 2001. *Large Mammal Restoration: Ecological and Sociological Challenges in the 21st Century*. Washington, DC: Island Press.

Mao, J. S., Boyce, M. S., Smith, D. W., Singer, F. J., Values, D. J., Vore, J. M., and Merrill, E. H. 2005. "Habitat Selection by Elk Before and After Wolf Reintroduction in Yellowstone National Park." *Journal of Wildlife Management* 69: 1691–1707.

National Park Service. 2015. "Beavers." Accessed April 6, 2015. http://www.nps.gov/yell/learn/nature/beavers.htm

Paine, R. T. 1966. "Food Web Complexity and Species Diversity." *The American Naturalist* 100: 65–75.

Terborgh, J., Lopez, L., Nunez, P., Rao, M., Shahabuddin, G., Orihuela, B., Riveros, M., Ascanio, R., Adler, G. H., Lambert, T. D., and Balas, L. 2001. "Ecological Meltdown in Predator-Free Forest Fragments." *Science* 294(5548): 1923–1926.

YouTube. 2014. "How Wolves Change Rivers." Accessed April 6, 2015. https://www.youtube.com/watch?v=ysa5OBhXz-Q

YouTube. 2007. "Silence of the Bees." Accessed April 7, 2015. https://www.youtube.com/watch?v=dIUo3STj6tw

L

Leather. *See* Non-Food Animal Products

Literature, Animals in

Narrative (story) exists in myth, fairy tales, legends, histories, and personal accounts. The appearance of animals in literature is culturally significant because literature reflects our understanding of ourselves and of society. For example, animals appear in satirical and political literature as vehicles for social commentary and change. Culture is also informed by our literature. Narrative meaning is produced and maintained though language and shared understandings in societies or cultural groups through traditions, conventions, norms, and standards. Stories not only tell us how we are, but they suggest how we *should* be.

Animals in literature have been used to guide human behavior in many ways. Bestiaries (animal encyclopedias popular in Medieval Europe) catalogued and illustrated real and imagined animals according to the human qualities they represented. Animal depictions and descriptions often inaccurately represented the actual animal in order to teach moral or religious lessons.

Aesop (620–564 BCE) used animals to dictate moral lessons in fables. The fable "The Hare and the Tortoise" teaches that keeping slow and steady will win a race, and the story of the shepherd boy who cried "wolf" teaches that those who habitually lie will not be believed when they tell the truth. Aesop's fables, and many other texts, anthropomorphize (ascribe human characteristics to) animals. Anthropomorphism of animals in literature serves to distance readers from subjects that may be difficult to address directly or to make life lessons easier to learn.

The ascription of human characteristics to animals can be positive or negative. Anthropomorphized animals feature prominently in the work of Rudyard Kipling, whose stories invoke stereotypes (oversimplified but widely accepted ideas) such as that of the deceptive snake in *The Jungle Book* (1894) and the independent and opportunistic cat in *The Cat Who Walked by Himself* (1902). Wolves are portrayed as intentionally cruel, snakes as deceptive, cats as evil or magical, and rats as criminal. Cruelty, deception, and the propensity toward evil or criminal behavior are not qualities inherent to the biological or social structure of nonhuman animals. Rather, they are human socially constructed qualities (qualities created by human society).

Human qualities assigned to animals in literature can transfer back to real animals and result in fears or beliefs that strain human-animal coexistence. For example, wolves were widely detested in Europe and were eradicated from many parts of Europe by the Late Middle Ages (ca. 1500 AD), but the wolf remained a part of western culture through stories. Wolves are commonly portrayed unfavorably in fairy tales and fables such as *The Three Little Pigs* (1886) and *Little Red Riding Hood* (1812) as deceiving, lustful, or crazy. These derogatory depictions have continued to increase heightened fear of real wolves, which inhibits conservation of wolves worldwide.

Hybrid animals such as the werewolf continue to be used to represent the animal qualities of a human that are socially undesirable, such as recklessness, gluttony, and aggressiveness. These qualities threaten the stability of orderly society and are characteristics and behaviors our moral lessons teach against. Humans *are* animals, but due to our perceived separation from the animal world, people can have difficulty accepting animal aspects of human nature.

Animals appear in literature metaphorically (comparatively using a figure of speech to say something *is* something else) to represent a human or humans. In these instances, animals may not be given human characteristics but are positioned literally to represent a human or humans. In many cultures animal metaphors are used to explain life and the workings of the universe. For example, biblical references to sheep are interpreted to be references to people who need to be guided to God. The reference to "a wolf in sheep's clothing" is a biblical reference to a false prophet (a person who falsely claims the divine gift of foretelling). The expression is used culturally to describe an untrustworthy person or organization. In the Bible, the lion and the snake also appear frequently. The snake represents evil or temptation. Adam's first wife, Lilith, became a succubus (a female demon often depicted with the body of a snake), and his second wife, Eve, was tempted by a snake in the Garden of Eden. Conversely, the biblical lion represents holiness and royalty. Royalty of the Ancient Near East and Africa kept lions in captivity, as did Greek and Roman rulers. The lion is the symbol of the Royal British empire. The cultural use of the lion to symbolize royalty has historically reinforced the position of royalty as close to God and above others in society.

The lion's link to royalty and godliness remains dominant in contemporary literature, such as the screenplay for Disney's *The Lion King* (1994), which mirrors biblical narratives. In the story, young lion Simba follows his destiny to assume his divine role as king of the land and all creatures. C. S. Lewis's *Chronicles of Narnia* (1949–1954) mirrors the life, crucifixion, and return of Christ and features talking lion Aslan, as a literary Christ figure.

While animals still figure in symbolically in literature, much contemporary literature focuses on exploration of what animals actually are, how they relate to people, and how they are situated in their environments. Books such as Jack London's

White Fang (1906) and Anna Sewell's *Black Beauty* (1877) are written from the perspective of the animal. They attempt to show how animals might interpret their relationships with humans and are noted for bringing attention to undesirable human behavior and human cruelty to animals.

Anita Hagy Ferguson

See also: Anthropomorphism; Art, Animals in; Eastern Religions, Animals in; Indigenous Religions, Animals in; Popular Media, Animals in; Social Construction; Wolves

Further Reading

Bruner, J. 1991. "The Narrative Construction of Reality." *Critical Inquiry* 18: 1–21.

Burke, C., and Coenhaver, J. G. 2004. "Animals as People in Children's Literature." *Language Arts* 81(3): 205–213.

Fireman, G. D., McVay, T. E., and Flannagan, O. J. 2003. *Narrative and Consciousness: Literature, Psychology, and the Brain.* New York: Oxford University Press.

Grimm, J. L., and Grimm, W. C. 2010 (1812). *Grimm's Fairy Tales.* Seattle: Pacific Publishing Studio.

Halliwell-Phillipps, O. 1886. *Nursery Rhymes of England.* London: F. Wayne and Co.

Zipes, J., ed. 1992. *Aesop's Fables.* New York: Signet Classics.

Livestock

"Livestock" is a composite term for terrestrial (land-based) animals domesticated to provide wool, eggs, milk, meat, and a number of other products (although some usages of the term exclude poultry). Today, livestock represents a small subset of all animal species but constitutes a large global population in terms of numbers of individual animals. This livestock population has enabled the production of vast quantities of animal-derived products but is also frequently identified as the source of environmental problems.

Among Earth's estimated 8–14 million animal species, no more than about 40 have been domesticated. Of these, a mere 14 provide 90 percent of the animal-derived human food supply. Only five species (cattle, sheep, pigs, goats, and chickens) are distributed in large numbers globally (International Livestock Research Institute 2007). The main reason so few species have been domesticated could probably be the nature of the animals themselves. To be suitable for domestication, animals should (in addition to being terrestrial) ideally have relatively broad herbivorous (plant-based) diets, a willingness to mate in captivity, a calm disposition, and form herds with well-developed hierarchies. Few species fulfill all these requirements.

Although livestock constitutes a tiny subset in terms of species, the global population of cattle, sheep, pigs, goats, and chickens alive at any given time is 25.5 billion. However, because many animals are younger than a year when slaughtered, around

65 billion are killed per year for food. With increasingly meat-intense diets and more thoroughly industrialized production processes in the livestock sector in wealthier countries, the livestock population has over the last 50 years grown at rates far surpassing that of Earth's *human* population. While the human population grew by 122 percent, from 3.2 billion to 7.1 billion, between 1963 and 2013, the combined population of cattle, sheep, pigs, goats, and chickens grew by 264 percent. Above all, this was spurred by a remarkable increase in the global chicken population, rising almost 400 percent, from 4.2 billion in 1963 to 20.9 billion in 2013 (FAO 2015). Although this increase in livestock numbers enabled an increase in livestock-derived products, the consumption of such products displays large regional differences. Per capita meat consumption, for example, ranges from well under 10 pounds (4.5 kg) per year in Bangladesh and India to over 200 pounds (91 kg) in the United States and Australia (OECD 2015).

Ever since some societies first domesticated livestock around 8000 BCE, the histories of these human societies have also been histories of livestock. Such histories include the increasingly intentional breeding of animals to meet human demands placed on them (more milk, meat, eggs, traction power, or aesthetic considerations), but also the fact that humans and human societies were affected by livestock they kept. For instance, with cattle domesticated, milk became a foodstuff consumed into adulthood. Hence, certain populations, particularly in Northern Europe, evolved to be able to digest lactose (milk sugar) as adults. Humans, in other words, also changed through human-cattle interactions.

Livestock is, moreover, a diverse category. Various species have their own particular biophysical characteristics, political-economic importance, and cultural connotations. Pigs are the source of much of the meat eaten in the United States, China, and Europe, and simultaneously considered inedible by large groups of Muslims and Jews around the world. "Holy" cows have become potent symbols for India, while beef is considered prime meat by many in North and South America. Between places and over time, cultural differences, technological developments, political-economic processes, and so on result in a wide range of different livestock practices and livestock landscapes.

To use a North American example, the 1873 invention of barbed wire reduced the amount of wood needed to build a fence. By lessening the costs of enclosing pasture, barbed wire "hasten[ed] the transition from prairie to pasture by further concentrating grazing in certain areas and by reducing the frequency of fires" (Cronon 1992, 221). Cattle and barbed wire, in short, produced another kind of landscape than cattle without barbed wire had. Before this, cattle had also radically transformed North American landscapes, with European cattle destroying pre-existing ecosystems. The colonization of this continent was, thus, not only a story of Native American land conquered by the French, British, Spanish, and so forth; it was simultaneously the story of environments reshaped by livestock.

While the current global trend in livestock is toward bigger populations of animals, often kept indoors, by national or even transnational companies, an estimated 30–40 million nomadic herders still follow their animals from pasture to pasture. Intensive "modern" practices in industrialized regions are thus complemented by more extensive "traditional" practices elsewhere. With large-scale industrialized operations, sometimes referred to as factory farming, criticized for how animals are treated, increased risk of disease, and waste generated, some producers in countries dominated by industrialized livestock practices have taken to also producing organic beef, milk, eggs, and so on. For example, organic labeling both in the United States and the European Union strives for higher animal welfare standards for living areas and food sources, along with reduced use of pesticides and antibiotics. Meanwhile, the resources used and emissions generated by the global livestock population have become prominent topics within discussions on sustainability. The livestock sector is responsible for 18 percent of greenhouse gas emissions, uses 70 percent of all agricultural land (for grazing and feed-production), and consumes 8 percent of all water used by humankind (Steinfeld et al. 2006). Hence, livestock have come to be regarded as both a source of products that are widely appreciated and also of local and global environmental problems of increasing concern.

Erik Jönsson

See also: Concentrated Animal Feeding Operation (CAFO); Domestication; Factory Farming; Husbandry; Meat Eating; Pastoralism; Pets; Slaughter; Zoonotic Diseases

Further Reading

Cronon, W. 1992. *Nature's Metropolis: Chicago and the Great West.* New York: W.W. Norton.

FAO. 2015. "Production/Live Animals." Accessed March 20, 2015. http://faostat3.fao.org/browse/Q/QA/E

Freidberg, S. 2009. *Fresh: A Perishable History.* Cambridge, MA and London: Belknap Harvard.

International Livestock Research Institute. 2007. "Safeguarding Livestock Diversity: The Time Is Now." Accessed March 20, 2015. https://cgspace.cgiar.org/bitstream/handle/10568/2479/AnnualRep2006_Safeguard.pdf;jsessionid=DF3A8C51266FF49A0D18474A0DFAFEB2?sequence=1

OECD. 2015. "Meat Consumption (Indicator)." Accessed July 3, 2015. https://data.oecd.org/agroutput/meat-consumption.htm

Steinfeld, H., Gerber, P., Wassenaar, T., Castel, V., Rosales, M., and de Haan, C. 2006. *Livestock's Long Shadow: Environmental Issues and Options*. Accessed August 27, 2015. http://www.fao.org/docrep/010/a0701e/a0701e00.HTM

Wilkie, R. M. 2010. *Livestock/Deadstock: Working with Farm Animals from Birth to Slaughter.* Philadelphia: Temple University Press.

M

Mad Cow Disease

Mad cow disease is the popular name for bovine spongiform encephalopathy (BSE). BSE is a type of transmissible spongiform encephalopathy (TSE), which are progressive and fatal neurological (nerve and brain) diseases. Although BSE is still not well understood, it is caused by a form of protein called a prion and transmitted through consuming body parts of an infected individual that are contaminated with the prions. It was first diagnosed in 1984 when several dairy cows in Britain began to appear strangely ill, and it is linked to Creutzfeldt-Jakob disease (CJD) in humans, making it a zoonotic disease (i.e., one that can be transmitted between different species).

TSEs are known in species other than cows, for instance, in sheep as the disease called "scrapie," also caused by prions. CJD is a human prion disease. Prions are a normal form of protein, and it is not yet known why some transform into the harmful version that attacks the central nervous system. Prions are living matter but are not living organisms in the same sense as other disease agents (called pathogens) such as bacteria. Prions cannot be killed by elements, such as very high temperatures, that can be used to destroy pathogens. Because TSEs affect the central nervous system, symptoms include changes in behavior, lack of muscle coordination, and difficulty moving.

The first cases of BSE were diagnosed after necropsies (dissection and examination of dead bodies) of the infected British cows were performed. In early 1988, the likely transmission agent for the disease was identified—feeding cows meat and bone meal (MBM) produced from slaughtered cows. This practice was banned that same year, but by that time thousands of new cases were occurring each week. The use of MBM had previously been considered safe for use in livestock feed because subjecting the material to high temperatures was known to kill typical pathogens. Also, with a typical pathogen, it may be possible to consume a smaller amount of contaminated matter and not become ill. With BSE, however, even a rather minute amount of matter can transmit the disease. Further, there can be a relatively long period of time (known as the incubation period), potentially several years, between when a TSE is contracted and when symptoms emerge. This long incubation period can make it practically impossible to know if an individual carries the disease when they are subclinical (not exhibiting symptoms). *The BSE Inquiry: The Report* (commissioned in 1998 by the British Parliament to investigate BSE-related events and government responses through March 1996) states that the U.K. government misunderstood these unique BSE characteristics. Because of this misunderstanding, thousands more animals became infected with BSE even after the 1988 ban.

A case can be made that U.S. farm policy in the early 1980s contributed to the occurrence of BSE in Britain. Technological advances in agriculture had led to enormous productivity increases in grain production and surpluses in industrialized nations. In the United States, then-President Ronald Reagan's administration implemented policies to encourage farmers to remove some crops from production. As a result, soy (a high-quality protein ingredient in livestock feed) prices increased and Britain's domestically produced MBM became a relatively cheaper and more attractive livestock protein source (Pielke 2008).

In March 1996, a new variant of CJD, called vCJD, was reported in the United Kingdom. According to the World Health Organization (WHO), vCJD "is strongly linked to exposure, probably through food, to BSE . . . [and] the hypothesis discussed during two 1996 WHO consultations [is] that the cluster of vCJD cases is due to the same agent that caused BSE in cattle" (WHO 2002). Just as there was a lack of understanding about disease transmission between cattle, so was there regarding transmission to humans. A large factor is that the disease's long incubation period obscured the connection between vCJD and BSE. Compounding this problem was the poor knowledge of the risks of consuming subclinical animals. To make matters worse, in the early years of BSE investigation there was evidently both a lack of clarity in, and a government misunderstanding of, a key BSE inquiry report about the importance of keeping cattle showing any signs of BSE out of the food supply in order to protect human health (Matravers et al. 2000). In 1989, the United Kingdom banned the use of cattle offal (organs and tissue) in human food, although beef continued to be declared safe to eat. It was discovered a number of years later that infected meat was still potentially going into the human food supply.

Between 1986 and 2002, there were more than 180,000 confirmed cases of BSE in Britain. In 1989, the first confirmed case outside of the United Kingdom occurred in Ireland. Since then, cases have been identified in other European countries, as well as Canada, Israel, Japan, and the United States (although numbering only in the thousands in total). In 1996, the European Union (EU) banned importation of British beef, and that same year, Britain killed cows most likely to have been affected. In the early 2000s, in the EU and elsewhere, testing for BSE began. In the United Kingdom, the incidence of BSE reached its peak in the mid-1990s and has steadily declined to almost zero in recent years. Outside the United Kingdom, annual case confirmations are now almost zero as well. Beginning in the 1990s, the EU and countries such as Australia, Canada, Japan, and the United States banned the inclusion of high-risk animal protein such as MBM in cattle feed.

Connie L. Johnston

See also: Factory Farming; Husbandry; Livestock; Meat Eating; Meat Packing; Slaughter; World Organisation for Animal Health (OIE); Zoonotic Diseases

Further Reading

Magdoff, F., Foster, J. B., and Buttel, F. H., eds. 2000. *Hungry for Profit: The Agribusiness Threat to Farmers, Food, and the Environment.* New York: Monthly Review Press.

Matravers, Lord Phillips of Worth, Bridgeman, J., and Ferguson-Smith, M. 2000. *The BSE Inquiry: The Report.* London: British House of Commons. Accessed October 16, 2015. http://www.bseinquiry.gov.uk/report/index.htm

Rhodes, R. 1997. *Deadly Feasts: Tracking the Secrets of a Terrifying New Plague.* New York: Simon & Schuster.

U.S. Food and Drug Administration. 2015. "All About BSE (Mad Cow Disease)." Accessed October 16, 2015. http://www.fda.gov/AnimalVeterinary/ResourcesforYou/AnimalHealthLiteracy/ucm136222.htm

World Health Organization (WHO). 2015. "Bovine Spongiform Encephalopathy (BSE)." Accessed October 15, 2015. http://www.who.int/zoonoses/diseases/bse/en

World Organisation for Animal Health . 2015. "Bovine Spongiform Encephalopathy (BSE)." Accessed October 16, 2015. http://www.oie.int/animal-health-in-the-world/official-disease-status/bse/list-of-bse-risk-status

Marine Mammal Parks

In recent years, marine mammal parks have been thrust into the international spotlight by high-profile events, mainstream documentaries, and endless media coverage. It is a controversial topic with passionate advocates on both sides. Many believe that these parks are cruel and outdated and need to be closed. However, there are generations of people with fond memories of feeding dolphins or getting splashed by Shamu, the killer whale.

Marine mammal parks are commercial theme parks that house marine mammals in tanks for the public to view, both in and out of shows. Killer whales (orcas), dolphins, beluga whales, and sea lions are among the most popular marine mammals kept in these parks. Modern marine mammal parks are a relatively new addition to the global amusement park industry. However, their origins can be traced back to the 1800s when beluga whales and dolphins were first captured and shipped to aquariums in the United States and Europe.

In the century that followed, aquariums used a system of trial and error to keep the animals alive. The 1950s saw the introduction of dolphin shows, but it was the introduction of the killer whale in the 1960s that really increased the popularity of these parks. The shows that these animals performed in were meant as entertainment, and this hybrid between aquariums and amusement parks is what we now know as marine mammal parks. Marine mammal parks focus more on entertainment than aquariums do. They have live animal shows featuring individual animal celebrities

People watch through glass as a killer whale swims by in a display tank at SeaWorld in San Diego on November 30, 2006. SeaWorld filed a lawsuit on December 29, 2015, challenging a California commission's ruling that bans the company from breeding captive killer whales at its San Diego park. The 2013 documentary *Blackfish* was instrumental in calling public attention to SeaWorld's practices. (AP Photo/Chris Park)

alongside traditional amusement park rides. Today, these parks now exist on every continent in the world except Antarctica and attract millions of visitors each year who are looking for a once-in-a-lifetime opportunity to see these impressive ocean animals up close.

Most parks also have education programs that teach guests about these animals' natural history and associated conservation concerns. Many of the parks also have programs that assist financially with conservation efforts. For example, SeaWorld, the world's largest marine mammal park corporation, has a conservation fund that donates thousands of dollars to marine mammal conservation and rescue.

In addition to their education and conservation efforts, marine mammal park advocates point to the high standard of care the animals receive while in captivity. In addition to being housed in state-of-the-art facilities, the animals are provided with "enrichment"—an industry term for activities that stimulate their brains. Enrichment can be as simple as hiding treats in different toys and allowing the animals to find them as they would in the wild. The industry claims that having high-quality food provided to them alleviates the stress and uncertainty of hunting for food as they would do in the wild. In most cases, the animals are not forced to perform. Only positive reinforcement is used. This means that when animals perform a task correctly, they are provided with a treat. If they perform a task incorrectly, they simply do not receive the treat. There is no further punishment. The parks claim that because the animals are provided with such excellent care, incidents with trainers are very rare.

The question remains if the public's interaction with these animals in captive settings is beneficial to overall education and conservation efforts. Opponents of marine mammal captivity cite conflicting studies about whether people actually retain information the parks provide. There are also concerns about the accuracy of the information given out. For example, SeaWorld has been documented informing guests that killer whales have a natural lifespan of about 25 years (Cowperthwaite 2013). However, the scientific community is in agreement that the average lifespan of a killer whale is actually between 40 to 80 years (Reeves 2002, 436–437). In captivity, most killer whales die in their teens, a stark contrast between industry claims and scientific studies.

The issue of the physical and mental health of these mammals is where the sides are most divisively split. In the wild, killer whales swim hundreds of miles every day in search of food. No matter how impressive the park, there is no way to replicate that behavior in captivity. Whales in captivity have experienced numerous health problems that have never been documented in the wild, such as chronic tooth decay, dorsal fin collapse (from lack of use), high infant mortality, and fatal infections.

The impressive intelligence of marine mammals—particularly killer whales and dolphins—leads many people to classify captivity as cruel. These animals have large brains that allow them to problem-solve, strategize, and create extensive social structures. For example, dolphins greet one another by exchanging their own distinctive whistles and remember the whistles of other dolphins for decades—like humans would a name. Wild killer whales have been documented teaching their young hunting methods that are passed down from generation to generation.

Opponents also see stress-induced aggression as a negative result of captivity. There have been three trainer deaths (by killer whales), one documented killer whale suicide, and one case of killer whale aggression that saw one whale kill another in front of a live audience. There is no documentation of a wild killer whale ever killing a human.

The public appears to be taking a stand on marine mammal captivity. The documentary *Blackfish*, which explored the treatment of SeaWorld's killer whales, was released in 2013 and the effect was tremendous. The issue has since become a topic of international discussion, and SeaWorld's corporate profits dropped 84 percent in the years that followed (Rhodan 2015).

Alicia McNorth

See also: Aquariums; Circuses; Cruelty; Dolphins; Intelligence; Stereotypic Behavior; Wildlife; Zoos

Further Reading

Cowperthwaite, G. 2013. *Blackfish*. Directed by Gabriela Cowperthwaite. New York: Magnolia Pictures.

Kirby, D. 2013. *Death at SeaWorld: Shamu and the Dark Side of Killer Whales in Captivity*. New York: St. Martin's Press.

Reeves, R. R., Stewart, B. S., Clapham, P. J., and Powell, J. A. 2002. *National Audubon Society Guide to Marine Mammals of the World*. New York: Knopf.

Rhodan, M. 2015. "Seaworld's Profits Drop 84% after *Blackfish* Documentary." August 6. Accessed January 10, 2016. http://time.com/3987998/seaworlds-profits-drop-84-after-blackfish-documentary

Rose, N. A. 2011. "Killer Controversy: Why Orcas Should No Longer Be Kept in Captivity." Accessed January 10, 2016. www.hsi.org/assets/pdfs/orca_white_paper.pdf

Ullrich, J. 2014. "On Dolphin Personhood: An Interview with Karsten Brensing." *Relations* 2.1, 123–131.

Meat Eating

Meat eating has been part of the human diet since prehistoric times and involves both economics and ethics. The degree to which humans have relied on meat-based proteins has fluctuated over time and place. Since the 20th century, the rate of meat consumption has been increasing and industrialized production facilitates rising global demand. Problems associated with the modern meat industry—concerns about human health, food safety, animal welfare, employee working conditions, and environmental damage—require additional research, careful consideration, and the attention of the consumer public.

Paleontologists (who study fossil animals and plants) and anthropologists suggest that hominins (a group that includes modern humans [*Homo sapiens*], extinct human species, and human ancestors) have eaten meat—the muscles, flesh, and organs of other animals—since the early Paleolithic Age (2.5 million to 10,000 years ago). Some scholars attribute the significance of meat eating to human evolution, associating the development of *Homo sapiens*' large brains with the fatty, high-energy, protein-rich food acquired through wild meats. Others describe how collective hunting established early social organization, such as bringing food to a home base and sharing it. Describing the adaptable digestive systems among early hominins, scholars point out that omnivorous diets (those including plants and meats) allowed for a more active and mobile existence. However, some researchers conclude that human predecessors were nearly all vegetarian. Throughout evolutionary history, some human diets may have consisted mostly of gathered nuts, berries, and fruits, while others may have consumed large amounts of meat; some may even have fed by scavenging meals left behind by larger predators. Overall, the scientific research on humans' ancestral diets suggests a range of possibilities and perspectives.

Meat consumption has varied according to physical geography, climate, technology, society, and culture. Environmental scientist Vaclav Smil (2013) suggests that meat's share of the overall food energy supply probably peaked at a time when hunters had plenty of "mega-herbivores" (large plant eaters like woolly mammoths) to hunt. As mega-herbivores became less plentiful, meat consumption fell, and societies shifted toward agriculture. Eventually, food provision by planting and harvesting crops led to permanent settlements, rising populations, and almost meatless diets. Smil concludes that famine and malnourishment increased in agricultural societies due to crop failure, population density, and diminishing numbers of wild game as a result of deforestation.

Agricultural societies rose during the Neolithic Age (or the "new" Stone Age, roughly 9000–3000 BCE). During this time, animals were domesticated for labor. Slowly, domesticated mammals (such as goats, sheep, pigs, and cattle) became the main supply of protein-rich food for the growing human population. Animal domestication also led to changes in the "appearance, functioning and productivity of organisms" (Smil 2013, 53). Thus, humans began to manipulate the characteristics of animals used for meat.

Since the growth of industrial, urban societies in the Western world (from the mid-18th to mid-20th centuries), meat consumption has continued to rise. Technological developments in agrochemicals, synthetic fertilizers, feed crops, food animal drugs, refrigeration, and genetic engineering fundamentally changed the way that meat is produced and distributed worldwide. Corporate consolidation and intensification of meat production led to large agribusiness, or "factory farming," which has replaced a majority of small, family-owned farms. While demand for meat in Europe and the United States has leveled off during the early 21st century, global demand continues to grow due to new markets in developing countries like China and Brazil. Rising disposable income, an international meat trade, and public access to inexpensive, processed meats have fostered this demand. The proliferation of fast-food restaurants like McDonald's and Kentucky Fried Chicken has further globalized the taste for meat. Nonetheless, cultural and religious prohibitions against eating certain meats endure. Some of these include avoiding pork in Judaism and Islam, beef in Hinduism, and horse in most of North America.

The industries and processes that supply the global demand for meat are challenged on multiple fronts. Nutrition experts have increasingly questioned the health aspects of meat eating. While it was once thought that only meat protein could provide certain amino acids, nutritionists now agree that such proteins can be obtained through legumes, grains, and vegetables. However, other nutrients, such as vitamin B-12, are not supplied through plant-based diets and must be obtained through fortified foods or supplements. Modern meat production may pose substantial risks to human health, food safety, animal welfare, and the environment. Human health risks associated with meat consumption include cardiovascular disease and cancer. There are also food safety concerns. Risk of foodborne bacterial pathogens (e.g., *E. coli*) may be elevated within the conditions of intensified food production. Overcrowding of animals and the presence of manure on the meat are some of the factors that lead to bacterial food contamination.

Finally, meat is an "environmentally expensive" food. Beef and pork are especially energy and land intensive. Environmental damage due to intensive livestock production includes the release of nitrogen compounds into the atmosphere (contributing to climate change), as well as significant water usage and contamination. Fertilizers, animal waste (e.g., manure), and antibiotic residue are the sources of many of these contaminants.

Questions about the sustainability of global meat production have ushered in new initiatives that may foreshadow the future of meat eating. In 2013, the first "lab-grown hamburger," created by growing strains of proteins in Petri dishes, was served. Others have begun to investigate the possibilities of insect-based foods (e.g., cricket flour) to meet global demands for high-quality protein.

Troy A. Martin

See also: Bovine Growth Hormone; Climate Change; Concentrated Animal Feeding Operation (CAFO); Domestication; Factory Farming; Humane Farming; Husbandry; Livestock; Mad Cow Disease; Meat Packing; Slaughter

Further Reading

Choi, C. Q. 2012. "Eating Meat Made Us Human, Suggests New Skull Fossil." Accessed June 30, 2015. http://www.livescience.com/23671-eating-meat-made-us-human.html

Dunn, R. 2012. "Human Ancestors Were Nearly All Vegetarians." Accessed June 30, 2015. http://blogs.scientificamerican.com/guest-blog/human-ancestors-were-nearly-allvege tarians/

Estabrook, B. 2010. "Where the Salmonella Really Came From." *The Atlantic.* Accessed June 30, 2015. http://www.theatlantic.com/health/archive/2010/09/where-the-salmonella -really-came-from/62585/

Isaac, G. 1978. "The Food-Sharing Behavior of Protohuman Hominids." In Isaac, G., and Leakey, R. E. F., eds., *Readings from Scientific American: Human Ancestors*, 110–123. San Francisco: W.H. Freeman and Company.

Physicians Committee for Responsible Medicine. "The Protein Myth." Accessed June 30, 2015. http://www.pcrm.org/sites/default/files/pdfs/health/faq_protein.pdf

Smil, V. 2013. *Should We Eat Meat? Evolution and Consequences of Modern Carnivory.* West Sussex, UK: John Wiley & Sons.

Steinfeld, H., Gerber, P., Wassenaar, T., Castel, V., Rosales, M., and de Haan, C. 2006. *Livestock's Long Shadow: Environmental Issues and Options*. Rome: Food and Agriculture Organization of the United Nations.

Stiftung, H. B., and Friends of the Earth Europe. 2014. "Meat Atlas: Facts and Figures about the Animals We Eat." Accessed June 30, 2015. http://www.boell.de/en/meat-atlas

Meat Packing

Meat packing describes the processing of the animal carcass for sale, including the handling of the animal (livestock) becoming meat (deadstock) in the abattoir (slaughterhouse), as well as to the practices of packaging, including the marketing of processed meat or cuts of meat. A meat packer's skill is in the commercial management of carcasses carrying different qualities of meat. Some parts of the carcass are more favored to eat than others, yet everything from the carcass needs selling, or at worst, disposal at minimal cost. Consequently, there are different challenges to processing chicken, cattle, salmon, or pig carcasses. The meat industry calls this "carcass utilization or balancing," a process where the skills of meat packing come to the fore, as an animal is becoming meat. Thus meat packers have direct involvement in how and why sentient (capable of experiencing positive and negative mental/emotional states) farm animals are selectively bred, live, and die in the way they do. When killed, their bodies are cut up and processed into edible products within a competitive marketplace of varied types of meat valued differently across meat-eating cultures across the globe.

Stressed animals at the point of death impair the quality and value of meat. In the abattoir, upside-down hanging carcasses move via conveyor belt to reach the hands of a butcher who, with machinery or knife, removes inedible outer layers. The work of the butcher carries its own dangers due to the industrial speed of carcass processing. Chickens are de-feathered via machinery, pig carcasses are tumbled through barrels of boiling water where brushes loosen and remove the bristles, and the skins of sheep and cattle are pulled off their bodies by the sheer strength of a butcher. Then the butcher prepares the edible animal for eating, ensuring the skin's appearance (for example, salmon sold whole for special occasion catering) meets required high aesthetic–quality standards. Once the outer skin is removed, carcasses that are now less recognizable as individual animals that lived in water or on land are well into the process of losing their animal identity. With inedible features removed, meatlike shapes and textures become obvious. Now carcass grading for meat quality and quantity takes place, along with food safety monitoring assessments.

Carcass grading determines the price farmers get for animals. For example, fat content and size of meat cuts are assessed from beef carcasses as caterers and restauranteurs who need beef steaks to fit plates prefer smaller carcasses, and gray, overmature salmon flesh is classed as low grade. Carcass conformation refers to fat and meat distribution across the bone structure. Dairy cows are cheaper, lean-muscled meat carcasses, producing meat suitable for low-fat diets. They are bred to maximize feed conversion to increased milk production, not to grow big muscles or put on body fat when fed a high-energy diet. Consequently, the cow that spends her life producing milk has an end-of-life carcass that carries lean-muscled meat. This meat is unsuitable for selling as quality beef-steak with marbling—seams of tasty fat, visible in the meat before cooking. Dairy cow meat is commonly used in burger manufacture where the consumer is not expecting an identifiable cut of meat. The industry produces different breeds with different conformations to meet the demands of meat packers' varied and assorted market of meat consumers. An extreme example is the Belgian Blue cow, selectively bred to have double muscles on the most highly valued parts of the carcass, who cannot give birth naturally as the broad-shouldered calf won't fit through the cow's narrow pelvis.

The work of the meat packer involves the complex task of finding a home in the global marketplace of meat consumption for all the parts of the carcass. Using the example of a chicken from the southwest of England, the white breast meat of this chicken, much in demand by Western consumers, and possibly the thighs, are sold in southwest U.K. supermarkets as locally grown chicken. What happens to the rest? Consumers in China and South Africa buy the feet and the beaks of these chickens to boil up to make stocks. During the summer outdoor grilling season, it is easier to find a home for the chicken drumsticks in the United Kingdom, but it is harder the rest of the year. The discarded carcass is shipped to Russia, where the remaining meat, known as the fifth quarter, is rendered from the bone to enter the human meat supply chain in a highly processed form, or alternatively to become animal feed.

Finally, the meat packer, through marketing, encourages the shopper to buy packs of meat that are typically harder to sell. For example, a meat processor innovates a stew-pack of cheaper cuts from the sheep carcass. This leads us to consider pricing. The overall price that the farmer gets from the meat packer is dependent on the variety and quality of cuts for which the meat packer can sell the carcass. Some parts of the carcass may be sold at discounted cost because it is actually more expensive to dispose of these parts of the carcass as waste rather than as food.

Emma Roe

See also: Factory Farming; Livestock; Meat Eating; Slaughter

Further Reading

Duncan, I. J. H. 2006. "The Changing Concept of Animal Sentience." *Applied Animal Behaviour Science* 100(1–2): 11–19.

Fernley-Whittingstall, H. 2004. *The River Cottage Meat Book.* London: Hodder and Stoughton.

Roe, E. 2010. "Ethics and the Non-Human. Sentient Matterings in the Meat Industry." In Anderson, B., and Harrison, P., eds., *Taking-Place: Non-Representational Theories and Geography,* 261–281. London: Ashgate.

Microbes

Microbes, short for and often referred to as microorganisms, are those organisms that are too small to see with the naked eye. The term "microbes" is derived from Greek, meaning small life. While the classification of microbes is complicated by the fact that many organisms are microscopic only during particular stages of their lives, it is commonly agreed that microbes include some of the oldest and most geographically widespread life forms on the planet. Fossils of microbes have been found dating back over 3.5 billion years. Extremely adaptable to myriad environmental conditions, microbial life thrives on every continent, in the deepest ocean trenches, under hundreds of meters of ice, in the hottest deserts, and even in our atmosphere. Microbes represent a diverse group of organisms including bacteria, archaea (a domain of single-celled microorganisms), most protozoa (a kingdom of single-celled, animal-like organisms), some fungi and algae, as well as certain animals. While many microbes are single-celled organisms, others, despite their small size, are rather complex multicellular life forms. Additionally, some scientists include viruses within the category of microbes, though others do not, as there is significant debate as to whether or not viruses are living organisms. Microbes are prevalent throughout the world, given that they are vital to the survival of humans and all living beings. Microbes play important roles in supporting human digestive and immune systems and contribute to the health of ecosystems, owing to their functional role in decomposition and nutrient recycling.

First discovered in 1676 by Dutch scientist Anton van Leeuwenhoek (1632–1723), the demonstrated existence of microbes was made possible by early microscopes. Jain (practitioners of an ancient Indian religion), Roman, Islamic, and Italian scholars, however, had speculated even earlier about microscopic organisms. Leeuwenhoek's demonstration of the existence of microbes eventually contributed to the development of the germ theory of disease, which attributes particular diseases to microbial causes. While the term microbes often carries a negative connotation, microbes are not necessarily harmful. Those microbes that do cause harm to their hosts (humans, animals, plants, etc., in and on which microbes live) are collectively termed pathogens. Other microbes, however, work as symbionts (organisms that work to the benefit of their hosts and in turn receive benefits from their hosts) to contribute to the survival of their host through assistance in digestion, production of amino acids, stimulation of the immune system, and by preventing pathogenic (harmful) microbes from taking hold. Microbes also assist in the fermentation of alcohol, vinegars, and dairy products, the production of vitamins and antibiotics, the composting of waste products, the decontamination of soil and water, and the recycling of nutrients back into the ecosystem.

In humans, microbes contribute to our survival by assisting in the development of our immune systems, by facilitating the digestion of our food, and by means of other mechanisms that are only now beginning to be uncovered. Research is under way mapping the human microbiome: the microbial diversity found within and on the human body. Such studies have found that over 90 percent of the cells that make up what we think of as the human body belong to microorganisms, a quantity that represents up to 2 kilograms (4.4 lb) of the average human's weight (NIH 2012). Current scientific thinking is that most microbes do not operate independently and instead are part of complex microbial ecosystems that function as a collective. According to this theory, imbalances in microbial ecosystems produce disease in humans and stress on the environment. Thus, recognition is growing around the importance of maintaining balanced microbial communities in and on human bodies to preserve health and within ecosystems for the purpose of preserving biodiversity.

Skye Naslund

See also: Biodiversity; Biotechnology; Mosquitoes; Nuisance Species; Research and Experimentation; Speciesism

Further Reading

Bingham, N. 2006. "Bees, Butterflies, and Bacteria: Biotechnology and the Politics of Nonhuman Friendship." *Environment and Planning A* 38: 483–498.

Crawford, D. 2007. *Deadly Companions: How Microbes Shaped Our History*. Oxford: Oxford University Press.

Dunn, R. 2011. *The Wild Life of Our Bodies: Predators, Parasites, and Partners That Shape Who We Are Today*. New York: HarperCollins Publishers.

Jabr, F. 2011. "They Like Your Guts." *Scientific American* 304: 22.

National Institutes of Health (NIH). 2012. "NIH Human Microbiome Project Defines Normal Bacterial Makeup of the Body." Accessed June 29, 2015. http://www.nih.gov/news/health/jun2012/nhgri-13.htm

Migration. *See* Biogeography; Tracking

Military Use of Animals

Throughout war's history, nonhuman animals have been primarily involved in five ways—as transportation, weapons, experimental subjects, casualties, and food. Today they are most often involved in three ways, as casualties in battle, experimental subjects, and food. Indeed, the industrialization of warfare would not be possible without the labor and slaughter of a multitude of nonhuman animal lives.

There has been vast use of nonhuman animals in wars. In Asia and Africa, elephants were used for transporting equipment and to attack enemy soldiers from 300 BCE to the mid-1900s. Also in Asia and Africa from the second century to the mid-1900s, monkeys were used to steal enemies' weapons, notes, pictures, and keys. Throughout ancient Egypt to the mid-1900s, poisonous snakes were used to protect important documents and to torture political prisoners through fear of being attacked by the snake.

Training birds to spy in modern warfare can be traced back to Germany's use of pigeons for aerial photography in WWI (1914–1918). The U.S. government deployed trained birds equipped with electronic transmitters to pick up conversations during the Cold War from approximately the 1950s through the 1980s. Also during the Cold War, the U.S. government trained ravens to open drawers and carry items of

A horse-drawn artillery wagon. Horses, mules, and other species such as elephants have been used extensively for transportation of military equipment and personnel, as well as in combat situations. (National Archives)

interest such as letters, keys, and pictures; marine animals for underwater defense work detecting bombs, placing bombs, and surveilling (keeping under surveillance) enemy ships through cameras placed on their bodies; and cats with devices surgically implanted in their inner ears to detect the location of and record conversations. Dogs have a long history of going to battle alongside humans in combat as soldiers, messengers, scouts, and skilled trackers. In World War II in the 1940s, dogs were trained by the Soviet military to carry mines under German tanks, where they would be detonated. Finally, a unique story of nonhuman animals being used in war is an orphaned bear donated to the Polish army during WWII. He was named Private Wojtek and traveled as a soldier carrying ammunition.

There have even been many ideas for uses of animals that have not been employed, such as, in the 1940s, the U.S. military's "bat-bombs," a canister filled with thousands of bats equipped with timed incendiary (fire-igniting) devices. The concept was that each bomb dropped would release a parachute before impact, slowing the bombshell to a safe speed for the bats to be released to land and nest in attics and eaves where each device would ignite fires in paper and the wooden structures. The British counterintelligence agency, the MI5, had plans in the 1970s to put a Canadian scientific discovery of gerbils' ability to detect the stress hormone adrenaline in sweat to use in potentially identifying spies or terrorists boarding planes or crossing borders.

Animals in war have frequently been represented in popular media. For example, the 2011 movie *War Horse* included a horse as the main character, focusing on his life, military bravery, and relationship with his owner (who later became a soldier). Another movie, based on a true story, is *Project X* (1987), and it is argued to have influenced animal advocacy more than any other movie related to nonhuman animals in war and the military. *Project X* is about the U.S. Air Force's testing of speed and oxygen tolerance on chimpanzees and how one human begins to care for the captive chimps and communicate with them using sign language, resulting in helping the chimpanzees gain freedom.

Today, most military research on nonhuman animals is done for four diverse purposes: 1) *medical testing* to discover cures and treatments for exposure to deadly and/or painful biological and chemical agents, burns, and gunshot wounds; 2) *psychological and behavioral studies*, which include restraining nonhuman animals on tables, chairs, military vehicles, and other devices and/or locking them in rooms and cages to measure and/or test for respiratory distress, hypothermia, speed tolerance, frostbite immunity, reactions to extreme sound and light exposure, food and water deprivation, and oxygen deficiency; 3) *weapons testing* by shooting at animals to determine the effect of shots from handguns and explosions from landmines; and 4) *resource possibilities for the military* by training nonhuman animals and insects to attack, defend, transport, or surveil with the assistance of technological devices such as micro-cameras, bombs, microphones, and guns attached to them. According to the U.S. Department of Defense, in 2007, 488,237 nonhuman animals were used not only for research but also for education and training of military personnel (New England Anti-Vivisection Society n.d.). As one example, until legal reform in 2013, the U.S. Army used nonhuman animals such as pigs as targets for the purpose of training soldiers on how to provide medical aid in a combat situation.

Currently, nonhuman animals play an array of roles in the military as their unique and superior capabilities are exploited. For example, researchers in the United States and in Europe have trained bees, rats, and elephants to detect minute traces of particular odors from explosives. Beginning in 2006, the U.S. Defense Advanced Research Projects Agency has been working with scientists to create "insect-cyborgs" and "micro air-vehicles" by modifying roaches, butterflies, caterpillars, and grasshoppers so that they can, as common insects, perform stealth surveillance for use as counterintelligence and national security against foreign and domestic terrorism. And in 2010, U.S. researchers were teaching crows to identify specific individuals through facial recognition.

Anthony J. Nocella II and John Lupinacci

See also: Body Modification; Cats; Chimpanzees; Cockroaches; Dogs; Dolphins; Elephants; Horses; Research and Experimentation

Further Reading

History Channel. n.d. *History of War Animals*. Accessed December 2, 2015. http://www.history.co.uk/studytopics/history-of-war-animals

Kaplan, J. 1987. *Project X*. Beverly Hills, CA: Starz/Anchor Bay.

Le Chêne, E. 2009. *Silent Heroes: The Bravery and Devotion of Animals in War*. London: Souvenir Press.

New England Anti-Vivisection Society. n.d. "Military Research." Accessed January 2, 2016. http://www.neavs.org/research/military

Nocella II, A. J., Salter, C., and Bentley, J. K. C. 2014. *Animals and War: Confronting the Military-Animal Industrial Complex*. New York: Lexington Books.

Shaw, Matthew. n.d. "Animals and War." Accessed December 10, 2015. http://www.bl.uk/world-war-one/articles/animals-and-war

Spielberg, S. 2011. *War Horse*. Universal City, CA: Dreamworks SKG.

Wadiwel, D. 2015. *The War Against Animals*. Leiden, Netherlands: Brill.

Milk. *See* Bovine Growth Hormone; Factory Farming; Livestock; Veganism; Vegetarianism

Moral Agency. *See* Personhood; Rights

Mosquitoes

The mosquito family, Culicidae, includes over 3,500 known species. Members of the family span tropical and temperate regions of the world; a few species live north of the Arctic Circle. It is estimated that as many as 1,000 species remain to be discovered and described. Mosquitoes play an important role in the ecology of transmissible disease.

Females lay their eggs in different types of water bodies. Some species prefer lake edges, others prefer flowing streams, others ephemeral puddles or human-made containers such as car tires, bird feeders, and bottle caps. Mosquito larvae (young) feed on aquatic microorganisms in the water bodies where they mature, playing an important role in lake, river, and pond food webs as a major food source for some types of small fish, which are in turn consumed by larger fish. These larger fish are important food for wading birds, reptiles, and humans.

In most mosquito species, males feed on plant juices while females use their piercing mouthparts, or proboscis, to draw blood meals from other animals. Blood meals support development of mosquito eggs. Mosquito species vary in their host species preferences; some prefer humans, others prefer other mammals or birds, and some are able to feed on multiple warm-blooded host species. Some species feed on reptiles, amphibians, fish, or the young of other mosquitoes.

Adult female mosquitoes locate animal hosts by means of a sensory complex that includes visual and olfactory (scent) cues along with heat detection. Chemical receptors on mosquitoes' antennae can detect compounds such as carbon dioxide and lactic acid, and these attract mosquitoes to hosts at close range. While biting, mosquitoes transfer small amounts of saliva to their hosts. The saliva contains a substance that provokes the itching response that many people and other animals experience at the site of a mosquito bite.

Mosquito bites may transfer parasites or viruses that cause diseases such as malaria, dengue fever, chikungunya, and viral encephalitides such as Zika and West Nile Virus, from one host to another. These microbes complete part of their life cycle in the mosquito's body. Microbe-mosquito relationships are often specific to mosquito genus; for example, only the genus *Anopheles* can vector (transfer) plasmodium, the single-celled microbe that causes malaria.

In 2015, there were over 200 million new cases of malaria, with nearly 450,000 deaths. The vast majority of deaths and cases (90 percent and 88 percent, respectively) occur in sub-Saharan African. Nearly three-quarters of malaria deaths in 2015 were among children under age five; pregnant women are also particularly susceptible (World Health Organization 2015). People infected with malaria experience bouts of fever, chills, aches, and fatigue that prevent them from attending work or school. Countries with high rates of malaria infection suffer losses of economic productivity and educational attainment that, combined with the high cost of malaria treatment, contribute to systemic poverty and economic stagnation.

Mosquitoes of the species *Aedes aegypti* and *Aedes albopictus* vector the virus that causes dengue fever, which has spread from nine countries in 1970 to more than 100 countries in 2015. The virus is most prevalent in tropical and semitropical regions, but local transmission has occurred in a few temperate-region countries. Nearly 100 million people develop clinical symptoms of dengue each year, and the virus has a fatality rate of 2.5 percent.

Human activities contribute to the prevalence of mosquitoes. Inadequate solid waste management—often a problem in low-income communities with poor sanitation infrastructure—leaves containers such as bottle caps and plastic bags that can serve as mosquito-breeding habitat. The global trade in used automobile tires helps spread the *Aedes* mosquitoes that vector dengue fever. Human-induced climate change is expanding the geographic range of both temperature-sensitive mosquito-borne parasites and vector mosquito species as the planet warms.

Traditional mosquito control often involved cultural and social adaptations. In East Africa, many communities settled at high altitudes where temperatures got too cold for mosquito species from nearby lowlands. Communities in Southeast Asia built houses on stilts above the flying range of the local *Anopheles* species. In the United States, the Tennessee Valley Authority, a public regional development agency, carefully regulated water levels and eliminated shoreline vegetation that provided mosquito habitat.

During World War II, American scientists developed the chemical DDT as a pesticide to control malaria mosquitoes in the South Pacific. After the war, DDT and related pesticides were used by farmers for crop pests, and by public health authorities attempting to control mosquito-borne diseases such as malaria and yellow fever. Mosquito populations first displayed resistance to DDT before 1950, and to replacement pesticides in subsequent years. DDT and related pesticides were soon found to persist in food webs and water supplies, threatening wildlife and humans with health issues. Many industrialized countries that were free from malaria, such as the United States, banned DDT in the 1970s. The 2001 Stockholm Convention on Persistent Organic Pollutants now regulates DDT worldwide, allowing its use only under specific conditions in countries where malaria remains a public health problem.

Countries where malaria is prevalent now use a combination of indoor residual spraying of pesticides, insecticide-treated bed nets, sanitation, and water management to reduce incidence of the disease. Communities affected by other mosquito-borne infections use similar interventions, along with broadcast pesticide applications. Mosquito populations worldwide continue to evolve resistance to pesticides, and diseases vectored by mosquitoes continue to emerge.

Dawn Biehler

See also: Wildlife Management; World Organisation for Animal Health (OIE); Zoonotic Diseases

Further Reading

Humphreys, M. 2001. *Malaria: Poverty, Race, and Public Health in the United States.* Baltimore: Johns Hopkins University Press.

Meade, M., and Emch, M. 2010. *Medical Geography*, 3rd ed. New York: Guilford Press.

Rueda, L. M. 2008. "Global Diversity of Mosquitoes (Insecta: Diptera: Culicidae) in Freshwater." *Hydrobiologia* 595: 477–487.

Secretariat of the Stockholm Convention. "Stockholm Convention." Accessed February 18, 2016. http://chm.pops.int/Home/tabid/2121/Default.aspx

Shaw, I., Robbins, P., and Jones, J.P., III. 2010. "A Bug's Life and the Spatial Ontologies of Mosquito Management." *Annals of the Association of American Geographers* 100: 373–392.

World Health Organization. 2015. *World Malaria Report 2015*. Geneva: WHO.

Multispecies Ethnography

Multispecies ethnography is a term used to describe the type of research conducted by anthropologists and social scientists who analyze the interactions between human and nonhuman beings. Traditional ethnographies have always included animals and plants, but they have tended to be depicted as resources, symbols, or as part of the natural landscape. Multispecies ethnographers seek to reverse this tradition by looking at animals and plants as beings whose lives and deaths shape human cultures and histories.

What is ethnography? Ethnography is a type of research originally employed by anthropologists who studied non-European cultures, but it is now also used in other disciplines such as geography, sociology, and media studies. Ethnography is a particular research method that generally involves spending considerable amounts of time with the populations being studied, including participating in and observing daily activities such as eating and relaxing. Ethnographers attempt to understand, at least partially, how research participants live their lives and assign meaning to their actions. In contrast with other research methods that are based on the researcher's point of view, ethnographical analyses seek to emphasize participants' perspectives, which are often very different from those of the researcher. The results of ethnographical research are typically presented in written form, which is called an ethnography. Therefore, multispecies ethnography can be defined as the use of ethnography to understand how different species construct different relationships among each other in specific times and geographies.

Multispecies ethnography as a practice has gained traction since the 2010 article, "The Emergence of Multispecies Ethnography," written by Eben Kirksey and Stefan Helmreich appeared in the journal *Cultural Anthropology* (in a special issue exclusively dedicated to multispecies ethnography). In their path-breaking paper, Kirksey and Helmreich (2010, 545) proposed multispecies ethnography as "a new genre of writing and mode of research" that studies how a "multitude of organisms' livelihoods shape and are shaped by political, economic, and cultural forces." In the same article, Kirksey and Helmreich (2010, 546) stated that multispecies ethnographers analyze "contact zones" where the division between nature and culture becomes blurred.

To better understand the emergence of multispecies ethnography, we must consider the ways in which human/nonhuman relationships are often discussed in the social sciences. As stated before, nonhuman beings are often treated as resources, symbols, or as part of the landscape, and because nonhuman beings are seen as part of the natural world, they are portrayed as lacking their own history. The fact that nonhuman beings, just like humans, have histories becomes clear when we think about certain animal and plant species. For example, we could analyze how cats and dogs became house pets and how we have interacted with these animals in different places and in different times to understand how human and animal histories are connected.

Multispecies ethnography is influenced by numerous thinkers and ideas but, in general, multispecies ethnographers question the claim that humans are exceptional, arguing that humans are only one of the many species in this planet. Multispecies ethnographies analyze culture as a shared product of human and nonhumans. For example, through ethnographic research in Ecuador, Eduardo Kohn (2007) found that the Runa (an indigenous, or native, people) see their dogs as agents, and take their dogs' dreams as a serious indication of what the future holds. For the Runa, Kohn tells us, the experiences of dogs are significant precisely because they are seen as being part of a larger human/dog community. In a similar manner, María Elena

García (2013) considers how Peru's "gastronomic revolution" is affecting indigenous and nonhuman bodies (specifically guinea pigs and alpacas). Through ethnographic research in indigenous communities and on animal farms, García compares what gastronomic experts and chefs say, and what indigenous communities actually experience. She concludes that the globalization of Peruvian cuisine has led to an increase in the slaughtering of guinea pigs and alpacas, while it has also contributed to the idea that because famous Peruvian chefs use indigenous foods, indigenous people are no longer discriminated against. In fact, García claims, indigenous communities are not benefiting from this global trend. For both Kohn and García, the objective of multispecies ethnography is to challenge how we think about the division between humans and nonhumans.

Multispecies ethnography has faced some opposition. Those who question the validity of the multispecies approach often mention some of the issues that can surface when conducting ethnography with nonhuman beings: How do we know what they are experiencing? And, how do we know that we know? Another challenge of multispecies ethnography is connected to the politics of representation and voice: Who can speak for whom? And, why? In this regard, many of the challenges that multispecies ethnographers face now have surrounded ethnography since its very inception and are part of the larger discussions around ethnographic authority: Can an individual claim to know a particular culture or human/nonhuman relationship based on personal experience? Despite these problems and discussions, multispecies ethnography is also raising important questions about what it means to be human and nonhuman. Multispecies ethnography provides important insights into the intricate, constant, and often overlooked relationships among human and nonhuman beings that construct the world we inhabit. To accomplish this, multispecies ethnographers rely on prolonged fieldwork visits in which they observe and participate in the daily interactions between human and nonhumans.

Ivan Sandoval-Cervantes

See also: Agency; Social Construction; Speciesism

Further Reading

García, M. E. 2013. "The Taste of Conquest: Colonialism, Cosmopolitics, and the Dark Side of Peru's Gastronomic Boom." *Journal of Latin American and Caribbean Anthropology* 18(3): 505–524.

Hartigan, J. 2015. *Aesop's Anthropology: A Multispecies Approach.* Minneapolis: Minnesota University Press.

Kirksey, E., and Helmreich, S. 2010. "The Emergence of Multispecies Ethnography." *Cultural Anthropology* 25(4): 545–576.

Kohn, E. 2007. "How Dogs Dream: Amazonian Nature and the Politics of Transspecies Engagement." *American Ethnologist* 34(1): 3–24.

Raffles, H. 2011. *Insectopedia.* New York: Pantheon Books.

N

Non-Food Animal Products

Humans have, for millennia, drawn on animals not only to provide much needed protein, but also to provide non-food materials such as hides, organs, bones, enzymes, glandular secretions, and products like silk and wool, from a broad spectrum of species. These products and their uses are fluid and shift through time to meet changing human wants and needs. Additionally, some groups or cultures may find that other groups and cultures have conflicting values with respect to acceptable forms of non-food animal products, highlighting issues of ethics, rights, and politics. For instance, using animal parts such as the esophagus of a lion or the sex organs of a honey badger as ingredients in medicine is seen as normal for some communities in Africa and Asia, but at the same time highlights potential conflict with the moral, ethical, and religious beliefs of other communities. Similarly, Western medicine draws on animal derivatives in anesthesiology, psychiatry, and orthopedic, plastic and general surgery, and therefore may violate the religious beliefs of certain groups that prohibit the consumption of specific animals.

While there will always be debates about acceptable uses of animal bodies for non-food products, their history reveals an incredible array of products and associated uses. Given this immense breadth, it is best to use general "types of use" as a means of classification. While both domesticated and wild animals are used for non-food products, at the most basic level, non-food animal products can be divided into two categories: utilitarian or decorative. Because some animal products are both utilitarian and decorative, such as a tooled leather coat, categorization is based on the primary use of an object. For instance, a leather coat is primarily used to protect the body from exposure to the elements, even though it may have designs or other embellishments. In contrast, a woven silk wall hanging is primarily valued for its aesthetic qualities, not its ability to insulate a room.

Some of the first utilitarian products derived from animals were leather clothes and bone tools. Scholars have suggested that human ancestors may have modified animal bones to become tools dating from 1.8 to 1 million years ago, making them some of the first examples of non-food animal products. In fact, bone tools are noted as having special characteristics that make them well suited for spinning wool, basketry, scraping and burnishing animal hides, and working ceramics. In addition to bone tools, prehistoric peoples are known to have regularly used animal hides and skins for a wide variety of applications, including clothing, footwear, and bags, as evidenced in the recovery of artifacts from archaeological sites around the world. A good example of this is seen in the oldest leather shoes, found in a cave in Armenia and possibly more than 5,000 years old. Moreover, tanning animal hides and skins to make leather is recognized as one of humanity's first manufacturing processes.

Other examples of utilitarian non-food animal products include wool and silk. While some associate wool strictly with sheep, it is important to recognize that other animals also produce a soft undercoat, such as Angora goats (mohair), Tibetan goats (Kashmir wool), camels, beavers, and rabbits. Probably the oldest textile produced by humans is felt, made from woolen fibers that have been compressed and matted together. The oldest evidence of felt-making dates from 6500–3000 BCE in Turkey, while another archaeological site in Siberia presents felt artifacts dating from the 5th century BCE. In addition to felt, woolen fibers were also spun into a thread or yarn and were then woven together to produce clothing and other functional items like bags. Some early examples of woven wool are seen in a nearly complete woolen tunic dating from the 8th century from northern Niger and some scraps of woolen textiles dating from the 11th–13th centuries from the Bandiagara escarpment in Mali.

Silk is also woven into an incredible array of products, including clothing and writing materials. Utilizing several species of silk worms, silk production started in China around the second century BCE and spread to surrounding areas over time. One of the oldest examples of silk clothing is a robe that dates back to 200–150 BCE, while one of the oldest manuscripts (written on a sheet of silk) dates to around 300 BCE. Found in another cache of Chinese artifacts is one of the oldest decorative examples of silk, a painting called "The Lady, the Dragon and the Phoenix."

Other examples of non-food animal products include the use of baleen whale oil, shark liver oil, and ambergris from sperm whales. From the 11th–20th centuries, whale oil rendered from the blubber of baleen whales was a highly valued commodity as it was used to fuel lamps, as an industrial lubricant, and in the manufacture of soap, varnish, and explosives. Shark liver oil, also known as squalene, had historically been used in the leather tanning industry, to oil textile machinery, and to seal the hulls of wooden boats. Currently, squalene is also highly valued in decorative applications such as deodorants, lipsticks, and other cosmetics. Due to high demand, both historically and contemporarily, several species of shark are now considered vulnerable by international conservation organizations. Similarly, ambergris (a bile duct secretion of sperm whale) has been valued since 1000 BCE as incense in Egypt. Subsequently, it has come to be used predominately in Europe and North America in perfume production, especially high-end perfumes.

Ian Edwards

See also: Bushmeat; Human-Wildlife Conflict; Indigenous Rights; Traditional Chinese Medicine (TCM); Whaling

Further Reading

Edwards, I. 2012. "The Social Life of Wild-Things: Negotiated Wildlife in Mali, West Africa." PhD dissertation. University of Oregon.

Edwards, I. 2003. "The Fetish Market and Animal Parts Trade of Mali, West Africa: An Ethnographic Investigation into Cultural Use and Significance." MA thesis. Oregon State University.

Eriksson, A., Burcharth, J., and Rosenberg, J. 2013. "Animal Derived Products May Conflict with Religious Patients' Beliefs." *BMC Medical Ethics* 14(48). Accessed March 24, 2015. http://www.biomedcentral.com/1472-6939/14/48

Read, S. 2013. "Ambergris and Early Modern Languages of Scent." *The Seventeenth Century* 28(2): 221–237.

Thomson, R. 2006. "The Nature and Properties of Leather." In Kite, M., and Thomson, R., eds., *Conservation of Leather and Related Materials.* Philadelphia: Elsevier Ltd.

Northern Spotted Owl

The northern spotted owl (*Strix occidentalis caurina*) is one of three spotted owl subspecies. Closely associated with the temperate rainforests of North America's Pacific Coast, it has become the focus of a heated debate over the politics and ethics of wildlife conservation.

The owl grows to around 45 cm (18 in) in length with a 90 cm (36 in) wingspan; its dark brown plumage with pale spots gives it its common name. The owl's range extends from the southwest of British Columbia, Canada, through Washington and Oregon as far as northern California in the United States.

Research on the owl's behavior suggests that it prefers to nest in old-growth forest. This is typically dominated by coniferous (cone-bearing) tree species such as Douglas fir and Sitka spruce, with some trees being centuries old and interspersed by dead and rotting standing trees. This environment provides suitable nesting sites for the birds and, as they are predatory, they are able to hunt for small animals living in the undergrowth. The owls generally form mating pairs for life with their offspring dispersing themselves widely upon maturity. This can aid owl conservation—dispersal increases its overall chances of survival if the habitat becomes degraded in one area—but it also means that the owls require a large area of suitable forest to breed successfully.

The northern spotted owl population is currently in decline throughout its range, with the main cause being logging. Since Europeans first settled in the area, nearly three-quarters of the old-growth forest has been either felled or converted to forestry plantations (from 40 million to 11.3 million acres). The owl is therefore listed

A northern spotted owl hunts in Muir Woods, California. Research shows that these owls prefer to nest in the old-growth forests of their range along the Pacific coast of North America. The logging industry of this region has significantly contributed to this species' decline, sparking conflict between industry and environmentalists. (Reese Ferrier/Dreamstime.com)

as "near threatened" on the International Union for Conservation of Nature (IUCN) Red List of threatened species. The species is viewed by conservationists as an "indicator species," meaning that its presence or absence gives an indication of the general health of the forest.

In the early 1980s, President Ronald Reagan's administration pursued a policy of converting the forest into a "fully managed" condition (i.e., with the remaining old-growth forest replaced by timber plantations), in order to satisfy global demand. In 1987, a coalition of environmental groups filed a lawsuit to protect the northern spotted owl. It was designated "threatened" under the U.S. Endangered Species Act in the summer of 1990. Following this, the U.S. Fish and Wildlife Service proposed in 1992 to set aside 5.4 million acres as conservation land for the owl—the Recovery Plan. At the same time, the Interior Department released its own plan, the Lujan Plan, which would protect 2.8 million acres of forest and override some of the requirements of the Endangered Species Act. Neither of these proposals was ever implemented. The Northwest Forest Plan, setting aside 5 million acres for habitat conservation, was finally approved in 1994. This failed to satisfy either environmentalists or the timber industry, as ultimately declines in the owl population would still occur. The Plan was predicted to lead to 28,000 job losses over a decade in a logging industry already facing unemployment (Foster 2002).

Throughout this time, environmental groups protested against logging. Some radical activists such as Earth First! practiced tree spiking—driving metal nails or rods into the trees to prevent logging—and even chained themselves to equipment. By 1996, environmentalists were using car blockades to try to halt logging. One group occupied a site in Willamette National Forest, Oregon, for over 11 months.

The threat of unemployment stoked tensions between forest workers and environmentalists. There are reports of workers proudly displaying bumper stickers with the slogan "I like spotted owls—fried." Logging bosses, it seems, were actively encouraging anti-environmental sentiments among their employees. Environmentalists were also sometimes unsympathetic to the workers' plight. *Forest Voice* magazine, published by the Oregon-based nongovernmental organization, the Native Forest Council, regularly argued that loggers could easily find work in other industries.

The controversy has been characterized as a classic case of jobs versus the environment. However, not everyone involved fits easily on one or the other side. In the opinion of one timber worker at a 1989 public hearing, the workers "have probably gained a respect for the forest and the land that few people will ever know. . . . Unless a person has actually sat quietly at a logging site and watched and listened, they cannot appreciate the amount of wildlife that is around" (Proctor 1996, 270–271). These opinions suggest that timber workers also express a great deal of concern for the environment.

Today, the northern spotted owl population continues to plummet, with an estimated overall population decline of 3 percent annually since 1990. In Washington State, the annual rate of decline is 7 percent. Estimates of the owl's population vary, but there are thought to be around 8,500 to 12,000 individuals throughout its range; in Canada the owl is almost extinct, with only around 30 breeding pairs left. As well as habitat destruction, the species faces new threats, including competition for food from the barred owl (*Strix varia*) and from drought and wildfires exacerbated by climate change.

The northern spotted owl can be considered a symbol of much wider issues concerning the ethics and ideology of environmentalism in North America amid a shifting political and economic climate. However, the example demonstrates that the biology of the owl itself also matters. The fact that the birds prefer old rotting trees, not younger trees grown on plantations, cannot be avoided.

Camilla Royle

See also: Animal Law; Biodiversity; Endangered Species; Extinction; Flagship Species; Habitat Loss; Indicator Species; Species; Wildlife Management

Further Reading

BirdLife International. 2013. "*Strix occidentalis*." The IUCN Red List of Threatened Species (version 2014.3). Accessed March 8, 2015. http://www.iucnredlist.org

Foster, J. B. 2002. "The Limits of Environmentalism without Class: Lessons from the Ancient Forest Struggle of the Pacific Northwest." In Foster, J. B., ed., *Ecology against Capitalism* 104–136. New York: Monthly Review Press.

Power, T. M. 1996. "The Wealth of Nature." *Forest Voice* 9(2). Accessed March 12, 2015. http://www.forestcouncil.org

Proctor, J. D. 1996. "Whose Nature? The Contested Moral Terrain of Ancient Forests." In Cronon, W., ed., *Uncommon Ground* 269–297. New York: W.W. Norton.

Nuisance Species

"Nuisance species" is a label for a species of wild animal or feral animals (domestic species who have returned to a wild state, such as stray cats) who are regarded as bothersome by humans. The label "pest" is sometimes used interchangeably. Reasons why animals might be considered a nuisance include the animal's perceived or real tendency to damage human property, to threaten the health or safety of humans or their domestic animals, to compete for human food or habitat, or to encroach in any way upon human interests. The nuisance species label represents one way that our terminology for animals shapes our moral obligations to them and makes certain animals seem more legitimate than others in a given space.

Nuisance species, and problems associated with them, vary widely from one context to another. In urban areas, nuisance species are likely to be generalists who can thrive in a variety of habitats and eat a varied diet. These animals are adept at using human-built environments. In North America, mice, rats, raccoons, squirrels, pigeons, geese, and bats are some species commonly branded as urban nuisances. Problems associated with urban animals are often related to concerns about property damage, about animals occupying spaces where they are not wanted (e.g., living in attics or under porches, hanging around golf courses or schools), about mess or disorder (e.g., digging up lawns, getting into garbage, defecating in public spaces), and about public health or safety issues (e.g., disease transmission or animal attacks).

In rural areas, nuisance species also include those that present a perceived economic threat to the livelihood of agricultural producers (e.g., animals that may damage crops or attack livestock). It is common practice for farmers to kill nuisance species as a means of protecting their assets, but the adverse effects of rural lethal control on some wildlife populations has brought the approach under scrutiny in recent decades. For example, aggressive poisoning campaigns over the last century have reduced prairie dogs to 2 percent of their former numbers in the United States, based on the belief that their burrow holes cause horse and cattle injury, that they compete with cattle for grass, that they spread disease, and that they destroy the environment, even though these claims have not been supported by scientific evidence (Hadidian et al. 2007).

A central problem with the term "nuisance species" is that it categorizes animals solely on the basis of their appeal to humans. Many have argued, in contrast, that animals have inherent value as beings in their own right, as well as important roles as members of their respective ecosystems. The phrase is often used today with quotes around the word "nuisance" to signal disagreement with the categorization of animals according to their utility to people. This type of classification is tied to the evolution of wildlife management in the United States. Early management agencies were motivated to increase populations of game species, which were of economic value because of their interest to hunters and trappers, and to extinguish animal populations that represented potential economic losses, which they branded nuisance species. In this utilitarian view, the lives of nuisance species have negative value—not only are they undesirable, it is considered a public good to remove them. Many nuisance species are considered "trash animals," a label that reveals their status as objects meant for disposal.

While changing attitudes toward the natural environment over the last 50 years have challenged a purely utilitarian perspective, laws relating to nuisance species continue to reflect historical attitudes. Nuisance species have few legal protections, and in many cases there are no formal processes by which an animal's status as a nuisance is determined. Canada's federal Fish and Wildlife Act, for example, states that residents may harass, capture, or kill an animal if they "believe on reasonable grounds" that the animal "is damaging or is about to damage" their property. Because many perceptions of animals as nuisances are based on fear, ignorance, or misunderstanding of a situation, statutes like this one, which depend upon personal discretion about whether an animal is "about to damage" property, clearly leave accused animals at a disadvantage.

A second problem with the nuisance species moniker is that it implies that, in a nuisance situation, *the animal itself* is the problem. An alternate view focuses on both human and animal behaviors, as well as human perceptions, as part of a perceived problem. For example, the expression "a fed bear is a dead bear" is often used to highlight the fact that feeding bears—and thus habituating them to human environments—is the reason why bears come too close to homes or campgrounds, a behavior that often results in their being killed. To highlight the importance of human factors in animal nuisance situations, the term "human-wildlife conflict" has replaced nuisance species in much academic and professional literature. Some have argued, however, that human-wildlife conflict is also not an ideal term because it implies a mutual antagonism between humans and animals. In reality, most so-called human-wildlife conflicts are simply humans perceiving particular animals as a threat or a conflict *between* people *about* wildlife (Peterson et al. 2010). Wildlife management organizations are increasingly recognizing the importance of addressing human-human conflicts by bringing together different stakeholders (e.g., conservationists, wildlife managers, landowners, hunters, and trappers) to work out mutually agreeable solutions.

Erin Luther

See also: Human-Wildlife Conflict; Species; Wildlife Management

Further Reading

Adams, C. E., and Lindsey, K. J. 2010. *Urban Wildlife Management*. Boca Raton, FL: CRC Press/Taylor & Francis Group.

Hadidian, J., Baird, M., Brasted, M., Nolfo-Clements, L., Pauli, D., and Simon, L. 2007. *Wild Neighbors: The Humane Approach to Living with Wildlife*. Washington, DC: Humane Society Press.

Hadidian, J., and Smith, S. 2001. "Urban Wildlife." In Salem, D. J., and Rowan, A. N., eds., *The State of the Animals 2001,* 165–182. Washington, DC: Humane Society Press.

Herda-Rapp, A., and Goedeke, T. L., eds. 2005. *Mad about Wildlife: Looking at Social Conflict over Wildlife*. Boston: Brill.

Nagy, K., and Johnson II, P. D., eds. 2013. *Trash Animals: How We Live with Nature's Filthy, Feral, Invasive, and Unwanted Species*. Minneapolis: University of Minnesota Press.

Peterson, M. N., Birckhead, J. L., Leong, K., Peterson, M. J., and Peterson, T. R. 2010. "Rearticulating the Myth of Human-Wildlife Conflict." *Conservation Letters* 3: 74–82.

O

OncoMouse

OncoMouse® is a transgenic mouse used in human medical research. It is transgenic (made of more than one species) because it has a human gene that promotes cancer embedded in its DNA. OncoMouse is unique because it is the first multicellular living being deemed patentable by the U.S. Patent and Trademark Office (USPTO). OncoMouse is key to understanding how present-day conflicts over animal ownership, research, and species integrity are playing out in society.

Animal research science has created a very particular relationship with mice by bringing them into the space of the laboratory as tools on par with other common pieces of equipment like microscopes and test tubes. Mice became researchers' animal of choice because they are easy to care for, easy to handle, and easy to manipulate through breeding due to their ability to reproduce so quickly. While these mice share a common wild ancestor—the house mouse (*Mus musculus*)—their genetic differences through domestication have made them their own species: *Mus domesticus*. Laboratory animals are purposefully bred, inbred, and genetically altered in order to conduct biological, biomedical, and chemical toxicity (how harmful a product is) research. The goal is not to breed animals for better physical characteristics (e.g., livestock) or temperament and appearance (e.g., pets), but to breed and develop abnormal and/or diseased animals as stand-ins for humans.

While strains of cancer-susceptible mice had been around since the early 1900s, they were created using traditional breeding methods. OncoMouse (from *onkos,* Greek for tumor) was different because it was produced through technological intervention. In the early 1980s, researchers Philip Leder (1934–) and Timothy Stewart (1952–) at Harvard University (using funding from DuPont Corporation) injected human genes that triggered cancer growth into fertilized mouse eggs and then used surrogate mouse mothers to carry the eggs to term. The resulting OncoMouse line of mice would then always manifest these human genes, even through traditional breeding. These mice were now highly susceptible to cancers and could be used to study potential cancer-fighting drugs and carcinogenic (cancer-causing) substances.

The decision to obtain and grant a patent for OncoMouse did not come out of the blue. Granting a patent is important because it 1) demonstrates the novelty of a creation, and 2) gives the patent owner exclusive rights of use and/or licensing for a 20-year period, during which time the owner can recoup the cost of research and make a profit. There were two previous patent disputes over whether living organisms were novel creations. In 1980, the U.S. Supreme Court ruled in *Diamond vs. Chakrabarty* that a genetically altered single-celled bacteria (in this case one that could digest oil) was patentable because it met the federal patent code specifications for new composition of matter. In 1987, the U.S. Board of Patent Appeals upheld the rejection of a patent for a genetically altered oyster due to a technicality unrelated to the organism itself; however, the appeal ruling, known as *Ex parte Allen*, stated the USPTO would now accept all nonnaturally occurring nonhuman organisms as patentable because they were novel creations and new compositions of matter. OncoMouse became patent no. 4,736,866 on April 12, 1988, thereby becoming the first patented multicellular living organism. Due to the fact that DuPont provided funding for OncoMouse, Harvard granted exclusive license/rights to the company. The patent has now expired, so researchers do not have to pay license fees, but the name OncoMouse is still a registered trademark (meaning no one else can use the name as their own).

Among the scientific community the only concerns that were raised about OncoMouse were related to the high fees and restrictive usage agreements that DuPont charged. There was no concern expressed over the ethics of patenting living beings, but the patent caused a global public debate about what it would mean to have intellectual property ownership of lifeforms and how far this ownership would go. Indeed, while the European Union and Japan changed their own patent laws to allow animal patents, countries like Canada, after their own Supreme Court case, did not. A main concern is that there is a slippery slope whereby scientists could make, and patent, such complex transgenic animals that they might arguably be human. In the United States, this would be in violation of both the 13th Amendment of the Constitution (banning slavery and therefore the ownership of a human being) and the parameters of the USPTO itself. To test patent law limits, in 2004 anti-life-patent advocates tried to patent an organism that would be a 50/50 human and mouse embryo combination (a "humouse"), but this was rejected by the USPTO on the grounds the animal would contain too much human material. U.S. patent law currently prohibits patenting a "human organism," but there has not yet been a definitive ruling on when a nonhuman organism might become a human organism via transgenic manipulations (USPTO 2015).

OncoMouse has also been used as an intellectual rather than physical model in social studies of science and society. For example, science philosopher Donna Haraway (1944–) has written extensively about the need for human society to pay attention to, and reflect on, the creation, meaning, and uses of these novel organisms. Through her concept of "shared suffering," she argues that such beings provide us with a chance to see them (and perhaps all animals) as something more than objects—as living subjects like humans that require attentive and humane treatment. She argues we must not close ourselves off from cultivating sensitivity, especially when OncoMouse is getting cancer so that humans may ultimately not.

Julie Urbanik

See also: Animals; Biotechnology; Ethics; Hybrid; Research and Experimentation; Sentience; Vivisection; Working Animals

Further Reading

Diamond v. Chakrabarty. 1980. 447 U.S. 303.

Ex Parte Allen. 1987. 2 U.S.P.Q.2d 1425, 1427 (Fed. Cir.).

Haraway, D. 1997. *Modest_Witness@Second_Millennium.FemaleMan©_Meets_OncoMouse™*. New York and London: Routledge.

Haraway, D. 2008. *When Species Meet.* Minneapolis: University of Minnesota Press.

Rader, K. 2004. *Making Mice: Standardizing Animals for American Biomedical Research, 1900–1955.* Princeton, NJ: Princeton University Press.

Raines, L. 1990. "Public Policy Aspects of Patenting Transgenic Animals." *Therigenology* 33(1): 129–149.

Rifkin, J. 1988. *The Biotech Century: Harnessing the Gene and Remaking the World.* New York: Jeremy P. Tarcher/Putnam.

Shorett, P. 2016. "Of Trangenic Mice and Men." Accessed January 5, 2016. http://www.councilforresponsiblegenetics.org/ViewPage.aspx?pageId=167

USPTO. 2015. "2105 Patentable Subject Matter—Living Subject Matter [R-07.2015]." Accessed February 8, 2016. http://www.uspto.gov/web/offices/pac/mpep/s2105.html

P

Pastoralism

Pastoralism originated during the "agricultural revolution" that occurred in human history around 10,000 years ago, when many people transitioned from food foraging lifeways involving the hunting and gathering of wild foods to food-producing ways of life centered on domesticated animals and plants. Pastoralism can be defined as the raising of herd animals—as large as or larger than goats and sheep—on "natural" pasture unimproved by human intervention (Salzman 2004). Humans and certain hoofed mammals (ungulates) that live off of natural resources such as grass, browse (leaves, twigs, and young shoots of trees or shrubs on which livestock feed), and lichens that grow in rangeland and pastures formed reciprocal relationships in which each side received benefits from the partnership. Humans have received far more benefits in the exchange than their large livestock partners. Domesticates such as cattle, horses, reindeer, camels, yaks, llamas, sheep, and goats provided their human partners with a variety of products including milk, meat, blood, bones, hides, wool, traction, transportation, and even symbolic objects for religious thought and ritual activities such as blood sacrifice. In return, pastoralists provided veterinary services, protection from predators, food and shelter, and companionship to the animals. Pastoralism today is on the decline as governments and military conflicts have reduced the territories of herding cultures.

The success of our species, *Homo sapiens*, on Earth is due, in part, to our relationships with herd animals. The human digestive system cannot process certain vegetative matter such as grass, but it can process the flesh of animals that do eat grass. Husbanding animals that eat plants that humans cannot provides an innovative solution to the problem of turning a disability into a resource. The pastoralist/animal relationship allowed humans to colonize vast areas of Earth, across a wide array of diverse environments. The relationship changed natural environments into landscapes shaped by human intentions and strategies in partnership with herd animals. Plains and savannah environments, from Central Asia to Africa, which receive at least the minimum annual rainfall to sustain grass and browse, support cattle and horse pastoralism. Hot and cold deserts, from the Sahara and Arabian deserts to the Arctic, with their prevailing sand or tundra environments, favor camel, sheep, or reindeer pastoralism. Mountain areas, from the Andes in South America to the Middle East and South Asia, sustain yak, llama, and goat pastoralism. Humans have therefore had a hand in the proliferation of diverse ungulate species exploiting different environments.

Domestication, the process of selecting for certain traits in the breeding process, has created an intensive relationship between human and nonhuman species.

The bond between pastoralist and animal is often very strong. The bond is one part emotions and another part economics. Spending days with a fast and reliable camel that takes its rider through a dangerous section of desert creates emotional ties to the animal. Having camels transport loads of merchandise to market makes them, in the eyes of their owner, economic animals—the lifeblood of the business. Some herders exalt the human/animal relationship and see themselves through the exchanges with a species of animal that they hold in high regard. They often self-identify as "the people of cattle," or "the reindeer people," and so on (Vitebsky 2005).

Keeping herd animals also means keeping up with the animals. In other words, herds on the open range must move from pasture to pasture, sometimes over a vast territory and across multiple ecological zones, in search of water and graze. For pastoralists, herd mobility is a natural part of life, even if they are mobile for relatively short periods of time. While herd mobility has been a key ecological and economic strategy of pastoralism, it is not its signature characteristic. In pastoralism, other activities such as farming, fishing, hunting, and trading may be as important, economically, as caring for the herd animals in the pastures. For example, some pastoralists in a very dry region of Madagascar have become "cactus pastoralists" by reducing herd mobility and devoting much of their labor to tending the plantations of prickly pear cactus that they feed to their cattle when the graze is dried up. Decreasing herd mobility puts more demands on human labor to provide fodder for the animals.

There have always been tensions in the relationship between pastoralists and their herds, often from outside forces. The advent of 19th-century colonialism, which involved militarily advanced, empire-building European countries taking control of resource-rich lands in Africa and South Asia, in turn greatly affected mobile peoples and their herd animals by dividing territories of open range. This resulted in a significant loss in pasture lands, which has constrained herd mobility. More recent efforts in economic development, in such places as Africa, have also reduced the territory of pastoralist peoples, turning grasslands and forests into farms and sometimes into urban spaces.

Jeffrey C. Kaufmann

See also: Animal Cultures; Animal Geography; Domestication; Habitat Loss; Human-Animal Bond; Husbandry; Livestock; Meat Eating; Military Use of Animals; Multispecies Ethnography; Species; Working Animals; Zoonotic Diseases

Further Reading

Humphrey, C., and Sneath, D. 1999. *The End of Nomadism? Society, State and the Environment in Inner Asia.* Durham, NC: Duke University Press.

Paine, R. 1994. *Herds of the Tundra: A Portrait of Saami Reindeer Pastoralism*. Washington, DC: Smithsonian Institution Press.

Salzman, P. C. 2004. *Pastoralists: Equality, Hierarchy, and the State*. Boulder, CO: Westview.

Schlee, G., and Shongolo, A. A. 2012. *Pastoralism and Politics in Northern Kenya and Southern Ethiopia*. London: James Currey.

Vitebsky, P. 2005. *The Reindeer People: Living with Animals and Spirits in Siberia*. Boston: Houghton Mifflin.

Wagner, A. 2013. *The Gaddi Beyond Pastoralism: Making Place in the Indian Himalayas*. London: Berghahn.

People for the Ethical Treatment of Animals (PETA)

People for the Ethical Treatment of Animals (PETA) is the largest animal rights organization in existence, with approximately 3 million members and 300 employees, and it is widely understood as playing a major role in popularizing the modern animal rights movement. Founded in 1980 by Ingrid Newkirk (1949–) and Alex Pacheco (1958–), the group strives for the elimination of animals as property for humans, and their exploitation for pleasure and profit by humans. This includes animals raised for food and clothing, used in research, bred as companion animals, confined in zoos, and used for entertainment in circuses. The organization is based in Norfolk, Virginia, and had a 2014 income of approximately $52 million. Pacheco left PETA in 1999, but Ingrid Newkirk remains the president. The group stages protests and conducts education, outreach, and media campaigns; all targeting the public, lawmakers, and corporations.

PETA first gained acclaim in 1981 when the group exposed research on 17 macaque monkeys at the Institute of Behavioral Research in Silver Spring, Maryland. The monkeys' afferent ganglia, the part of the brain that allowed them to feel their arms, were cut in an attempt to see if the monkeys could relearn how to use their arms. Going undercover for five months, Pacheco worked as a volunteer researcher in the lab, documenting monkeys living with open, rotting wounds, neurotic behavior, expired food, and feces in cages. With this documentation, and soliciting the help of five expert witnesses, Pacheco approached police on September 8, 1981 (Pacheco 1985). Three days later, police engaged in the first-ever raid on a research facility, confiscating the monkeys and charging researcher Edward Taub (1931–) with 17 counts of animal cruelty. He was convicted on six charges, all of which were overturned on appeal.

The group's philosophy is largely derived from Australian philosopher Peter Singer's book *Animal Liberation* (1975), which makes a philosophical case for considering the interests of animals without regard to species. PETA's goal is, ultimately, absolute—"animals are not ours to eat, wear, experiment on, use for entertainment, or exploit in any way" (PETA 2015). Although the group's goal remains animal

Members of People for the Ethical Treatment of Animals (PETA) protest against mad cow disease in Washington in 2004. PETA's protests cover the spectrum of animal use and exploitation. The organization's methods frequently include very visible, eye-catching demonstrations that capture public and/or media attention. (AP Photo/Ted S. Warren)

liberation, it also campaigns for industry and legislative reforms. These reforms would not abolish the use of animals, but instead improve the conditions of animals that are used or exploited by humans. Because PETA employs both approaches, the organization has been criticized by not only animal welfare activists, many of whom do not oppose the human use of animals but believe they should be well-treated, but also more hardline elements of the animal rights/liberation movement who believe reforms and welfare to be counterproductive to the ultimate goal of ending animal exploitation.

The group has affiliate programs all over the world—operating in the United Kingdom, France, Germany, the Netherlands, Australia, India, and PETA Asia-Pacific, based in the Philippines. At their home office in Norfolk, Virginia, the group operates a mobile low-cost spay/neuter clinic and delivers, via truck, free PETA-built dog houses and straw bedding, as well as free educational services, largely to low-income neighborhoods. The organization operates a prominent youth division, peta2, which partners with popular musicians and taps into youth movements—particularly the punk subculture—to promote animal rights and vegetarianism. In this vein, peta2 organizes events on college campuses and reaches out to attendees at the traveling music festival Warped Tour. Other demographic-specific initiatives include PETA Kids, PETA Latino, and Jesus People for Animals.

One of PETA's primary methods of outreach is through celebrity endorsements. An extensive list of high-profile actors, comedians, musicians, and other figures have participated in PETA public service announcements in video and print. Actors Alec Baldwin, Pamela Anderson, and James Cromwell; musicians Justin Bieber, Paul McCartney, and Morrissey; and comedian Bill Maher are some of PETA's most noted public supporters.

A chief component of PETA's outreach revolves around media coverage. Utilizing street theatre (such as publicly showering to highlight the amount of water used to produce meat), civil disobedience in the form of commandeering microphones at events, footage from their undercover investigations of factory farms or animal laboratories such as the one described above, and other tactics that can be traced back to the U.S. and European social movements of the 1960s, the group is known for its use of billboards, social media, news coverage, and viral videos (spread on social media) as conduits to spread its talking points.

PETA's tactics have, however, been criticized by feminist organizations and some other animal rights groups for its depiction of nude or nearly nude women in protests and advertising campaigns. In 2012, PETA released an ad featuring imagery of a bruised woman in a neck brace, whose injuries were attributed to the increased sexual vigor of her boyfriend due to his vegan diet. Feminist and anti-domestic violence groups criticized the ad on the grounds that its imagery trivializes domestic violence. Their 2003 "Holocaust on Your Plate" exhibit, which compared the systematic and large-scale confinement, mutilation, and killing of animals on factory farms to Nazi concentration camps, was denounced by the Anti-Defamation League (CNN 2003), a Jewish advocacy group, prompting PETA President Ingrid Newkirk to end the campaign and issue an apology. The organization nevertheless defended the campaign, pointing out that concentration camps were modeled after slaughterhouses, and quoting the Jewish author Isaac Bashevis Singer, who said that "in relation to animals, all men were Nazis" (Singer 1972, 257). Subsequent campaigns have drawn parallels with other forms of human exploitation and oppression.

Drew Robert Winter

See also: Advocacy; Ethics; Humane Education; Rights; Veganism; Welfare

Further Reading

CNN. 2003. "Group Blasts PETA 'Holocaust' Project." February 28. Accessed June 16, 2015. http://www.cnn.com/2003/US/Northeast/02/28/peta.holocaust

Pacheco, A., with Francione, A. 1985. "The Silver Spring Monkeys." In Singer, P., ed., *In Defense of Animals,* 135–147. New York: Basil Blackwell.

PETA. 2015. "All about PETA." Accessed June 16, 2015. http://www.peta.org/about-peta/learn-about-peta

Singer, I. B. 1972. *Enemies, A Love Story*. New York: Farrar, Strauss and Giroux.

Singer, P. 1975. *Animal Liberation*. New York: HarperCollins.

Personhood

Personhood can be thought of in three interrelated ways. First, in many cultures the word "person" in everyday language signifies a human being. Second, the idea that a person is only human has strong links with the status of having moral rights to ethical treatment. Finally, this status of having moral rights is connected to legal rights. The definition of "person" as solely a human being, and humans as the only bearers of moral and legal rights, is being challenged by animal advocates in many ways.

Although subject to ongoing debate, in the context of Western philosophy there are several widely accepted key aspects of personhood, the first of which is having a "self." A self is understood as a psychological entity that can change over one's life but (for the most part) remains consistent over time and space. In other words, a self has a connected past, present, and future and remains identifiable in different places. Another aspect of self that many argue is essential to personhood is self-awareness (also known as self-consciousness). This means that an individual is conscious of his/her own existence and aware that he/she exists separately from other individuals. A final key aspect is the ability to act with intention (also known as agency), that is, to be conscious of one's actions and, at some level, make decisions about those actions. Although some human beings (e.g., infants and many mentally disabled people) may not have all three of these qualities, most humans do. The famous cognitive (related to brain functions and the mind) scientists Sigmund Freud (1856–1939) and Jacques Lacan (1901–1981) developed influential theories of humans' sense of self, but today scientists are studying these aspects in other species.

Challenges to the idea that only humans can be persons are based on claims that many nonhuman species can also be self-conscious agents, and a number of scientific studies have shown that many animals, including chimpanzees, dolphins, elephants, gorillas, and whales, do have these capacities. Research on whales has shown that life and learning patterns once thought to be uniquely human, such as language dialects and transmission of culture, also exist in these animals who evolved, like humans, in complex social groups (Whitehead and Rendall 2014). A study of (captive) gorillas' eye contact with human caretakers demonstrated that these animals can differentiate between self and other and also between inanimate objects (such as tools) and subjects with agency (Gomez 1990). Most, if not all, of the claims for personhood for nonhuman species are based on studies that demonstrate that they have the intelligence necessary for the key aspects of selfhood.

Accepting animals as persons changes our obligations toward them. Scientific studies showing that at least some animals can have "selves" at one level make only a moral argument—that humans *should* view at least some nonhumans as persons and treat them accordingly. However, a moral obligation is not always the same as a legal obligation. It is only when notions of moral obligations are codified (written down into law) that those defined as persons gain moral and legal rights. Existing legal rights associated with personhood include the right not to be owned as property, to not being treated in a manner that causes harm, and, frequently, also the right to life.

Although many people may view certain or all animals as persons in a moral sense, and many laws do give nonhumans a variety of legal protections, in most sets of written laws worldwide, nonhuman animals are legally seen as property and not as persons. In other words, they do not have legal personhood and the related rights. This is why, for example, animals can be held captive in zoos and raised and killed for food or used for medical experiments. In terms of human persons, it would be an unthinkable violation of their rights, for example, to hold them captive and put on display as nonhumans are in zoos.

There are limited examples of legal personhood status for animals. For example, in 2013 India granted this status to dolphins, citing their high intelligence, and in 2014 Argentina deemed Sandra, a captive orangutan, a nonhuman person, based on her similarity to humans. In addition, the organization *The Great Ape Project* has sought to gain global legal recognition of all of the four nonhuman great ape species (bonobos, chimpanzees, gorillas, and orangutans) as persons. The lawyer Steven Wise (1952–) has been at the forefront of arguing for establishing the legal personhood of great apes in the United States. Advocates have also stated that scientifically, cetaceans (biological classification of marine mammals that includes dolphins, porpoises, and whales) and even a number of farmed species meet the requirement for personhood. These advocates argue that while the legal concept of person has historically been applied only to humans, it is also true that not even all humans (e.g., slaves in the United States and Europe) have always been legally defined as persons. This precedent of expanding the legal definition provides a moral and legal justification for doing so now and is supported by scientific evidence.

Finally, many humans share their lives and become very familiar with nonhuman individuals (e.g., household companions such as dogs and cats or working companions such as horses). Close relationships with animals can challenge the idea of only humans as persons because physical proximity to and emotional closeness with animals allows humans to experience first hand other species as distinct individuals who exemplify the aspects of personhood.

Connie L. Johnston

See also: Advocacy; Agency; Animal Cultures; Animal Law; Chimpanzees; Communication and Language; Dolphins; Elephants; Emotions, Animal; Ethics; Evolution; Great Apes; Intelligence; Multispecies Ethnography; Pets; Rights; Sentience; Speciesism

Further Reading

Cavalieri, P. 2008. "Whales as Persons." In Armstrong, S. G., and Botzler, R. G., eds., *The Animal Ethics Reader*, 2nd ed., 204–210. New York: Routledge.

Cavalieri, P., and Singer, P., eds. 1994. *The Great Ape Project: Equality Beyond Humanity*. New York: St. Martin's Press.

Farm Sanctuary. 2015. "Someone, Not Something." Accessed October 23, 2015. http://www.farmsanctuary.org/learn/someone-not-something/about-the-someone-not-something-project/

Gomez, J. C. 1990. "The Emergence of Intentional Communication as a Problem-Solving Strategy in the Gorilla." In Parker, S. T., and Gibson, K. R., eds., *"Language" and Intelligence in Monkeys and Apes: Comparative Developmental Perspectives*, 333–355. Cambridge, UK: Cambridge University Press.

Nonhuman Rights Project. Accessed October 23, 2015. http://www.nonhumanrightsproject.org

Whitehead, H., and Rendell, L. 2014. *The Cultural Lives of Whales and Dolphins.* Chicago: University of Chicago Press.

Wise, S. M. 2000. *Rattling the Cage: Toward Legal Rights for Animals.* New York: Basic Books.

Pests. *See* Nuisance Species

Pet Overpopulation. *See* Pets; Shelters and Sanctuaries; Spay and Neuter

Pets

The scale of pet keeping is remarkable. In 2012, according to the Humane Society of the United States, 60 percent of American households included at least one pet. Pets are also big business: During that year it is estimated that $50 billion was spent on these 160 million animals. At the same time, it is obvious that many pet animals are often poorly treated, and animal welfare organizations have rightly drawn attention to a range of problems associated with pet keeping—to the point at which many activists suggest that pet keeping should be banned. The keeping of pets is thus one of the most significant, but also one of the most complex, of all human-animal relationships.

Even trying to define pets and pet keeping is difficult. One straightforward answer has been suggested: A pet is an individual animal with a proper name, one that is kept in the house, and one that is never eaten. It is easy to think of counterexamples, though, as the most common animals kept as pets are fish (not usually individually named), and some pets like horses live separately from us. And while the importance of domesticity has been prominent in much recent writing, with the history of pets being linked to the emergence of the middle-class family home, to a culture that increasingly considered cruelty to animals unacceptable, and one that prized a

No one is really sure when the first pets came to be, but many animal studies researchers believe their increasing importance in our lives is the direct result of society's move toward industrialization and urbanization. Pets not only provide loving companionship, but they are also a way to connect to a wildness everyday modern life lacks. (Alena Ozerova/ Dreamstime.com)

sentimental attachment to the natural world, this is certainly too simple. Pets and pet keeping are clearly a very ancient phenomenon, and evidence for pets can be found in classical Greece and Rome, in ancient Egypt and China, and indeed in virtually every culture in the world. The further back in history we go, the harder it is to be sure about why people kept pets, but there can be no question that some people loved their animals and grieved for them after their death. We have to be careful about taking contemporary pet keeping as unique.

It has been argued that the modern culture of pet keeping is different, however, not only in terms of scale but also because it has become more normal and acceptable (in many societies) for large numbers of people to keep animals in their homes solely for the pleasure of their company. Keeping pets is not uncontroversial, but it is an extreme position now to say that people should be banned from keeping pets. In the past, though, pets were suspect (in Western Europe, at least) for a number of reasons: They were viewed by many as a luxury, a frivolity, and a diversion from proper objects of care, such as children or spouses. Perhaps only from the late 1700s, in places like Britain, did caring for pets not only become acceptable but even something that marked you as a good person. At the same time, pets may also have become easier to define as societies became more modern. To take Britain as an example again, for a long time the word "pet" was applied to people as well as animals: Children, for instance, could be "petted" and indulged, as could slaves and servants. It is only much more recently that "pet" became a term more or less exclusively used for animals. Furthermore, it became increasingly important to distinguish between "useful" and "useless" animals (usefulness has been associated with "working" dogs, excluding the modern pet kept for pleasure and companionship alone), to see pets as property (pets, unlike children, say, are defined through being "owned"), and to contrast the pet with the stray animal (the proper pet has a proper home, unlike the "homeless" street dog, for example). All these characteristics help make up the modern "pet."

As noted, however, even if pet keeping has become more widespread and more normal, many controversies over keeping pets remain. Today, pedigree pets and designer breeds are widely known for health problems. The modern craze for pets is responsible for the intensive volume breeding of animals—animals that may be kept in the most inappropriate of conditions: physically and psychologically maltreated, natural behaviors cruelly curbed or denied, and irresponsibly dumped or destroyed when no longer wanted or considered too costly or inconvenient to keep. Exotic pets, popular as an alternative to the more common animals, are captured from the wild and globally traded. Worst of all, perhaps, the attention lavished on pets has sometimes been contrasted with the fate of the other animals that we value only for purposes of food. Some animal activists (though they are in the minority) take the position that pet keeping is unacceptable, excepting only caring for shelter animals that would otherwise be euthanized.

So what, ultimately, does "pet love" mean? Why are we so attached to our pets? Some scholars have suggested that pet keeping allows human beings to connect with the natural world that is otherwise distant. Others see pets as compensations or substitutes for human companionship. Others still think of pets as subsidiary humans whom we dominate even when we demonstrate affection toward them. More positively, though, many have argued that there is something special and indeed worthwhile in our relationship with domestic animals, in our ability to communicate with them, and vice versa. Often enough the term "companion animal" or "companion species" is preferred, but whatever words we use, many scholars have argued that because pets are physically and symbolically close to us, they may even help to make us "human."

Philip Howell

See also: Animal Geography; Breed Specific Legislation; Cats; Cruelty; Designer Breeds; Dogs; Domestication; Emotions, Animal; Empathy; Exotic Pets; Human-Animal Bond; Puppy Mills; Shelters and Sanctuaries; Spay and Neuter; Welfare

Further Reading

Fudge, E. 2008. *Pets*. Stocksfield, UK: Acumen.

Haraway, D. 2003. *The Companion Species Manifesto: Dogs, People, and Significant Otherness*. Chicago: Prickly Paradigm Press.

Shell, M. 1986. "The Family Pet." *Representations* 15: 121–153.

Tuan, Y-F. 1984. *Dominance and Affection: The Making of Pets*. New Haven, CT: Yale University Press.

Pit Bulls

In Western society, especially the United States, we are used to thinking of dogs as a human's best friend. Pit bulls, however, are a group of dogs who are often perceived very differently. Their association with dogfighting, their physical appearance, and their links to criminal groups have combined to give these dogs a bad reputation. Pit bulls are an important case study for Human-Animal Studies because they are simultaneously part of the category of pet dog, yet they are also extremely feared and legally banned in many areas.

While many people think there is just one pit bull breed, they are actually a group of dogs—like retrievers or spaniels. The three most common are the American Staffordshire terrier, the American pit bull terrier, and the Staffordshire bull terrier. Due to variations in body size, however, it is often difficult to tell pit bulls from each other and even from other breeds like boxers and English bulldogs. General characteristics include a muscular body on a compact frame, athleticism, a large/square head, short hair, easy-going disposition, and loyalty to humans. Contrary to popular conceptions, pit bulls' jaws do not lock; they are simply bred for, and equipped with, strong jaws, allowing them to bite and hold.

It is generally accepted that the pit bull originated from the mastiff, a large working/herding dog. Mastiff-like dogs were used in England during the Middle Ages for baiting bulls (and later bears) in pits. The aim of these "fights" was not to kill, but to see which animal could last longest—the bull won if it could not be "pinned" (made immobile for a short time), and the dog won if it could "pin" the bull by latching onto its nose. Breeding new dogs, called bulldogs or pit bulls, to have more strength in the front of their bodies enabled them to latch on better and to help them not break their backs when shaken by the bull/bear. Baiting and fighting animals was outlawed in England in 1835, but many people went underground and continued the practice by fighting dogs with each other.

English immigrants brought these dogs, and fighting, to the United States in the early 1800s. Up until the early part of the 20th century, dog fights were a common event entire families would attend. Despite their use in fighting, their loyalty to humans made them the quintessential American dog. They were used to advertise patriotism during WWI (1914–1918), and a pit bull was the sidekick to the children in the *Little Rascals* (a comedy movie series about children that ran from 1922–1944). Sergeant Stubby was a pit bull and the first dog to be decorated by the U.S. military for his service in WWI. The United Kennel Club was formed in 1898 to register what they recognized as the American pit bull terrier. The American Kennel Club (AKC) officially recognized the American Staffordshire terrier in 1936 but has never recognized the American pit bull terrier because of its links to fighting—which at the time was becoming illegal (and therefore going underground) in many states.

It wasn't until the 1980s that pit bulls gained a reputation for human aggressiveness among the public at large. This was a decade that saw several cases of lethal pit bull attacks on humans (adults and children), the rise of visual music culture where pit bulls became symbols of "street credibility," and the rise of violent drug gangs that not only fought dogs but trained them for protection. A 1987 *Sports Illustrated* magazine cover titled "Beware of this dog," showing a pit bull baring its teeth, exemplified growing public concern. These events led to the passing of breed specific legislation (BSL) banning pit bulls and many related "bully" breeds in communities of all sizes.

Today, dog fighting is a felony in all 50 U.S. states, but, like England, it continues underground in both urban and rural areas. A famous recent case involved the American football quarterback Michael Vick, who was arrested in 2007. Not only was he illegally fighting dogs and gambling on them, but he admitted to killing numerous "inferior" dogs. He was required to pay almost $1 million for the care of the 54 dogs found on his property. Some of those dogs have been adopted out and others, due to injuries or inability to be rehabilitated, were euthanized.

According to the American Temperament Test Society (ATTS), the only breed that ranks higher than pit bulls in good temperament are Labrador retrievers. In fact, this easygoing nature was bred into them because of fighting. Handlers of the fighting dogs needed to be able to enter the pits and separate dogs, so any dog that showed aggression toward humans was removed from fighting and no longer bred. Because of their dispositions they have increasingly been used as therapy dogs.

While pit bulls are commonly thought to bite/attack more than other dogs, studies of breeds and dog bites have not revealed this to be the case (AVMA 2015b). In fact, toy breeds, spaniels, and collies are more prone to bite, but because these breeds are not as large and/or strong, the medical impact of their bites/attacks is much less than dogs like pit bulls. The studies showed a stronger link to how the dogs were raised, the context of the bite/attack (was a child teasing the animal or was the human known/unknown to the dog), and whether the owner/guardian was around.

Julie Urbanik

See also: Breed Specific Legislation; Dogs; Fighting for Human Entertainment; Pets

Further Reading

AVMA. 2015a. "Infographic: Dog Bites by the Numbers." Accessed January 12, 2016. https://www.avma.org/Events/pethealth/Pages/Infographic-Dog-Bites-Numbers.aspx

AVMA. 2015b. "Dog Bite Risk and Prevention: The Role of Breed." Accessed January 12, 2016. https://www.avma.org/KB/Resources/LiteratureReviews/Pages/The-Role-of-Breed-in-Dog-Bite-Risk-and-Prevention.aspx

Colby, J. 1936. *The American Pitbull Terrier.* Sacramento, CA: The News Publishing Company.

Pitbulls.org. n.d. "A Brief History of the American Pit Bull Terrier." Accessed January 12, 2016. http://pitbulls.org/article/brief-history-american-pit-bull-terrier

Stratton, R. 1983. *The World of the American Pit Bull Terrier.* Neptune City, NJ: TFH Publications, Inc.

Poaching

Poaching is a crime that has various overlapping but surprisingly similar definitions. These all relate to the illegal removing, stealing, or extracting of natural resources such as game, wildlife, or fish. Put simply, poaching is the *illegal* taking of wildlife, and this constitutes a large threat to biodiversity and wild animals. In most cases, poaching requires the animals to be killed, for example for their meat, skins, or ivory. A global dissatisfaction with poaching has led to an increase of antipoaching initiatives to save a variety of animals from extinction. In the meantime, however, poaching itself also seems to be on the rise.

Large conservation nongovernmental organizations (NGOs) such as the World Wildlife Fund (WWF) and Conservation International (CI) have a strong voice in defining the global legal frameworks like the Convention on the International Trade in Endangered Species (CITES). Therefore, they also shape what is legal and what is criminal, based on moral judgments and shaped by political dynamics. Currently, most of the conservation NGOs try to overcome poaching at the local level where it happens and not by educating the wealthier consumers who purchase many of these wildlife products. However, in today's world of inequality, human poverty *and* wealth are strong drivers of poaching.

It is important to make a distinction between subsistence and commercial poaching. Since the mid-1800s, when Yosemite National Park was established in the United States, many more wildlife reserves and national parks have been established around the world in such a way that local people's subsistence activities, such as hunting local species for food, have been outlawed. This is, in part, a legacy of European colonialism (1500s–1900s) and a time when nonlocal people were in control and made political and land-use decisions that benefited themselves instead of the cultures already living in their traditional homelands. In fact, some of the world's best-known nature areas that have been set up for the protection of animals and enjoyment by tourists have created new, and problematic, regulations for the local people. For example, the Bushmen of the Central Kalahari Game Reserve in Botswana are being criminalized as poachers, even though they have been hunting for subsistence for centuries. In March 2015, Botswana hosted the second conference of the United for Wildlife initiative, a consortium of some of the world's leading conservation NGOs. The Duke of Cambridge, Prince William, who is an eager sports hunter, supported the event, which was aimed at tackling poaching and illegal wildlife trade. Just before the conference, however, the Botswanan government had enacted a law prohibiting all hunting in the country except for regulated, legal trophy hunting for wealthy tourists. Despite attempts by the Bushmen to ask assistance from the Duke, United for Wildlife continues its collaboration with the government of Botswana that outlaws the Bushmen's subsistence hunting and transforms it into poaching (SI 2015).

In contrast to subsistence poaching, commercial poaching is the type that is generally connected with larger criminal syndicates. For example, it also takes place in our oceans where illegal, unregulated, and unreported fishing is still a regular event in Europe. And in the Galápagos Marine Reserve, commercial, illegal shark fishing has led to a big decline of shark populations to meet the growing international demand for shark products. Among the best-known animals that are targeted for commercial poaching is the rhino. In South Africa's Kruger National Park, the poaching of rhinos has increased since 2008, after many years in which the population had been stable. Most of the demand comes from Asia, especially China and Vietnam, where demand is increasing based on the assumed healing properties of the horn (including the idea that it can cure cancer) and its status as a luxury item; some people simply put the horn on display in their houses. This demand leads to commercial poaching, and therefore the location and act of poaching itself are often only a small part in a global economy in which the poached animals are valuable. People all over the world, especially in wealthier countries, often purchase products such as medicines, clothes, food (e.g., cockles [a particular type of edible saltwater clam] or caviar) and jewelry that contain parts of wild animal bodies that have been obtained through poaching and illegal wildlife trading.

Various antipoaching strategies have so far been tried. For example, in South Africa there have been discussions about the legalization of ivory and rhino horn. Even the public has become more involved; tourists can now use a smartphone app (Wildlife Witness), developed by the Trade Records and Analysis of Flora and Fauna in Commerce (TRAFFIC), to report any suspicious activities when traveling in Southeast Asia. But today the most dominant antipoaching strategy is militarization: Governments, NGOs, and private security companies, often in cooperation, search for poachers using military technologies and equipment (e.g., drones, weaponry, and recently also the introduction of tracker and sniffing dogs). This has often created small war zones in wildlife areas, and the increase of violence has not only led to an increase in poached animals but also to the killing of many poachers and park rangers who work in antipoaching units.

Stasja Koot

See also: Black Market Animal Trade; Ecotourism; Ivory Trade; Trophy Hunting; Wildlife

Further Reading

Carr, L. A., Stier, A. C., Fietz, K., Montero, I., Gallagher, A. J., and Bruno, J. F. 2013. "Illegal Shark Fishing in the Galápagos Marine Reserve." *Marine Policy* 39: 317–321.

Duffy, R. 2010. *Nature Crime: How We're Getting Conservation Wrong*. New Haven, CT: Yale University Press.

Elof, C., and Lemieux, A. M. 2014. "Rhino Poaching in Kruger National Park, South Africa: Aligning Analysis, Technology and Prevention." In Lemieux, A. M., ed., *Situational Prevention of Poaching*, 18–43. London: Routledge.

Lemieux, A. M. 2014. "Introduction." In Lemieux, A. M., ed., *Situational Prevention of Poaching*, 1–17. London: Routledge.

Lunstrum, E. 2014. "Green Militarization: Anti-Poaching Efforts and the Spatial Contours of Kruger National Park." *Annals of the Association of American Geographers* 104(4): 816–832.

Stanciu, S., and Feher, A. 2010. "Combating Illegal Fishing." *Scientific Papers: Animal Science and Biotechnologies* 43(2): 56–60.

Survival International. n.d. Accessed May 17, 2015. http://www.survivalinternational.org

White, N. 2014. "The 'White Gold of Jihad': Violence, Legitimisation and Contestation in Anti-Poaching Strategies." *Journal of Political Ecology* 21: 452–474.

Polar Bears

Native to the oceans, seas, and land masses found within or near the Arctic Circle, the polar bear is perhaps the most emblematic animal of the north polar region. Named *Ursus maritimus*, or "maritime bear," these bears are generally classified as a marine species because adults live primarily on sea ice. Distributed across portions of Russia, Norway, Greenland, the United States, and Canada, these bears face significant threats to their survival from anthropogenic (human-caused) activities, including climate change, habitat loss, pollution, and conflicts with human enterprise in the region.

Polar bears have evolved to occupy a narrower ecological niche than other bear species and are well adapted to surviving Arctic conditions, possessing a translucent, water-repellant coat and an insulating layer of body fat. Capable of swimming hundreds of miles, the polar bear is the only marine mammal adapted to also move efficiently in terrestrial environments. Weighing 775–1,300 lbs (350–590 kg) and measuring from 7–10 ft (2–3 m) in length, polar bears are the largest mammalian predators that spend at least part of their time hunting on land. The only other land

Scientists weigh a polar bear that has been tranquilized. Polar bears are considered a flagship species, or icon, for global climate change. Many conservationists fear that melting Arctic ice will lead to their extinction in the near future. (Corel).

carnivore that rivals the polar bear in size is the Kodiak bear, a subspecies of the brown (grizzly) bear (*Ursus horribilis*) found in southwest Alaska.

While many bear species are omnivores, consuming both plant and animal matter, polar bears are carnivorous, eating only meat. Throughout most of its range, the polar bear's diet is primarily dependent on ringed (*Pusa hispida*) and bearded seals (*Erignathus barbatus*). Polar bears hunt at the edges of ice packs, catching seals as they surface from underwater to breathe. However, polar bears do not depend solely on seals and are opportunists with a diverse prey base. They consume aquatic species like blue mussels and green sea urchin, and scavenge marine mammals including walrus and whales that wash up on shore. Polar bears also prey on terrestrial (land-based) species including muskox, reindeer, caribou, birds and their eggs, and rodents. Although terrestrial species do not possess the large amounts of calorically rich fat that polar bears require in their diet, predation on land-dwelling animals can increase the odds of a bear's survival during the warmer months when the absence of ice prevents them from hunting seals.

Unlike brown bears, male and nonbreeding female polar bears are active year-round and do not hibernate in the winter. Polar bears most frequently utilize the edges of pack ice, as these sites are ideal for hunting seals. Following seasonal changes in sea ice distribution, bears may migrate thousands of miles in search of prey. Adult polar bears are also dependent on pack ice on the continental shelf for mating and raising young.

Polar bears occupy a unique and important place among indigenous societies throughout the Arctic region. Polar bear hunts have long been imbued with spiritual and cultural meaning while fulfilling material needs within these communities. This traditional subsistence, or small-scale, hunting intended to meet the needs of a community did not significantly impact polar bear populations. However, more technologically advanced methods of hunting significantly increased the number of polar bears killed in nonsubsistence trophy hunts, prompting the development of international regulations on polar bear hunting by the mid-20th century.

In 1973, the International Agreement on the Conservation of Polar Bears, signed by the five countries inhabited by polar bears, mandated cooperation on research and conservation efforts throughout the polar bears' range. Signatories agreed to restrict commercial and recreational hunting while still allowing for subsistence hunting by local groups utilizing traditional methods and technologies. In the United States, polar bears have been protected from hunting since 1972 by the Marine Mammal Protection Act. In 2008, the U.S. Fish and Wildlife Service listed the polar bear as a threatened species under the Endangered Species Act, citing ongoing concerns regarding the species' population status and viability due to loss of sea ice habitat.

The World Conservation Union (IUCN) estimates the total polar bear population worldwide to be 20,000–25,000 individuals (Schliebe et al. 2008). These population estimates have increased over the past 50 years, a positive change attributed to hunting controls. However, the IUCN classifies the polar bear as a vulnerable species, identifying 8 of the 19 polar bear subpopulations as being in decline. Polar bears face significant threats to their survival, including climate change, habitat loss, pollution, and conflicts with human economic enterprise in the region, such as shipping and oil and gas development. The IUCN, U.S. Geological Survey, and many polar bear specialists identify climate change, specifically warming trends and subsequent loss of ice cover, to be the most significant threat to polar bears, impacting their hunting strategies, migration, reproduction, and survival. The outlook for these bears is grim; diminishing summer sea ice due to global climate change is forecasted to have sustained and significant impact on the world's polar bear population, resulting in the loss of over 65 percent by 2050 (USGS 2007).

Sharon Wilcox

See also: Climate Change; Endangered Species; Extinction

Further Reading

Atwood, T. C., Marcot, B. G., Douglas, D. C., Amstrup, S. C., Rode, K. D., Durner, G. M., and Bromaghin, J. F. 2015. *Evaluating and Ranking Threats to the Long-Term Persistence of Polar Bears*. Accessed September 1, 2015. http://dx.doi.org/10.3133/ofr20141254

Derocher, A. E. 2012. *Polar Bears: A Complete Guide to Their Biology and Behavior.* Baltimore, MD: Johns Hopkins University Press.

Norwegian Polar Institute. 2015. "Polar Bear Survey 2015." Accessed September 1, 2015. http://www.npolar.no/en

Schliebe, S., Wiig, Ø., Derocher, A., and Lunn, N. 2008. "*Ursus maritimus*: The IUCN Red List of Threatened Species." Accessed September 1, 2015. http://www.iucnredlist.org

Stirling, I. 2011. *Polar Bears: A Natural History of a Threatened Species*. Markham, Ontario: Fitzhenry & Whiteside.

U.S. Department of the Interior, U.S. Geological Survey. 2009. "Polar Bear Finding Webpage." Accessed September 1, 2015. http://www.usgs.gov/newsroom/special/polar_bears/

U.S. Fish and Wildlife Service, Marine Mammals Management. 2008. "Final Rule Listing the Polar Bear as a Threatened Species Under the Endangered Species Act." Accessed September 1, 2015. http://www.fws.gov/alaska/fisheries/mmm/polarbear/esa.htm

Popular Media, Animals in

Animals are used as symbols and species representatives in many forms of media and popular culture. There are television programs, documentaries, and films about them, advertising that employs them to stand in for human emotions and activities, as well as real dogs and cats who are used to advertise products pets use, such as food and toys. They are available as playthings such as stuffed toys and appear in cartoons, comic strips, on greeting cards, travel brochures, in political propaganda, animated films, computer-generated programs, and even as the names of software, such as Apple computer operating systems (Snow Leopard™, Cheetah™, and Panther™). With origins in special cultural traditions, the representational use of animals makes sense to a wide range of people, even if the animal has no clear connection to the story told or product sold.

The first animals to appear in symbolic form were those that ancient humans pecked into rock art, painted onto cave walls, and carved into tools and decorative objects. Drawings and paintings of animals were later made on ancient woodblock prints, papyrus, etchings, paintings, and as jewelry.

The spoken word is another early form of the inclusion of animals in global myth, folklore, and legend. Western culture is full of stories that employ animals for metaphorical purposes, such as Aesop's fables, which were originally spoken word

Mickey Mouse greets visitors to Disney World during its opening year, 1971. Disney is famous for personifying animals throughout the company's various forms of media, and many Disney characters are known worldwide. (Photofest)

stories later written down and eventually published in books. Anthropomorphized animals (those given human emotions, appearances, and behaviors) are found throughout the world. In American and European culture, the stork is said to deliver babies, the raven is a harbinger of both good and evil, and doves signify peace and hope. Many cultures employ animals to teach children moral lessons, and mythology helps explain difficult questions about human and animal origins, relationships, and appropriate behavior. In Mali, the Bambara people tell a tale of an antelope who taught them agriculture. In Eastern faiths and religions such as Buddhism, the lion protects entrances to temples. Hindu mythology is replete with magical monkeys such as Hanuman.

How an animal figures in myth is typically how he or she is reproduced in media. The development of portable media and the printing press made animal representations available throughout the world, including natural histories of animals, children's storybooks, newspapers, and magazines. Electronic forms of communication such as movies are intimately connected to animals. The first animal to appear in motion pictures was a galloping horse named Occident, captured in Eadweard Muybridge's (1830–1904) 1877 American film *The Horse in Motion.* Beginning with *Seal Island* (1948), Walt Disney (1901–1966) created a formula for the production of wildlife documentaries, turning wildlife into personalities. In fact, the Disney formula (giving animals identifiable personalities, emphasizing life struggles, situating stories in family life, employing stop-motion and time-lapse photography techniques, and narrating through music) set the standard to which most animal-related documentaries still subscribe today. Disney's global impact in popular culture includes theme parks with animal adventures, characters (e.g., Donald Duck, Mickey Mouse, and Pluto [Mickey's dog]), and blockbuster movies such as *Bambi* (1942), *The Jungle Book* (1967), *The Lion King* (1994), and *Finding Nemo* (2003).

Radio programs featured animals and animal sound effects from its earliest days. For example, barking dogs serve as warnings of intruders or welcoming family members. Television programs, particularly those that target children, typically feature animals. The U.S. television program *Lassie* (1954–1974), about the relationship between a young boy (Timmy) and his faithful canine companion, was an early entry in bringing animals into the home, as was *Mr. Ed* (1958–1966), about a talking horse. In the United States, the popularity of nature-based television programs grew from weekly programs such as *Wild Kingdom* (1963–1988; 2002–present) and occasional specials from National Geographic to, by the 1990s, cable channels, such as Animal Planet, dedicated specifically to animals. Today, the National Geographic, Discovery, and Disney channels export animal-related television programming around the world. In the United Kingdom, the BBC similarly created and distributed nature documentaries, beginning with *Wildlife in Australia* (1954), naturalist Sir David Attenborough's (1926–) nature series from the 1960s onward, and topic-specific films, such as *Animal Super Parents* (2015). Television programs include *Wildlife on One* (1977–2005), *Birding with Bill Oddie* (2005–present), *Big Cat Diary* (1996–2008), and *Deadly 60* (2009–present).

Animals are widely used as brand images and in advertising as tools for drawing attention, adding humor, or conveying complex cultural understandings. For example, Kellogg's cereal brand Frosted Flakes® has, since 1951, used Tony the Tiger as a spokes-animal for the product. Old English sheepdogs, known as the Dulux dogs, are familiar for advertising this paint brand, and United Biscuits employs adorable puppies to build affection toward the brand.

The Internet hosts a wide variety of animals and animal-related issues. Due to the speed and global delivery, it has been an effective tool for distribution of animal advocacy information such as undercover videos of cows being beaten after falling on the slaughterhouse floor and the conditions inside mass production poultry buildings. While the Web is an important source for factual information about animals, the most popular content is sentimental and funny animal videos that also find their way to Facebook and other social media sites. Grumpy Cat (the blue-eyed cat with an underbite), the toothless and tiny cat Lil Bub, Boo the celebrity loved

Pomeranian, and lions returned to the wild reunited with former human caretakers are popular fare. The most popular Internet animals are cats, whose videos on YouTube receive more views than any other category.

Debra Merskin

See also: Advertising, Animals in; Animals; Anthropomorphism

Further Reading

Baker, S. 2001. *Picturing the Beast: Animals, Identity, and Representation.* Chicago: University of Illinois Press.

King, M. J. 1996. "The Audience in the Wilderness." *The Journal of Popular Film & Television,* 24(2): 60–69.

Phillips, B. J. 1996. "Advertising and the Cultural Meaning of Animals." *Advances in Consumer Research* 23: 354–360.

Wells, P. 2009. *The Animated Bestiary: Animals, Cartoons, and Culture.* New Brunswick, NY: Rutgers.

Puppy Mills

What have come to be known as "puppy mills" are essentially intensive breeding facilities that arose after World War II (1939–1945). Prior to the advent of puppy mills, dogs were bred in more traditional social structures with closer relationships to other dogs and humans. Puppy mills are often seen as the largely hidden and problematic side of humans' increasing love for pets, and dogs in particular.

Today puppy mills are known to exist in Australia, East Asia, Europe, and North America. In the United States, puppy mill operations are concentrated in Missouri and Minnesota and other parts of the Midwest. There is also a high concentration of puppy mills among Amish and Mennonite (two religious sects) breeders, particularly in Pennsylvania. In Asia, South Korea specializes in so-called "teacup" dogs, breeding for a genetic variant of dwarfism so that they will remain tiny into adulthood, while in eastern European countries such as Poland, breeders frequently concentrate upon established small dog breeds such as dachshunds and cocker spaniels.

The conditions under which dogs are bred and raised in puppy mills, which sometimes contain over 1,000 dogs, typically involve overcrowding; unsanitary conditions; the prevalence of certain congenital (inherited) illness; restrictions upon movement and access to clean air, water, and light; lack of socialization; and a disruption of traditional puppy raising by humans and other dogs, resulting in behavioral issues. After the puppies are born and weaned (taken off their mother's milk), they are then purchased by brokers such as the Hunte Corporation (one of the largest) at auctions or directly from breeders. From there they are transported, frequently in large tractor-trailer trucks, to pet stores and put up for sale.

Puppy mills in the United States are regulated by the U.S. Department of Agriculture's Animal Plant Health Inspection Service (USDA APHIS), under rules established through the Animal Welfare Act (AWA) and by such state welfare or anti-cruelty laws as may apply in the jurisdiction where the puppy mill is located. Any breeder possessing five or more breeding females who makes sales of $500 or more per year is required to obtain a USDA Class B breeder's license, which entails periodic inspections. Although puppy mills are inspected, APHIS agents do not report state law cruelty violations and have been criticized for overlooking violations and offering violation avoidance counseling instead. These concerns are detailed in a 2010 report from the USDA Inspector General titled "Animal and Plant Health Inspection Service Animal Care Program Inspections of Problematic Dealers." The report was drafted with input from the Companion Animal Protection Society and the Humane Society of the United States, and notes many weaknesses in the regulatory structure.

Many U.S. municipalities, such as Los Angeles, San Diego, New York City, Sarasota (Florida), and Chicago, have enacted "Retail Pet Store Ordinances" which ban the sales of dogs obtained from puppy mills, in response to growing public awareness of puppy mill conditions. The New York City Retail Pet Store Ordinance is unique because it contains the world's first ban on puppies sourced from brokers, such as the Hunte Corporation. Often these ordinances are challenged in the courts based upon claims that the term "puppy mill" is not defined by the USDA or in American legal texts, or that such ordinances are unconstitutional. However, in *Avenson v. Zegart* (577 F. Supp. 958 (D. Minnesota 1984)), a U.S. District Court in Minnesota defined "puppy mill" as "a dog-breeding operation in which the health of the dogs is disregarded in order to maintain low overhead and maximize profits." More recently, the district court in Kansas in *Martinelli v. Petland, Inc.* (No. 10-407-RDR (D. Kan. Oct. 7, 2010)), citing *Avenson*, applied the definition "a dog breeding operation in which the health of the dogs is disregarded in order to maintain a low overhead and maximize profits." All of the foregoing point to a growing acceptance of a generally agreed-upon meaning of "puppy mill" within the law.

At the consumer level, puppies offered for sale in retail pet stores are almost always obtained from puppy mills and are usually misrepresented to consumers as originating from "local breeders" or "hobby breeders." In many cases consumers are told that puppy mills do not exist because of USDA regulation. These statements are untrue and have led to state consumer fraud statute enforcement actions as well as private lawsuits based upon misrepresentations. The consumer will often not comprehend the industrial puppy mill component of production in a purchase based upon impulse and emotion, or will disregard such knowledge in favor of instant gratification from a consumer purchase. Until the consumer demand for puppies matches a sense of human-animal social responsibility for the conditions of production, puppy mills will continue to exist.

John T. Maher

See also: Animal Law; Designer Breeds; Dogs; Pets; Welfare

Further Reading

Fumarola, A. J. 1998. "With Best Friends Like Us Who Needs Enemies—The Phenomenon of the Puppy Mill, the Failure of Legal Regimes to Manage It, and the Positive Prospects of Animal Rights." *Buffalo Environmental Law Journal* 6: 253.

Hinds, M. 1993. "Amish at Heart of 'Puppy Mill' Debate." *New York Times*. September 20. Accessed January 30, 2016. http://www.nytimes.com/1993/09/20/us/amish-at-heart-of-puppy-mill-debate.html?pagewanted=all

McMillan, F. D., Duffy, D. L., and Serpell, J. A. 2011. "Mental Health of Dogs Formerly Used as Breeding Stock in Commercial Breeding Establishments." *Applied Animal Behaviour Science* 135(1): 86–94.

USDA. 2010. *Animal and Plant Health Inspection Service Animal Care Program: Inspections of Problematic Dealers*. Accessed January 30, 2016. http://www.usda.gov/oig/webdocs/33002-4-SF.pdf

R

Race and Animals

The social realities of race and human-animal relations are complex. While the two phenomena have distinct histories, race and human-animal relations have often been connected. Some racial groups have been demonized as animalistic or less than human. Other groups have defined their racial identity based on how they treat animals. And some animals have been treated negatively because of their association with certain racial groups. While race and human-animal relations are often debated as separate issues, it is important to understand how they intersect with one another.

Race, while often defined in relation to biological features, is not biologically fixed. Rather, what it means to be "white" or "Asian" changes over time and is determined by social processes, such as historical lineages of culture and kinship, one's experiences of racism or racial privilege, or dominant ideas about skin color in one's society. Racial categories, however, are often seen as "natural" and unchanging. This obscures the unequal social relations that produce racial difference in the first place, as well as the racist inequalities that persist among different racial groups. For this reason, race is a social construction, but it also has very real consequences.

"Animal" *is* a term based in biology, however, and refers to various organisms grouped within the animal kingdom. Though humans are also within the animal kingdom and humans share many characteristics with nonhuman animals—mortality, for instance—humans are also often thought to be distinct from other animals and higher up in a hierarchy of living beings. For this reason, "animal" also has an everyday meaning of referring to that which is the opposite of human. This meaning of animal as less-than-human can be attached to nonhuman animals. For instance, a person might insult an ape as being "just a stupid beast." But this meaning can also be applied to other humans. For instance, someone might say "that criminal is an animal."

Once one understands that both what it means to be of a certain racial category *and* what it means to be an "animal" (in the everyday sense) are determined by social processes that change over time and space, we can begin to understand how race and human-animal relations become connected.

One way this happens is when a human social group is racialized—meaning its racial identity is defined—based on how that group's relationships with animals differ from the larger social norm. For instance, in 1990s Miami, Caribbean immigrants were portrayed in the popular press as "backward"—a euphemism for calling a group racially inferior—because they practiced Santeria, an Afro-Caribbean religion that can involve the ritual slaughter of dogs and other animals (Wolch and Emel 1998). Similarly, in post-Apartheid South Africa, suburban whites who resented the integration of black South Africans into their previously all-white neighborhoods accused blacks of being "primitive" and "uncivilized"—also euphemisms for racial inferiority—because of their religious slaughter of cows (Ballard 2010). The white South Africans defined their whiteness based on an imagination that modern farming techniques were more humane. In both examples, racial superiority was defined according to how a certain social group related to nonhuman animals.

While these examples demonstrate how racial status becomes defined based on a group's relationships with animals, in other instances racism operates by systematically placing racially oppressed groups in positions that require certain human-animal relationships over others. For instance, following the ratification of the North American Free Trade Agreement (NAFTA), much of the pig slaughter industry relocated from the United States to Mexico, where poorer Mexicans were paid low wages to work in unsafe conditions of animal slaughter. This division of labor perpetuated a racial hierarchy, in which non-white workers in Mexico were forced into human-animal relations of violence while the majority-white executive boards of U.S. animal slaughter companies—and American consumers—avoided such violent interactions.

Racial inferiority can also be created through the animalization of certain social groups. "Animalization" here means the assigning of certain characteristics associated with "inferior" animals to certain human groups or the more straightforward use of animal names as racial slurs. For instance, there exists a long history within the United States of denigrating people of color as animal or beastly, and specifically associating black people with "dumb" and "savage" apes. Jackie Robinson, the first African American player to desegregate Major League Baseball, was often taunted with monkey gestures. Barack and Michelle Obama, the first black U.S. president and First Lady, have been compared with monkeys. For instance, Univision TV host Rodner Figueroa was fired after saying Michelle Obama should have been cast in the film *Planet of the Apes*. This racist animalization of people of color has a long history based in the European colonization of African peoples and has traveled beyond just the United States. For instance, at a 2014 soccer game in Spain, black Brazilian player Dani Alves had a banana thrown at him. This racist association has even extended to the depiction of animals themselves. Great apes such as gorillas and chimpanzees have been racialized as black and threatening, as in the Hollywood film *King Kong*, wherein a monstrous African gorilla kidnaps a white woman and terrorizes "civilized" New York City. Animals can also become associated with racist imaginations in more roundabout ways. For instance, in the United States, the breed of dog known as a pit bull has often been portrayed as a vicious fighting dog associated with racist stereotypes about black violence and urban black communities.

William L. McKeithen

See also: Animals; Great Apes; Pit Bulls; Social Construction

Further Reading

Ballard, R. 2010. "'Slaughter in the Suburbs': Livestock Slaughter and Race in Post-Apartheid Cities." *Ethnic and Racial Studies* 33(6): 1069–1087.

Delgado, R., and Stefancic, J. 2001. *Critical Race Theory: An Introduction.* New York: New York University Press.

Frayer, L. 2014. "Spain Fines Team of Racist, Banana-Throwing Fan, but Is It Enough?" *National Public Radio.* Accessed March 18, 2015. http://www.npr.org/sections/codeswitch/2014/05/09/310990212/spain-fines-team-of-racist-banana-throwing-fan-but-is-it-enough

Hutchinson, E. O. "Nothing New in the Ape Crack about Michelle Obama." Accessed March 18, 2015. *The Huffington Post.* http://www.huffingtonpost.com/earl-ofari-hutchinson/nothing-new-in-the-ape-crack-about-michelle-obama_b_6869650.html

Tarver, E. C. 2013. "The Dangerous Individual('s) Dog: Race, Criminality and the Pit Bull." *Culture, Theory and Critique* 2: 1–13.

Wolch, J. R, and Emel, J. 1998. *Animal Geographies: Place, Politics, and Identity in the Nature-Culture Borderlands.* London and New York: Verso.

Research and Experimentation

For most scientists, the gold standard for making scientific breakthroughs is research and experimentation performed on animals. Many medications and devices in use today were tested on animals to ensure their safety and efficacy, but as mounting evidence supports that animals feel physical pain and experience emotions, animal welfare concerns are increasing and support for alternatives is rising. Regardless of personal beliefs, one cannot understand the relationship between humans and animals without acknowledging the use of animals by humans in ways that are controversial. Animal research and experimentation falls squarely under this domain.

"Animal research" is often used interchangeably with the terms "animal experimentation," "animal testing," and "in vivo testing." However, there are clear differences between these terms.

Animal research is the umbrella term that covers the use of animals in studies looking to discover new information that is intended to benefit society; it may be done in the field or the lab, and may be either observational or invasive. For example, watching how gorillas socialize in the wild is observational fieldwork, whereas removing different areas of a rat's brain to see how this affects the rat is invasive laboratory work. Thus, the term animal research is generic and may be applied to

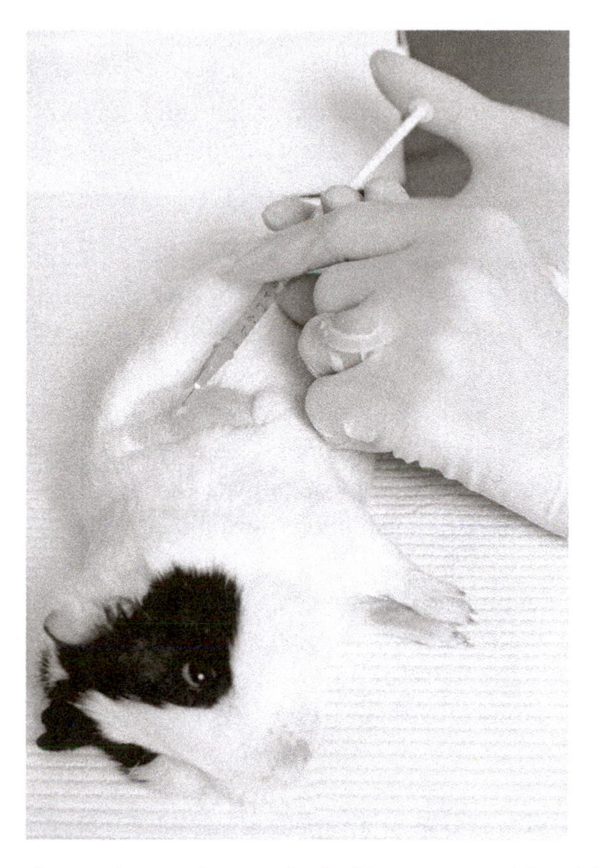

Animal research and experimentation can be highly controversial and both proponents and opponents of this way of using animals have ample evidence to show when animals have helped advance scientific understanding and when they have not. (National Cancer Institute)

a wide variety of animal uses for the purpose of expanding knowledge in scientific and sociological senses.

Animal experimentation and animal testing are subsets of animal research. Animal experimentation is often used in the early phases of research to investigate a new hypothesis. Animal testing is often used in the later stages of research, when a new product has been created and needs to be used to show it accomplishes its intended outcome.

Animal experimentation may be done to advance human health but is also frequently performed to satisfy curiosity, with no apparent benefit to society, and, along with animal testing, may inflict pain and suffering. Thus, both terms often carry a more negative connotation and are usually used when studies are considered controversial.

"In vivo" is Latin for "within the living" and refers to testing on a living and fully functioning organism (plant, animal, or human). Scientifically, this refers to the testing of such things as medications (e.g., Prozac), medical devices (e.g., pacemakers), or novel therapies (e.g., stem cell therapy). "Vivisection" is the cutting open of a living being while it still alive; specifically, it refers to procedures performed without any pain medications or anesthesia. "Dissection" is the cutting open and/or dismemberment of a living being after its death or while it is anesthetized to study its anatomy and/or physiology.

Animal research and experimentation are documented from the fourth century BCE, when Aristotle (384–322 BCE) recorded his dissection and vivisection of animals. As the Europeans' desire to expand knowledge grew, animal experimentation increased during the 1600s. The famous mathematician and philosopher René Descartes (1596–1650) absolved humans of concern with animals as sentient beings with his theory that "animals are machines" (Descartes 1649) and therefore did not experience pain. This Cartesian view of animals was widely accepted and used to support vivisection at medical institutions during the 1700s:

> [The students] administered beatings to dogs with perfect indifference . . . [t]hey said that the animals were clocks; that the cries they emitted when struck, were only the noise of a little spring which has been touched, but that the whole body was without feeling. (Fontaine 1738, as cited in Fudge 2006)

By the 1800s, the exploration of a variety of medical conditions led to unabated animal use. In 1876, Britain responded to an outcry from citizens opposed to vivisection by passing the controversial Cruelty to Animals Act, becoming the first nation to specifically regulate animal experimentation.

Many European countries followed Britain's lead, but the United States did not pass its first federal law addressing animal experimentation until the Animal Welfare Act (AWA) of 1966. Although Congress had considered previous animal welfare legislation, they did not pass the law until two separate articles published in reputable magazines highlighted the abusive handling of dogs; it was the subsequent public demand for regulation that forced Congress to finally act.

The original AWA set minimum standards for the handling, sale, and transport of live dogs, cats, monkeys and apes, guinea pigs, hamsters, and rabbits held by animal dealers or pre-research situations in laboratories; it also specifically addressed the use of dogs and cats in dealer and laboratory settings. It did not regulate animal treatment during the actual research and excluded all animals not specifically mentioned above. Subsequent amendments to the AWA addressed a variety of perceived shortcomings in the original Act. Perhaps the most pertinent here is the 2002 amendment, which updated the definition of animal to align with current regulations but still excluding from protection birds, rats, and mice bred for use in research.

There is ongoing debate about the value of animal research. Some believe it has been invaluable in helping move human health forward, while others believe the vast majority of animal research has resulted in abject failure. The debate will likely continue for many years to come, as both sides have data to support their positions.

Katherine Fogelberg

See also: Human-Animal Bond; Institutional Animal Care and Use Committees (IACUCs); Vivisection

Further Reading

Adams, B., and Larsen, J. 2015. "Legislative History of the Animal Welfare Act: Introduction." Accessed March 5, 2015. http://awic.nal.usda.gov/legislative-history-animal-welfare-act/intro

Akhtar, A. 2012. *Animals And Public Health: Why Treating Animals Better Is Critical to Human Welfare.* New York: Palgrave Macmillan.

Descartes, R. 2011. "Animals Are Machines." *Journal of Cosmology*, 14: np. Reprinted from *Passions of the Soul* (1649).

Fudge, E. 2006. *Brutal Reasoning: Animals, Rationality, and Humanity in Early Modern England.* New York: Cornell University Press.

National Institutes of Health. 2010. "Animal Welfare and Scientific Research: 1985 to 2010." Accessed April 11, 2014. http://grants.nih.gov/grants/olaw/seminar/docs/Booklet_AWSR.pdf

Roberts, I., Kwan, I., Evans, P., Haig, S. 2002. "Does Animal Experimentation Inform Human Healthcare? Observations from a Systematic Review of International Animal Experiments on Fluid Resuscitation." *BMJ* 324: 47–476.

Rights

The concept of "rights" has its roots in Western philosophical traditions. The application of this concept to nonhuman animals is a relatively recent phenomenon, however. Distinctions can be drawn between the concept of moral and legal rights and between animal rights and welfare. Many if not most animal rights advocates consider freedom from human-caused death, pain, suffering, and captivity to be basic rights for animals.

Approaches to ethics that are based on the idea of intrinsic rights—or rights that exist within an individual—are "non-consequentialist," meaning they do not rely on an evaluation of the consequences of a particular action to determine whether it is morally acceptable. The German philosopher Immanuel Kant (1724–1804) wrote extensively on rights, influencing modern views on justice and morality in Western societies and globally, for example in the *Universal Declaration of Human Rights*, adopted by the United Nations in 1948, and in beliefs about the basic rights of individual humans, for example the right to "Life, Liberty, and the pursuit of Happiness" stated in the U.S. *Declaration of Independence*. With limited exceptions, a non-consequentialist approach respecting an individual's basic human rights trumps any benefits that might be gained by others by violating those rights.

The U.S. philosopher Tom Regan (1938–) wrote a detailed argument for non-consequentialist rights for animals in *The Case for Animal Rights* (1983). Using evidence from evolutionary biology and animal behavior to show similarities between humans and other animals, Regan asserts that, like humans, certain animals maintain a mental and physical identity over time; consciously experience both good and bad events during their lives; have biological, social, and psychological interests; and exhibit what is known as preference autonomy, or the ". . . ability to act in pursuit of their goals" (p. 116). Regan calls individuals with these qualities "subjects-of-a-life" (p. 264). All such human and animal subjects benefit from the ability to successfully pursue their interests/goals, which may be as simple as the avoidance of pain or hunger. They are also harmed by elimination of/reductions in this ability to pursue their interests. Regan argues that it does not matter that the humans' and other species' interests may differ in complexity. Each individual is seeking to satisfy interests relative to his/her individual capacities and needs.

Regan further asserts that individuals of different species equally have inherent value (or value in and of themselves and not based on their usefulness to others) and therefore have a valid claim to respectful treatment. This valid claim is what is known as a *right* and exists for all individuals with inherent value. This right to respectful treatment is, according to Regan, a *basic moral right*, whether that animal be human or not. The claim can only be made against humans, however, because most humans consciously understand when they are causing harm and typically have the capacity to choose whether or not to do so. Therefore, animals do not have rights in their relations with each other. For example, a zebra attacked by a lion experiences great harm, but the zebra does not have a claim to respectful treatment from the lion because lions are not thought to understand the harm they cause or the concepts of rights and morality.

The concept of sentience—the ability to have conscious awareness of one's experiences—is key to Regan's argument and for many animal rights advocates. A human or nonhuman animal has to be conscious of good and bad experiences for rights to matter. Also key are relevant interests, meaning that different individuals/species will have different life interests. For example, learning to read would not be an interest for a squirrel but it is an important interest for a human and critical to a good life.

Moral rights can be contrasted with legal rights, which vary considerably worldwide. Although many people may strongly believe in or support a particular moral right (e.g., that chimpanzees have a right to not be used in medical experiments), such rights are not enforceable unless backed by law. Although there are anti-cruelty and welfare laws that govern animals' treatment, a major concern for rights advocates is that animals are, with limited exceptions, legally considered to be property around the world, which means their inherent rights as "subjects-of-a-life" are not being taken into account. Anti-cruelty and welfare laws, while promoting kindness and compassion, do not question animals' status as beings that can be owned and/or used for humans' benefit.

A variety of individuals and organizations are working to get individuals of certain species legally recognized as persons with rights, like humans, as opposed to property. For example, the U.S. attorney Steven Wise (1952–) has worked on behalf of great apes such as chimpanzees, and his not-for-profit *Nonhuman Rights Project's* mission is to gain legal rights for various nonhuman species, such as elephants. Governments such as the Balearic Islands, who in 2007 granted great apes the right to life, liberty, and protection from torture, have also taken action. Organizations such as the Animal Liberation Front (ALF) and People for the Ethical Treatment of Animals (PETA), two of the highest profile animal rights organizations, work (often aggressively) to sway public opinion toward recognizing rights. Although some animal rights advocates have taken the more radical position that humans should not interfere in nonhumans' lives at all (including having pets), many take a more measured approach, focusing on animals' legal status as property; their captivity in zoos, laboratories, and farms; and the harm they endure being used for human benefit.

Connie L. Johnston

See also: Advocacy; Agency; Animal Law; Animal Liberation Front (ALF); Animals; Chimpanzees; Elephants; Emotions, Animal; Ethics; Evolution; Humans; Indigenous Rights; Intelligence; Personhood; People for the Ethical Treatment of Animals (PETA); Sentience; Species; Speciesism; Vivisection; Welfare

Further Reading

DeGrazia, D. 2002. *Animal Rights: A Very Short Introduction.* New York: Oxford University Press.

Feder, K. L., and Park, M. A. 1990. "Animal Rights: An Evolutionary Perspective." *The Humanist* 50(4): 5–7, 44.

Francione, G. L. 1996. *Rain without Thunder: The Ideology of the Animal Rights Movement.* Philadelphia: Temple University Press.

Kant, I. 1976. "Duties to Animals and Spirits." Translated by Louis Infield. In Regan, T. and Singer, P., eds., *Animal Rights and Human Obligations*, 122–123. Englewood Cliffs, NJ: Prentice-Hall.

Nonhuman Rights Project. Accessed October 21, 2015. http://www.nonhumanrightsproject.org

Regan, T. 1983. *The Case for Animal Rights.* Berkeley and Los Angeles: University of California Press.

Wise, S. M. 2000. *Rattling the Cage: Toward Legal Rights for Animals.* New York: Basic Books.

S

Sacred Cow. *See* Eastern Religions, Animals in

Sentience

Sentience is a critical concept in human-animal relations because it is frequently utilized as a basic threshold that determines the level of ethical consideration, and resulting treatment, given to a nonhuman species. Which species are sentient—and whether nonhumans can even be sentient—has been a matter of debate in western science for centuries, although today there is more widespread agreement that many species are sentient. For members of the general public, however, the notion of animal sentience has been less controversial.

Sentience is defined as the cognitive (mental) ability to have subjective awareness, or perception, of one's experiences. Sentience is frequently used synonymously with "consciousness." Here "conscious" is not the same as "awake," but indicates that one knows what is happening to oneself. The concept of sentience goes beyond just physical experiences like pain, however, and includes emotions from the more basic feeling of fear to more complex feelings like happiness. A distinct but related term is "sapience," which includes the additional capacities of being able to evaluate and remember actions and consequences and to assess risk.

For centuries, Western science took the position that humans were the only beings capable of sentience. The roots of these beliefs about nonhuman sentience can largely be traced back to the French philosopher-scientist René Descartes (1596–1650), who did invasive experiments on live, unanesthetized animals such as dogs, saying that they were like inanimate objects and therefore were not conscious of pain. A more recent example is that, for a number of years, many animal and cognitive scientists and biologists argued that fish are not sentient. They claimed, instead, that fish move away from a source of pain (e.g., a sharp object) not because they *consciously* experience the pain, but because the brain causes them to simply *react* to this negative stimulus by moving away to avoid harm (the way you might pull your hand away from something hot before you actually feel pain from the heat). Whether one believes that some or all nonhuman animals are sentient is clearly important—if you believe that an animal, like a chair or a rock, cannot consciously experience something negative like pain, then there are no real ethical constraints on how that animal should be treated. An increasing number of scientists from fields such as animal welfare, ethology (the study of animal behavior), and neurology (the study of the brain and nervous system) now claim that there is no need for irrefutable proof of nonhuman sentience. They base their claims on significant observational evidence, but they further draw on neuroscience and evolutionary biology, both of which show numerous similarities between humans and many other species. These scientists claim that all vertebrates (animals with a spinal column) have neurological pain systems, which indicates that they have the cognitive capacity to consciously experience pain. This means that not only animals such as dogs, horses, and dolphins can feel pain, but that fish can as well. In addition, studies are showing that some invertebrates, like cephalopods (octopus and squid) and spiders, have more sophisticated cognitive abilities than previously thought. However, many nonscientists, based on their everyday interactions, do believe that many other species are sentient. For example, people who have dogs or cats frequently express certainty that their companions are consciously aware of their experiences and show evidence of this through their tail movements, facial expressions, and barks, growls, and meows.

The terms "sentience"/"consciousness" are distinct from "self-consciousness," or the awareness of oneself as an individual separate from other individuals. An animal can be consciously aware of their surroundings and experiences but may not recognize that they are a separate being. Besides humans, we currently have evidence from the "mirror test" that chimpanzees, dolphins, and elephants have self-consciousness. With chimpanzees, for example, a colorful sticker is applied to the forehead while away from a mirror. When the chimpanzees then see themselves with the sticker in the mirror, they have been observed using their hands to remove it. Because they are reacting to their own reflection, this indicates that they recognize themselves in the mirror. However, because we do not know their thoughts, the chimpanzees' actions are *strong indicators*, but not proof, of self-consciousness.

If animals can consciously have negative experiences, then they also have the capacity to suffer. Viewing animals as sentient has a vital impact on societal attitudes, policy, and laws that protect them. In recent official documents, several political entities (e.g., the European Union [2007], France [2015], the Canadian province of Quebec [2015]) have stated that at least some nonhumans deserve a certain level of humane treatment and/or legal protections because they are sentient. A major step forward in the animal science community arose in 2012, with the signing of *The Cambridge Declaration on Consciousness* by an international group of neuroscientists. The Declaration states that, despite limitations in animals' (and humans') abilities to communicate internal feelings/perceptions, the behavioral and neurological evidence that exists overwhelmingly indicates that humans are not alone in their capacity for consciousness and that other sentient animals include "all mammals and birds, and many other creatures, including octopuses" (p. 2).

Connie L. Johnston

See also: Advocacy; Agency; Animal Law; Chimpanzees; Dogs; Dolphins; Elephants; Emotions, Animal; Ethology; Intelligence; Welfare

Further Reading

Bekoff, M. 2016. "The Science of Sentience: An Interview about Animal Feelings." *Psychology Today.* Accessed January 25, 2016. https://www.psychologytoday.com/blog/animal-emotions/201601/the-science-sentience-interview-about-animal-feelings

Broom, D. M. 2014. *Sentience and Animal Welfare*. Wallingford, UK: CABI.

Cambridge Declaration on Consciousness. 2012. Cambridge, UK. Accessed January 25, 2016. http://www.fcmconference.org/img/CambridgeDeclarationOnConsciousness.pdf

Canadian Press. 2015. "Quebec Passes Animal Protection Law." *The Star*, December 4. Accessed January 25, 2016. http://www.thestar.com/news/canada/2015/12/04/quebec-passes-animal-protection-law.html

European Union. 2007. Treaty of Lisbon. Accessed January 25, 2016. http://eur-lex.europa.eu/legal-content/EN/TXT/?uri=uriserv:OJ.C_.2007.306.01.0001.01.ENG

Fletcher, A. B. 1987. "Editorial: Pain in the Neonate." *The New England Journal of Medicine* 317(21): 1347–1348.

Kirkwood, J. K. 2006. "The Distribution of the Capacity for Sentience in the Animal Kingdom." In Turner, J., and D'Silva, J., eds., *Animals, Ethics and Trade: The Challenge of Animal Sentience*, 12–26. Petersfield, UK: Compassion in World Farming Trust.

Neumann, J-M. 2015. "The Legal Status of Animals in the French Civil Code." *Global Journal of Animal Law* 1: 1–13.

Stevan, H. 2016. "Animal Sentience: The Other-Minds Problem." *Animal Sentience: An Interdisciplinary Journal on Animal Feeling* 1: 1–11. Accessed January 25, 2016. http://animalstudiesrepository.org/animsent/vol1/iss1/1/

Service Animals

Service animals are trained to assist people with a wide variety of disabilities in order to help them achieve specific tasks or independence. The most well-known type of service animals are guide dogs and guide miniature horses that are specifically trained to assist people with severe visual impairments. However, service animals also assist people with other physical and psychosocial disabilities. A person with significant hearing loss partners with hearing or signal dogs that alert them to sounds like an alarm or a baby crying. Psychiatric service dogs assist individuals in detecting and lessening the effects of psychiatric episodes. Sensory signal dogs alert the handler to distracting repetitive movements so the person is able to stop the movement. Diabetic and seizure alert dogs predict and warn the person of oncoming episodes. Seizure response dogs guard their handler or go for help during a seizure.

Emotional support animals and therapy animals are often confused with service animals, but they are distinct categories. Emotional support or comfort animals are pets that are not trained to provide a specific service or perform a specific task. They provide therapeutic benefits to individuals with mental or psychiatric disabilities. An example is an emotional support dog whose presence provides comfort to its handler in public settings, whereas a service dog is trained to sweep the site for triggers of the handler's psychiatric episodes. Therapy animals partner with their handlers to provide therapeutic interventions. A key distinguishing factor between service animals and emotional support and therapy animals is that neither emotional support animals nor therapy animals are trained to perform a specific task for a person with disabilities.

Service animals have assisted humans throughout history. There are depictions of animals guiding individuals who appear to be blind as far back as ancient Rome on a fresco in the ruins of Herculaneum dating from the 1st century CE. There are similar scenes in a 13th-century Chinese scroll painting and 16th-century European woodcuts, paintings, and engravings. More recently, the Paris Hospital for the Blind began training dogs to serve as guides as early as the 1780s. And in 1819, the Institute of the Training of the Blind in Vienna published manuals for coaching guide dogs. Following World War I, the German doctor Gerhard Stalling retrained war dogs to work as service dogs for blind veterans. This began an effort in Europe and the United States to improve access to service dogs, and in 1929 and 1934 the first American and British guide dog schools were founded, respectively.

The growing use of service animals in the United States led to their legal protection under the 1991 Americans with Disabilities Act (ADA). The ADA provides protection for people with service animals and mandates that all public facilities allow people with service dogs into any area of the building where other people are permitted. The dog must be housetrained, under the verbal or physical control of the handler, and vaccinated in accordance with state and local laws. In the United States, service animals are defined as any dog "that is individually trained to do work or perform tasks for the benefit of an individual with a disability" with a separate provision for miniature horses (ADA Requirements 2011). Under the ADA, a person is not required to provide proof of a disability or proof that the animal is a trained service animal. The lack of required identification of an animal as a service animal leaves room for abuse of the law, as seen by the rise of people bringing their pets into businesses under the guise of being a service animal and unaccredited service animal certifications online.

In response to concerns about the public's physical and health safety because of other species being characterized as service animals and the lack of power of business owners to respond to these concerns, the ADA was revised in 2011 to only include dogs and miniature horses. Proponents of the exclusion of other species, such as birds, pigs, lizards, snakes, and monkeys, note concerns to human health through injury and communicable diseases, as well as animal welfare concerns with regard to training and care. Those who argue in support of their inclusion as service animals cite their careful training, frequent veterinarian visits, and unique set of skills as justification for their inclusion as ADA service animals.

The distinction between service and emotional support animals results in the exclusion of emotional support animals from nondiscrimination protection of the ADA. For example, some schools have denied students' requests to bring to school emotional support animals that provide important supportive and coping tools. However, the Fair Housing Act and the Air Carriers Access Act does provide broader protection to people who use emotional support animals. The Fair Housing Act states that housing providers must allow emotional support animals into homes and that additional fees or housing restrictions are illegal. The Air Carrier Access Act allows animals to accompany passengers on commercial airlines, though the size and species may be regulated.

Ethical considerations for service animals include managing the expectations and workload of the animal, providing for physical and psychological needs, and preparing for separation from the handler through retirement or death. The handler's awareness of these issues will help both handler and animal prepare for and mitigate them. The service animal is an important member of the handler's life and cared for with frequent veterinarian visits and time "off duty."

Andy VanderLinde

See also: Animal Assisted Activities; Animal Assisted Therapy

Further Reading

ADA Requirements. 2011. "Service Animals." Accessed May 12, 2015. http://www.ada.gov/service_animals_2010.htm

Ascarelli, M. 2010. *Independent Vision: Dorothy Harrison Eustis and the Story of the Seeing Eye*. West Lafayette, IN: Purdue University Press.

Assistance Dogs International, Inc. 2015. "Home." Accessed May 12, 2015. http://www.assistancedogsinternational.org

Brennan, J. 2014. *Service Animals and Emotional Support Animals*. Nguyen, V., ed. Houston, TX: Southwest ADA Center.

History. 2011. "Assistance Dogs: Learning New Tricks for Centuries." Accessed May 12, 2015. http://www.history.com/news/assistance-dogs-learning-new-tricks-for-centuries

Hornsby, A. 2000. *Helping Hounds*. Lydney, UK: Ringpress Books.

Huss, R. J. 2010. "Why Context Matters: Defining Service Animals under Federal Law." *Pepperdine Law Review* 37(4): 1163–1216.

International Guide Dog Federation. 2015. "Home." Accessed May 14, 2015. http://www.igdf.org.uk/

Wenthold, N., and Savage, T. A. 2007. "Ethical Issues with Service Animals." *Topics in Stroke Rehabilitation* 14(2): 68–74.

Shamanism. *See* Indigenous Religions, Animals in

Sharks

The earliest sharks appeared more than 400 million years ago, and presently it is estimated that there are more than 500 surviving species of sharks. Sharks are highly adaptable fish and can be found in all the seas of the world. Although the vast majority of shark species are found in salt waters, there are a few, such as river sharks, that are able to survive in fresh waters, too. The diversity of sharks is such that the smallest known species of sharks, the dwarf lanternshark, is only about 6 inches long, whereas the most well known of sharks, the great white shark, can easily reach more than 20 feet in length. The largest species of sharks, the whale shark, can grow up to 40 feet. Humans' fascination with sharks has seen us engaging with this ancient animal in two main ways, for entertainment and for food. In recent years, shark culling (lethal removal) has also become a controversial subject.

Western popular culture has a sustained interest in sharks. In 1975, Steven Spielberg directed *Jaws*, a movie about a small coastal town terrorized by a man-eating great white. The movie itself is based on a best-selling novel by Peter Benchley (who has since expressed regret that the shark in the movie was portrayed as an aggressor instead of a victim). Since then, Hollywood has churned out no fewer than 50 shark-themed films. While most of these, with titles like *Shark Attack* (1999) and *Sharknado* (2013), are dismissed by critics, they adopt the common theme of killer sharks attacking humans indiscriminately. In reality, according to the Florida Museum of Natural History, fatal attacks of humans are extremely rare, with fewer than 10 documented annually in recent years. Moreover, very few shark species (such as the great white and tiger shark) are known to attack humans unprovoked.

Despite the low incidence of fatal shark attacks, in January 2014, the state government of Western Australia initiated a controversial shark culling program in response to the seven fatal shark attacks between 2010 and 2013. The policy has attracted a storm of objections by various animal welfare groups. By the end of 2014, the state government announced a cessation of the program, citing the uncertainty over the effectiveness of the program in reducing shark attacks as the main reason. The vocal objections to the culling program show that sharks have a special place in human-animal relationships, despite their persistent negative portrayal in popular culture.

Humans' fascination with sharks has grown with better knowledge of their behavior and physiology. For instance, studies have shown that many shark species possess high-order problem-solving skills and complex sociality. A group of sharks off the coast of South Africa has been observed cooperating to bring a beached dead whale into deeper waters so that they could share it for food. Great white sharks also practice a unique hunting technique called "breaching" that involves the shark approaching the surface of the water at such a high speed that it is propelled partially or completely out of the water. This enables them to effectively hunt for prey such as seals.

Sharks have been around for millions of years and are found throughout the world's oceans. Although fearsome in appearance and frequently portrayed negatively in popular media, sharks very rarely attack humans. (Steven Melanson/Dreamstime.com)

An increasingly popular way to interact with sharks is through "cage diving." While this affords close and personal encounters with sharks, there are criticisms nonetheless. For example, to draw sharks to the cages, blood is deliberately released in the water, creating an unnatural situation where sharks follow the scent to the cages without actually having prey to consume. Some shark conservationists also believe that this practice (called "chumming") will alter the innate ability of sharks to hunt effectively by associating human presence with feeding.

By far the most devastating impact humans have on sharks is through consumption. While consuming shark meat is common in many cultures, killing sharks to consume their fins is the most controversial issue today. The consumption of shark fins was first documented in China during the Song Dynasty (960–1279 CE). Since the 1600s, shark fin has become a high-status food, only enjoyed by the privileged class. Today, the consumption of shark fins remains popular in Chinese societies across the world, and the fins can cost up to $800 per pound.

Although touted as a delicacy and often used as an ingredient in soups, shark fins are tasteless and are more prized for their texture. Claims of the high nutritional value of shark fins have been consistently debunked. The main argument against shark finning is the cruel manner in which sharks are captured, as they have their fins cut off and are thrown back into the sea (because the meat on their bodies is virtually worthless) to drown. The demand for shark fins has also raised concerns about the sustainability of this form of consumption. While fins can be harvested from any shark species, the industry tends to focus narrowly on about 14, ranging from the blue shark to the scalloped hammerhead. In the 2000s, the total number of sharks killed annually through the fin trade is estimated to be between 26 to 73 million. (The large variance in the estimate is due to the prevalence of illegal catches and under-reporting.) However, since 2010, there have been increasing signs that things are changing. In 2013, for example, the Chinese government directed all its officials to not serve shark fin in official banquets. With many hotel chains in Asia removing shark fin from their menu, it may only be a matter of time until shark fin as a culinary choice will fall out of favor.

Harvey Neo

See also: Dolphins; Human-Wildlife Conflict; Popular Media, Animals in

Further Reading

Clarke, S., Milner-Gulland, E., and Bjørndal, T. 2007. "Social, Economic, and Regulatory Drivers of the Shark Fin Trade." *Marine Resource Economics* 22: 305–327.

Florida Museum of Natural History. Accessed December 19, 2015. https://www.flmnh.ufl.edu

Jacques, P. 2010. "The Social Oceanography of Top Oceanic Predators and the Decline of Sharks: A Call for a New Field." *Progress in Oceanography* 86: 192–203.

Lovgren, S. 2005. "*Jaws* at 30: Film Stoked Fear, Study of Great White Sharks." Accessed November 18, 2015. http://news.nationalgeographic.com/news/2005/06/0615_050615_jawssharks.html

Philpott, R. 2002. "Why Sharks May Have Nothing to Fear Than Fear Itself: An Analysis of the Effect of Human Attitudes on the Conservation of the Great White Shark." *Colorado Journal of International Environmental Law and Policy* 13: 445–472.

Shelters and Sanctuaries

Animals in shelters include unwanted pets, stray (former) pets, and abused/neglected animals (pet and non-pet). Animal shelters can be open-access—typically taxpayer (publicly) funded, providing animal control services, and required to accept all animals—or closed access—privately funded and able to reject animals for space and/or adoptability considerations. Unlike animal shelters, whose focus is on rehoming animals, animal sanctuaries provide a permanent home for the rescued animals, although they may also offer adoption for some. Also unlike shelters, sanctuaries can keep a variety of animals and often specialize in a certain type or species; for example, farmed animals, injured wildlife, or senior dogs.

There had been a tradition in England, dating to the Middle Ages, of having village or town shelters, called "pounds" (because they impounded animals), which were used to hold animals who had wandered away from their homes until they were bought back by their owners. Livestock were typically redeemed, but cats and dogs, who had less monetary value, were not and were generally killed by the "poundsman." These pounds were found in colonial America as well. Pounds, which evolved into animal control agencies, were set up to protect the public, not to care for animals, and for years worked at cross-purposes with the anti-cruelty societies.

While people have obviously been rescuing stray animals for hundreds of years, organized animal shelters are relatively recent, dating to the early 19th century in England. Here, in 1824, a group of animal advocates formed the first Society for the Prevention of Cruelty to Animals (SPCA) in the world, later renamed the Royal Society for the Prevention of Cruelty to Animals (RSPCA).

In the United States, Henry Bergh (1813–1888) formed the first formal animal welfare group in 1866—the American Society for the Prevention of Cruelty to Animals (ASPCA). This group was established not to rescue animals but to protect specific animals like New York City carriage horses and to fight animal cruelty. While initially the ASPCA worked on anti-cruelty campaigns, the organization (which began handling New York City's animal control services in 1894) became the model for many of the country's shelters (many of which use SPCA in their names today). The first private animal shelter was the Women's Humane Society, founded in 1869, which opened its first animal shelter in Philadelphia in 1912.

It wasn't until the 1970s that animal shelters began to take a role in educating the public on humane animal care and to tackle the problem of unwanted dogs and cats by offering both spay/neuter services (to reduce breeding) and veterinary care. Before that time, veterinarians were rarely involved in animal shelters, and euthanasia methods were often inhumane.

Because many animals in animal shelters never get adopted, the need for a permanent, non-euthanasia solution began to emerge. Private individuals, usually animal rescuers, began offering their homes and land for the permanent care of unadopted or unadoptable animals, leading to the rise in animal sanctuaries. Many of these were (and still are) unincorporated and are simply one individual's response to caring for the animals in a given community. Others are incorporated as nonprofits (in the United States) or, outside the United States, nongovernmental organizations (NGOs), and take donations from the public to support their activities. In the United States, one of the first such sanctuaries was the Wildlife Way Station, founded in 1965 to provide permanent homes for wild animals. Outside of the United States, sanctuaries have been operating for much longer. India, for example, is home to the Bori and Senchal Wildlife Sanctuaries, founded in 1865 and 1915, respectively.

Other animal sanctuaries, like Farm Sanctuary and Animal Place, founded in 1986 and 1989, respectively, specialize in farmed animals. These sanctuaries offer not only a permanent home to animals who would have otherwise been killed and eaten, but also educate the public about factory farming and meat-free diets. Still other sanctuaries focus on a particular animal species. Examples include the Lone Pine Koala Sanctuary in Australia, the Donkey Sanctuary in England, and the Sepilok Orangutan Sanctuary in Indonesia.

Besides sanctuaries, which care for unadoptable animals, another approach to the companion animal overpopulation problem comes from the no-kill movement, which began in India and calls for ending the killing of healthy animals. Although shelters throughout the United States (and elsewhere) are trying to get close to this ideal, the no-kill concept has also created divisions within the animal welfare community, pitting no-kill shelters against open-access shelters. Because most no-kill shelters are privately run and closed access, these shelters can refuse to accept animals; this option is not available to taxpayer-funded shelters, which must take in all animals, regardless of space availability or animals' adoptability. These shelters are then blamed for their often high euthanasia rates even when much of the problem is out of their control.

Animal shelters and sanctuaries can be found around the world today, although nations that are less financially well-off, which often have greater problems with stray animals, tend to have fewer groups with fewer resources. For instance, in Latin America, where free-roaming street dogs are common, the typical response to the problem was often shooting, poisoning, or otherwise killing the dogs when their numbers grew too high, tourists complained, or there were reports of dogs harming people. Today, local animal shelters are starting to implement spay/neuter as a partial solution and are creating education campaigns aimed at encouraging the public to adopt street dogs rather than to purchase purebred animals from pet stores.

Margo DeMello

See also: Breed Specific Legislation; Cats; Cruelty; Dogs; Hoarding; Humane Education; Pets; Puppy Mills; Spay and Neuter

Further Reading

Glen, S., and Moore, M. T. 2001. *Best Friends: The True Story of the World's Most Beloved Animal Sanctuary.* New York: Kensington Books.

Harbolt, T. L. 2003. *Bridging the Bond: The Cultural Construction of the Shelter Pet.* West Lafayette, IN: Purdue University Press.

Leigh, D., and Geyer, M. 2003. *One at a Time: A Week in an American Animal Shelter.* Santa Cruz, CA: No Voice Unheard.

Winograd, N. J. 2007. *Redemption: The Myth of Pet Overpopulation and the No Kill Revolution in America.* Los Angeles: Almaden Books.

Silk. *See* Non-Food Animal Products

Skins. *See* Non-Food Animal Products

Slaughter

Raising and killing animals for food has long been, and continues to be, one of the most significant social forms of human-animal relations. The practice of killing animals for food production is called slaughter. By the beginning of the 21st century, humans were slaughtering well over a billion cattle, sheep, and pigs and over 60 billion chickens globally each year. Ninety-nine percent of all domesticated animals are commodities in animal agriculture (Williams and DeMello 2007). Methods of killing food animals are of increasing public concern in Europe and other industrialized countries because of rising interest in humane treatment.

The practices of slaughter vary worldwide, but in most countries there are basic rules to protect the welfare of animals, and the human food supply, at the time of killing. The World Organisation for Animal Health (OIE) slaughter welfare standard sets basic minimum standards that every member country should follow. Even though not imposed by law in all countries, they are the only global, science-based standards on animal welfare agreed upon by OIE members. These standards have an emphasis on the animals' subjective experiences and include recommendations about their handling. For example, animals should be handled in such a way as to avoid harm or injury, and stress and pain at the time of killing should be minimized by correct use of stunning (i.e., those methods, either mechanical or electric, that help to ensure that animals do not feel pain by rendering them unconscious prior to their being slaughtered).

Traditional, non-mechanized methods of slaughter typically involve slitting the animals' throats or cutting off their heads (used primarily with birds). These means continue to be used worldwide in non-industrial settings that slaughter only a few animals at a time. In industrialized animal agriculture, the focus is on efficient slaughtering of a high volume of animals and is highly mechanized, although with human involvement. Animals are stunned first, and then a shackle is placed around an ankle. They are then hoisted upside down onto a hanging conveyor belt and moved past a person who will slit their throats, after which they will "bleed out."

Stunning is a technical process that, if done correctly, induces immediate unconsciousness and inability to feel physical sensation, so that slaughter can be performed without fear, anxiety, pain, suffering, and distress. Many people who eat meat, especially in industrialized countries, have expressed concern over animals' potential for pain and fear at the time of slaughter. Two common stunning methods are "captive bolt" (the mechanical "shooting" of a steel bolt into a cow's, sheep's, or pig's brain) and electrical (passing an electric current through a cow's, sheep's, pig's, or bird's brain). In North America, the European Union, Australia, and many other countries, stunning of animals before slaughter is legally required. However, what is known as ritual slaughter is exempted from these regulations in order to allow members of religious groups to meet the requirements of their religions with respect to slaughter.

Ritual slaughter is the killing of animals for food performed according to the requirements of either the Jewish or Muslim religions. Jewish slaughter is called *shechitah* (kosher slaughtering), and Muslim slaughter is called *halal*. Stunning prior to *shechitah* is prohibited because, according to Jewish religious rules, the animals need to be alive and in good health at the time of death, and stunning methods are considered to cause injuries that render them *trefah* (animals unfit to eat). Moreover, only a Jew specially trained can perform *shechitah*. He is required to study for a number of years and is examined in the laws of *shechitah*, animal anatomy, and disease. The animal is slaughtered by slitting its throat with a razor-sharp knife. Jewish religious authorities argue that, when the cut is done correctly, the animal appears not to feel it; however, this claim is highly contested, for example by many animal advocates and those in animal professions, such as veterinarians.

Halal slaughter is defined as the process of killing an animal according to Islamic law. The process must be carried out by a trained Muslim and begins by invocation of Allah. The instruments for slaughter must be sharp to ensure the most stress-free and quick cut of the throat possible. *Halal* slaughter requires that the animal is alive at the time of killing, and therefore it does not allow any method of stunning.

The United States is one of the countries that has enacted legislation (the Humane Slaughter Acts of 1958 and 1978) to address public concerns about animal suffering. These Acts also allow for the protection of Jewish and Muslim slaughter practices, including ritual slaughter as one of two humane methods (the other being stunning). In Europe, practices of religious slaughter without stunning have been protected since 1928, and the 2013 European Union (EU) regulation on the protection of the welfare of animals at the time of killing (EU 2009/1099) maintains the exemption from stunning for religious slaughter. However, these practices have become particularly contested in the last decade, especially in conjunction with the increased number of Muslim immigrants in several EU countries and with the great increase in the demand for *halal* meat. The ritual slaughter exception is highly contested, and several European countries (e.g., Poland, Denmark) have asked for a ban on religious slaughter without stunning. How animals should be slaughtered, therefore, raises not only questions of animal welfare but also ethnic and cultural traditions, prejudices, and conflicts.

Mara Miele

See also: Animal Law; Eastern Religions, Animals in; Factory Farming; Meat Eating; Meat Packing; Western Religions, Animals in; World Organisation for Animal Health (OIE)

Further Reading

Cudworth, E. 2008. "'Most Farmers Prefer Blondes'—Dynamics of Anthroparchy in Animals' Becoming Meat." *The Journal for Critical Animal Studies* 6(1): 32–45.

Evans, A., and Miele, M. 2012. "Between Food and Flesh: How Animals Are Made to Matter (and Not to Matter) within Food Consumption Practices." *Environment and Planning D: Society and Space* 30(2): 298–314.

Lever, J., and Miele, M. 2012. "The Growth of the Halal Meat Markets in Europe: An Exploration of the Supply Side Theory of Religion." *Journal of Rural Studies* 28(4): 528–537.

Steinfeld, H., Gerber, P. Wassenaar, T., Castel, V., Rosales, M., and De Haan, C. 2006. *Livestock's Long Shadow*. Rome: UN Food and Agriculture Organization (FAO).

Velarde, A., and Dalmau, A. 2012. "Animal Welfare Assessment at Slaughter in Europe: Moving from Inputs to Outputs." *Meat Science* 92(3): 244–251.

Velarde, A., Rodriguez, P., Fuentes, C., Llonch, P., von Holleben, K., von Wenzlawowicz, M., Anil, H., Miele, M., Cenci Goga, B., Lambooij, B., Zivotofsky, A., Gregory, N., Bergeaud-Blackler, F., and Dalmau, A. 2010. *Improving Animal Welfare during Religious Slaughter, Recommendations for Good Practice*. Accessed December 13, 2015. http://www.dialrel.eu/images/recom-light.pdf

Vialles, N. 1994. *Animal to Edible*. Cambridge, UK: Cambridge University Press.

Williams, E. E., and DeMello, M. 2007. *Why Animals Matter: The Case for Animal Protection*. New York: Prometheus Books.

Social Construction

Social construction is a key concept for Human-Animal Studies (HAS) because it reveals that human attitudes and practices toward animals are not permanent and unchanging, but rather related to the specific contexts of time, place, and culture. This concept provides a framework for HAS scholars to 1) study and uncover the contexts of human-animal relations in different cultures, and 2) empower societies to actively reflect upon, and sometimes challenge and change, practices and relations that may no longer be relevant or may now be seen to be unethical.

Intellectuals have discussed and debated how societies form their understandings of reality for centuries; it is agreed, however, that the 1966 publication of *The Social Construction of Reality* by sociologists Peter Berger (1929–) and Thomas Luckmann (1927–) formalized one way societies do this through the concept of social construction. Rather than seeing knowledge as completely objective, or social conventions (such as marriage being only between a man and a woman) as universally accepted, they argued that social life/practices and the way we perceive the world are subjective and come from within a social group. Therefore, social construction can be defined as "any category, condition, or thing that exists or is understood to have certain characteristics because people socially agree that it does" (Robbins et al. 2010, 288).

So how does the concept of social construction apply to animals and human-animal relations? Consider how we often talk about "humans and animals." HAS scholars, and indeed most people, use this phrasing, but why? It conveys that humans are different from animals, yet humans are biological animals, just like dogs and cats.

What are the consequences of using "humans and animals"? In the history of human societies, this social construction of humans as not-animals has led to what is termed an "othering" of animals. Making animals "other" than humans has allowed us to treat them differently. For example, humans find it unacceptable to eat other humans, but they do eat other animals, although societies differ as to which animals are acceptable to eat. In the United States it is not considered acceptable to eat cats, dogs, or horses because they are constructed as pets and therefore "closer" to humans. Yet many Europeans enjoy horse meat, and many east Asians enjoy cat and dog meat. While HAS scholars also discuss "humans and animals," because this phrasing is the norm and the most easily understood by the public, they do so with an awareness of its problematic nature.

One of the most often cited examples of social construction is the concept of wilderness. In terms of physical landscape, many people think of this as an area that does not have any evidence of human interference. In the United States, national parks are seen as protected wild areas, but is a national park really wilderness? What about the roads, the human-made boundaries, and the often overlooked history of native peoples' uses of these places? What we believe to be the wilderness of a national park is really a human social construction.

To see how these social constructions of wilderness intersect with a specific animal, we can turn to a population of "wild" horses in the Ozark National Scenic Riverway of southern Missouri, managed by the National Park Service (NPS). Most people consider wild animals to be those that are not under human control, normally living on their own and making their own decisions. In this case the horses are thought to be the descendants of animals abandoned during the Great Depression (1929–1939). They were living in the area self-sufficiently when the park opened in 1964, but in the 1970s the government decided that parks should contain only native wild animals and classified the horses as a non-natural destructive disturbance to the park's native ecosystem. The horses went from being seen as "wild" to being seen as "not-wild" and the government made plans to remove them. Local people argued that the horses *were* wild and natural to the park. Ultimately, the horse advocates won, but it remains a good example of how categories applied to animals are often fluid and that there is no definite line between objective and subjective ideas of "wild."

Some scholars have argued that there are problems with a social constructivist approach. They are concerned it is a slippery slope that, taken as far as it could go, would mean denying an objective reality. For example, is a horse not really a horse? Social constructivists do not deny objective reality, but instead emphasize the importance of revealing social and historical context. They respond by openly recognizing there is a balance that must be made between objective facts and subjective social knowledge. An example is primatologist Jane Goodall's (1934–) discovery in the 1960s that chimpanzees used tools. Previous to her discovery, science had constructed humans as the only animals capable of tool use, saying it was a fundamental separation between humans and other animals. Goodall's work used objectively real evidence to debunk a subjective scientific construct.

Another way to address concerns about denying objective reality is through the concept of co-production (Robbins et al. 2010). This concept argues that social knowledge creation is not a one-way street shaped only by humans. Instead, co-production recognizes that humans, animals, and the environment are always influencing each other, which, in turn, reshapes human constructs. In this way, the horses and the chimpanzees, because of their actions, helped to reshape how humans understood the "wild" and the uniqueness of the human species.

Julie Urbanik

See also: Animal Geography; Horses; Human-Animal Studies; Humans; Personhood; Speciesism; Wildlife

Further Reading

Berger, P., and Luckman, T. 1966. *The Social Construction of Reality.* Garden City, NY: Doubleday.

DeMello, M. 2012. *Animals and Society: An Introduction to Human-Animal Studies.* New York: Columbia University Press.

Rikoon, S., and Albee, R. 1998. " 'Wild-and-Free, Leave-'Em-Be': Wild Horses and the Struggle over Nature in the Missouri Ozarks." *Journal of Folklore Research* 35(3): 203–222.

Robbins, P., Hintz, J., and Moore, S. 2010. *Environment and Society.* Malden, MA, and Oxford, UK: Wiley-Blackwell.

Taylor, N. 2013. *Humans, Animals, and Society: An Introduction to Human-Animal Studies.* New York: Lantern Books.

Urbanik, J. 2012. *Placing Animals: An Introduction to the Geography of Human-Animal Relations.* Lanham, MD: Rowman and Littlefield.

Spay and Neuter

"Spay" and "neuter" refer to the permanent removal or loss of function of reproductive organs. Spay is female-specific. Neuter applies to both sexes, but particularly in the United States, it is generally associated with males. In males, removal of the testicles is also called castration. The procedures may also be called sterilization or desexing; these terms, too, apply to both sexes. Animals who have *not* been spayed/neutered may be referred to as intact, entire, or whole. Spaying/neutering is the most common method to prevent reproduction of dogs and cats; it is also practiced in smaller pet species such as rabbits, rats, and ferrets. Male cattle, goats, horses, pigs, and sheep are also commonly neutered (castrated), and female cattle sometimes spayed.

Every year millions of unwanted dogs and cats are dropped off at shelters around the United States, and many end up being euthanized. Spay/neuter programs, such as the one pictured above, are seen as a key method for reducing pet overpopulation. (AP Photo/Eric Risberg)

Among companion animal species (specifically dogs and cats), spaying most often entails ovariohysterectomy (surgical removal of the uterus and ovaries). Neutering in males most often involves orchiectomy (surgical removal of the testes), although vasectomy is (rarely) used; the latter prevents reproduction while preserving testosterone. Male animals raised for food are alternately sterilized through orchiectomy; banding, which restricts blood flow to the scrotum and testes; or the Burdizzo clamp, which breaks the spermatic cord and blood vessels connected to the testicles. Some producers spay female cattle by ovariectomy (removal of ovaries).

Use of general or local anesthesia and/or analgesia (pain control) varies across species. Although exceptions exist, general anesthesia is now standard for male and female cats and dogs undergoing surgical sterilization. For species of animals raised for food, in some countries non-veterinary personnel may perform the aforementioned procedures on conscious animals without pain control. Cost, labor intensiveness, and access to appropriate pharmaceuticals have all been cited as reasons for not using anesthesia or analgesia. Failure to use anesthesia/analgesia is significant cause for concern about animal welfare.

There is an emerging field of pharmaceutical-based non-surgical fertility control for nonhuman animal species. Multiple approaches target various aspects of the reproductive system in a variety of ways. Permanent injectable sterilants using zinc gluconate and calcium chloride exist for male dogs and cats. Long-acting male and female contraceptives have been studied in both companion animal species and wildlife species (including but not limited to wild canids [dogs] and felids [cats], deer, elephants, wild boar, wild horses, and prairie dogs). Limited products are registered for contraception.

Why spay/neuter? Particularly in dogs and cats, the most fundamental reason is to prevent unwanted reproduction. The practice is credited for significantly decreasing euthanasia in American shelters; the American Society for the Prevention of Cruelty to Animals (ASPCA) estimates that cats and dogs euthanized in shelters dropped from 20 million in 1970 to approximately 3 million in 2011, largely due to increases in spaying (particularly before a first pregnancy) and neutering. Sterilization has also been widely adopted as a method to humanely control free-roaming or feral cat and dog populations worldwide and in many studies has been found to be more effective long term and socially acceptable than culling (killing). Since the 1990s, the reduction in shelter euthanasia has also paralleled the popularization of "pediatric" spay/neuter, which has enabled animal shelters to place only sterilized pets into the community and reach free-roaming cats at a younger age.

Additional factors may influence spay/neuter decision-making, especially for companion animals. The first is health benefits and risks. For females, documented health benefits of traditional spay (ovariohysterectomy) include, but are not limited to, eliminating risk of ovarian cancer, uterine cancer, and pyometra (potentially fatal uterine infection) and reducing risk of mammary cancer, particularly when performed before the first estrus. For males, surgical neutering eliminates risk of testicular cancer and adverse physical consequences of fighting between intact males. At the same time, research has found that surgical spaying and neutering, particularly when performed at an early age, may correlate with a higher risk of certain health problems in certain breeds of dogs (not cats). This has been attributed at least in part to absence of sex hormones. It is critical to emphasize, however, that research findings are limited in scope and are breed specific.

Behavior is another motivation for spaying/neutering many species. Reduction of sex hormones can reduce sex hormone–related behaviors that people consider problematic. Neutering in males is often performed based on the assumption that it will reduce aggression toward one another and toward people; it may also reduce behaviors such as roaming, sexual mounting, and urine marking, and make animals easier to handle. It is important to recognize that the relationship between sex hormones and behavior vary dramatically by species and individual, however, and that behaviors *not* driven by sex hormones will not change with neutering.

Studies have shown that cultural, social, regional, and economic factors may influence peoples' attitudes toward spaying/neutering, particularly dogs and cats. This can lead to dramatic variations in rates of spaying/neutering by country, region, and community. The sex of the pet owner has also been found to correlate with attitudes toward neutering male companion animals, in particular, with men more resistant than women. The reason for obtaining an animal (e.g., companionship, work, protection) can also affect decision-making. Accounting for these factors is critical to understanding attitudes and behaviors toward sterilization, particularly as part of efforts to increase spay/neuter numbers for population control. New approaches to sterilization (e.g., nonsurgical, preserving hormones and/or reproductive organs) have the potential to address factors that may currently prompt resistance to spaying/neutering.

Valerie Benka

See also: Animals; Cats; Dogs; Ethics; Humane Farming; Livestock; Pets; Welfare

Further Reading

Alliance for Contraception in Cats & Dogs. 2013. *Contraception and Fertility Control in Dogs and Cats*. Accessed August 1, 2015. http://www.acc-d.org/resource-library/e-book

American Veterinary Medical Association. 2011. "Literature Review on the Welfare Implications of Ovariectomy in Cattle." Accessed August 1, 2015. https://www.avma.org/KB/Resources/LiteratureReviews/Documents/ovariectomy_cattle_bgnd.pdf

American Veterinary Medical Association. 2013. "Literature Review on the Welfare Implications of Swine Castration." Accessed August 1, 2015. https://www.avma.org/KB/Resources/LiteratureReviews/Documents/castration-cattle-bgnd.pdf

American Veterinary Medical Association. 2014. "Literature Review on the Welfare Implications of Cattle Castration." Accessed August 1, 2015. https://www.avma.org/KB/Resources/LiteratureReviews/Documents/swine_castration_bgnd.pdf

Kustritz, M. V. R. 2007. "Determining the Optimal Age for Gonadectomy of Dogs and Cats." *Journal of the American Veterinary Medical Association* 231(11): 1665–1675.

Tasker, L. 2007. *Stray Animal Control Practices (Europe): A Report into the Strategies for Controlling Stray Dog and Cat Populations Adopted in Thirty-One Countries.* London and West Sussex, UK: WSPA & RSPCA International.

Species

A species is the smallest unit of the eight major taxonomic categories (biological classifications)—domain, kingdom, phylum, class, order, family, genus, and species. From the time of Classical Greece, the concept of species has shifted from a physiological one (based on physical characteristics) to an evolutionary one (based on shared genetic material). Regardless, throughout modern history, significant weight has been placed on the category of species by scientists, conservationists, and lay people alike. Today, there are an estimated 8.74 million species of organisms on Earth, an estimated 7.77 million of which are animals. Of those, however, only 953,434 have been described and cataloged by (predominantly) Western scientists (Wall 2011).

Species classification dates back to the Greek philosopher Aristotle (384–322 BCE), who classified approximately 500 species of animals into categories that loosely correspond to the modern categories of vertebrate (having a spinal column) and invertebrate. Aristotle is attributed with having introduced the terms *genus* (derived from "general") and *species* (derived from "specific"). Aristotle's system remained largely unchanged until the 18th century when Carl Linnaeus (1707–1778), a Swedish botanist, zoologist, and physician, greatly expanded and modified Aristotle's classification and developed the modern naming system.

In this modern system, organisms are given binomials, or two-part names, that include both the genus and the species. For example, the tiger's binomial is *Panthera tigris*, meaning that it is a member of the *Panthera* genus (along with lions, jaguars, and leopards), and the *tigris* species.

Prior to the 1859 publication of *On the Origin of Species* by British naturalist Charles Darwin (1809–1882), in Western society species were commonly thought of as distinct groups of organisms in which any two individuals could produce fertile offspring. Many people believed that species could be traced back to the creation of Earth as described in the book of Genesis in the Bible. Darwin's theory of evolution through natural selection demonstrated how organisms could change dramatically over very long time spans—much longer than the few thousand years that were assumed to have passed since divine creation in the Christian worldview. Darwin's theory moved away from the idea that species are unchanging and defined by a particular form based on organisms' physical characteristics (i.e., typological) to how organisms have evolved over time and reflect shared ancestry. While Linneaus and others thought of species as unchangeable, Darwin popularized the idea that species do not have clear boundaries. In the last few decades, the improvement in DNA sequencing has largely solidified evolutionary (also called phylogenetic) classification.

Given the historical shift in the definition of "species," confusion sometimes arises. Many species have relatively clear-cut boundaries. The giant panda (*Ailuropoda melanoleuca*), for example, is a very clear-cut species given the geographical separation between it and other genetically related members of the bear family, and the genetic/size differences between it and similarly named and geographically closer red panda (*Ailurus fulgens*). Others are difficult to define, however. Called the "species problem," this difficulty arises when a species complex, or group of related species, is so closely related that the boundaries between them blur. Such problems often arise when in the past a species had been defined based on observable morphological (physical) similarities, but more recent genetic analyses show the group to be made up of more than one closely related species. For example, the African forest and African bush elephants had been classified as a single species (the African elephant) due to similar physical appearance. Now, as a result of recent DNA testing, they are considered to be two distinct species that diverged genetically between 2 and 7 million years ago. Given that the two species can produce viable offspring when they mate, some scientists refuse to recognize them as distinct species, but physical evidence showing different sizes, diets, and habitats, and genetic evidence showing only limited hybridization (cross-breeding between two species) in the wild along with significant genetic differences, suggest two species, an important distinction for the conservation of both unique populations and their differing genetic material.

Modern ecology and biodiversity research adopts the species as the fundamental taxon, or unit of measure, at which most analysis takes place. For example, significant conservation efforts are based around the International Union for Conservation of Nature (IUCN) Red List, which contains over 5,000 species of plants and animals classified as vulnerable, endangered, or critically endangered. Such species are often protected by national laws that forbid hunting, control land development, and create preserves. A growing recognition of the importance of the genetic diversity contained in subspecies (a taxonomic classification subordinate to species) is now strengthening conservation efforts at that scale as well.

While many humans interact directly with animals as individuals, humans often interact with and conceptualize animals as species. Frequently, humans engage with a single member as a representative of the species as a whole. For example, when one thinks of a lion, rarely does one picture a specific lion, but more often imagines a generic lion as a representative of all. Such conceptualization occurs most often with those species that are most distant from humans. For example, pets are often thought of as individuals, whereas wild animals, fish, and insects are more generally conceptualized as groups at the species level. Zoo animals, in particular, are displayed as representative specimens of species that many humans will never encounter in the wild, forgoing their individual identities in favor of their equation with the species as a whole (Malamud 1998).

Skye Naslund

See also: Animals; Biodiversity; Extinction; Speciesism; Taxonomy

Further Reading

Arthur, W. 2014. *Evolving Animals: The Story of Our Kingdom.* Cambridge, UK: Cambridge University Press.

Darwin, C. 1859 [1993]. *On the Origin of Species.* New York: Random House.

International Union for Conservation of Nature and Natural Resources. 2015. *The IUCN Red List of Threatened Species.* Accessed January 15, 2016. http://www.iucnredlist.org/

Linnaeus, C. 1758. *Systema Naturae, Per Regna Tria Naturæ, Secundum Classes, Ordines, Genera, Species, cum Characteribus, Differentiis, Synonymis, Locis Vol. 1*, 10th ed. Accessed January 15, 2016. http://www.biodiversitylibrary.org/item/10277#page/3/mode/1up

Malamud, R. 1998. *Reading Zoos: Representations of Animals and Captivity.* London: MacMillan Press Ltd.

Nijhuis, M. 2012. "Which Species Will Live?" *Scientific American* 307(2): 74–79.

Ruse M., ed. 2013. *The Cambridge Encyclopedia of Darwin and Evolutionary Thought.* New York: Cambridge University Press.

Wall, T. 2011. "8.74 Million Species on Earth." Discovery News. Accessed January 15, 2016. http://news.discovery.com/earth/plants/874-million-species-on-earth-110823.htm

Speciesism

The concept of speciesism functions in the same way that the concepts of racism and sexism do—to highlight arbitrary discrimination and unjust treatment of a group. The term was made popular by the Australian philosopher Peter Singer (1946–). Many animal advocates and ethical philosophers utilize the idea of speciesism in forming their animal advocacy arguments. At its core, the concept of speciesism challenges many human ideas and social institutions.

Speciesism indicates that (unfair) preference is given to humans over all other species and that certain nonhumans are viewed more or less favorably by humans than others. Although popularized by Peter Singer in his book *Animal Liberation* (1975), the term "speciesism" actually comes from the British psychologist Richard Ryder (1940–). In the 2012 documentary *The Superior Human?* Ryder stated that "[i]n 1970, I coined the term 'speciesism' to describe the prejudice against other species, and to draw the analogy with other prejudices like racism and sexism." Racism and sexism work the same way in that preference is given based solely on race or sex. These prejudices are not seen only in individuals' acts but in broader societal structures as well. For example, for centuries in the United States and Europe, non-whites and women were seen as inferior to white males, and economic, political, and social structures were developed that made them legal property, either bought and sold as slaves, in the case of non-whites, or being under complete legal control of a husband, in the case of white women. Much like the white males of this earlier time, humans overall today largely view themselves as superior to all other animals, and preference is given to human interests.

This presumption of superiority is often based in religious beliefs about the divinely determined order of life and/or on what is seen to be humans' higher intelligence. However, religious scholars such as Paul Waldau (1950–) challenge these religious foundations, as has the author Matthew Scully (1959–), who has asserted that the Christian virtues of kindness and compassion are too frequently overlooked. Others, such as Singer and Ryder, as well as animal advocates, claim that it is arbitrary, and therefore speciesist, to use human intelligence as the sole or main criterion for making ethical judgements. Using scientific evidence, they argue that many animals experience pain, stress, and suffering in similar ways to humans and that high intelligence is not relevant to these experiences. The American philosopher James Rachels (1941–2003), for example, drew on evolutionary biology, which shows physiological continuity between humans and other animals. These various claims against a special category for humans challenge a number of long-standing cultural and religious beliefs, as well as social structures (e.g., legal systems) that consistently give nonhuman animals significantly inferior status and in this way are similar to the challenges posed by slavery abolitionists and advocates for women's rights.

The most common manifestation of speciesism relates to favoring human interests, simply because they are human, without accounting for the *type* of interest involved. For example, it is speciesist to give preference to a human's interest in killing an elephant for a trophy because it is reasonable to argue that the elephant's, and his or her social groups', interest in life and maintaining the stable group are greater than the human's interest in trophy hunting for entertainment or prestige. With regard to respecting life in general, Singer has argued that it is speciesist to draw the boundary at the human without considering the relevant factor of an individuals' potential quality of life. For example, a normal adult human, because of higher mental capabilities, may have the possibility for a richer mental life than many other animals. However, the mental lives of some animals, for example chimpanzees, dogs, or elephants, are undeniably richer than the lives of some humans, such as an individual with severe brain damage. To hold the lives of humans more valuable, just because they are human, than the lives of all other animals is to arbitrarily discriminate in favor of one's own species.

These issues can be summed up by saying that the concept of speciesism does not mean that all interests are equal or that all lives are of equal value. It does mean that in making ethical judgments, one should consider the type/magnitude of interest (e.g., entertainment versus life itself) and the relevant characteristics of the individuals involved. For example, high intelligence matters for composing a symphony but not for suffering from a painful laboratory experiment.

Some scholars disagree with the idea that showing preference for one's own species is ethically wrong. For example, the American philosopher Michael Bradie agrees that evolutionary biology shows that nothing rigidly separates humans from other species, but he argues that this does not prove that speciesism is wrong. Bradie claims that evolutionary theory indicates that we humans, as social beings, are biologically disposed through evolution to give preferential treatment to our close relatives. This does not mean that we cannot extend our moral circle to include other species, but it is less arbitrary than the term speciesism indicates.

Finally, some use the term speciesism to also include preference given to certain nonhuman species. For example, the website veganism.com states that when humans favor companion species, such as dogs and cats, over species that they eat, such as cows and pigs, this is speciesism because no *relevant* differences can be identified in terms of intelligence and capacities for suffering.

Connie L. Johnston

See also: Advocacy; Animal Liberation Front (ALF); Animals; Chimpanzees; Cruelty; Dogs; Ethics; Evolution; Humans; Intelligence; Rights; Sentience; Social Construction; Species

Further Reading

Bradie, M. 1994. *The Secret Chain: Evolution and Ethics*. Albany, NY: State University of New York Press.

McAnallen, S. 2012. *The Superior Human?* Cairns, Australia: Ultraventus Film.

Rachels, J. 1990. *Created from Animals: The Moral Implications of Darwinism*. New York: Oxford University Press.

Scully, M. 2003. *Dominion: The Power of Man, the Suffering of Animals, and the Call to Mercy*. London: St. Martin's Griffin.

Singer, P. 1976. *Animal Liberation*. New York: Harper Collins.

Waldau, P. 2003. *The Specter of Speciesism: Buddhist and Christian Views of Animals*. Oxford Scholarship Online.

Species Survival Plans. *See* Endangered Species

Stereotypic Behavior

Animals sometimes display repetitive behaviors that do not have obvious function or purpose and are not common in natural living. These behaviors are called stereotypy (Greek; *stereo*: solid, without change; *typus*: type). Even if a behavior is abnormal, it cannot be considered as a stereotypy if it lacks repetitive nature and has obvious functions. Studies have related stereotypies with captivity and lack of environmental enrichment (providing something extra so that the environment is not barren) and thus with the welfare quality of the animals. Often, stereotypy in animals is compared to obsessive-compulsive disorder in human beings.

Stereotypies are prevalent in a wide range of species, including domesticated animals, captive wild animals, and lab animals. Commercially raised farm animals display stereotypies such as belly nosing (repeatedly rubbing a nose onto another's stomach), sham-chewing (chewing air), and bar biting in pigs and tongue-rolling, self-suckling, and rolling-eyes in calves. Stereotypic behaviors in chickens include pacing and feather-pecking (one hen pecking feathers of other hens in a group). Self-biting and repetitive movement are also common stereotypies in captive primates. Companion animals display stereotypies like circling, pacing (walking to and fro), and self-biting in dogs, and weaving (repetitive swaying side-to-side), stall-walking (walking rapidly around a stall), wind-sucking (sucking air into the wind pipe without having any solid object in the mouth), and cribbing (biting a stall door or fence and sucking in air while having a solid object in the mouth) in horses. Barbering (excessive hair loss by pulling) is a common stereotypy in lab rodents (rats and mice).

To complicate the matter, what apparently seems to be normal behavior in one species might be abnormal in others. For example, ruminants (animals that have a four-chambered stomach and chew regurgitated, semidigested food) display chewing behavior even when they are not provided with food. This is a normal behavior because they swallow their food first and chew the cud (a portion of food that returns to the mouth from the stomach for second digestion in ruminants) later on. However, pigs also display a similar behavior of chewing, but it is considered a stereotypy. This is because pigs are mono-gastric (one chambered stomach) and do not chew cud. When they display chewing behavior, it is just air so the behavior is called sham-chewing.

Studies have shown that stereotypic behaviors are more prevalent in captive environmental conditions where animals are not able to display their natural behavior, when they have a barren environment, or cannot interact much with other animals of the same species and lack natural light, sounds, and other stimuli. For these reasons, stereotypies are believed to be the result of suboptimal environmental conditions and poor welfare standards.

Animals in natural conditions display certain behaviors. For example, birds are used to laying eggs in nest-sites. Industrially farmed laying chickens, however, are housed indoors where they do not have access to nest-sites to lay eggs. In such a case, birds start developing pacing behavior, which ultimately develops into a stereotypy. Laying hens are also raised under controlled lights. Sometimes, due to an imbalance in their internal physiology as a result of lack of sunlight, laying hens start pecking feathers, which later develops as a stereotypy. Similar issues arise with industrially farmed pigs. Pigs display rooting behavior in the wild. Sows (female pigs) in confined concrete stalls do not have substrate (straw or soil) to root in. Sows usually fight to maintain a hierarchy when they are in nature, which they are not able to do while individually housed. They then start chewing metal bars out of frustration, which turns into a stereotypy. In the case of monkeys used in scientific research, their natural behaviors of jumping and climbing trees are unavailable to them in their cages; so out of frustration they start displaying stereotypies like pacing and self-injury.

There could be a synergistic (additive) effect of multiple factors to prolong or increase the rate of a certain stereotypy. Once animals start displaying the behaviors, they become habit and the animals continue displaying the behaviors even after they go back into their natural environment. Self-harm and injury can lead to losses in animals raised for economic purposes and animals used in research.

Different approaches have been taken to minimize stereotypies. These include social and environmental enrichment. Social enrichment can be done by housing animals in pairs or groups if they are social animals. It can be also achieved by frequent interaction with humans. Environmental enrichment can be achieved by providing more individual space, toys, and interactive tools, depending on species (e.g., scratch posts for cats, ropes and rooting materials like straw and soil for pigs); modifying the physical environment (light duration and intensity, sound, air quality); and altering food quality (e.g., inclusion of high-fiber diet in cows). The types and extent of stereotypies may also depend on how the animals were raised in their early life, the breed and genetics of the animal, as well as individual differences. The approach should be to provide a natural environment as much as possible such that animals can display their normal behavior.

Avi Sapkota

See also: Cruelty; Ethics; IACUCs; Welfare

Further Reading

Broom, D. M. 1983. "Stereotypies as Animal Welfare Indicators." *Current Topics in Veterinary Medicine and Animal Science* 23: 81–87.

Colgoni, A. 2005. "Abnormal repetitive behavior in captive animals." Accessed August 4, 2015. http://www.aps.uoguelph.ca/~gmason/StereotypicAnimalBehaviour/library.shtml

Garner, J. P. 2005. "Stereotypies and Other Abnormal Repetitive Behaviors: Potential Impact on Validity, Reliability, and Replicability of Scientific Outcomes." *ILAR Journal* 46(2): 106–117.

Marriner, L. M., and Drickamer, L. C. 1994. "Factors Influencing Stereotyped Behavior of Primates in a Zoo." *Zoo Biology* 13(3): 267–275.

Mason, G. J. 1991. "Stereotypies: A Critical Review." *Animal Behaviour* 41: 1015–1037.

Nora, P. n.d. "Towards an Understanding of Stereotypic Behaviour in Laboratory Macaques." Accessed August 4, 2015. http://www.awionline.org/lab_animals/biblio/at-phil.htm

T

Tagging. *See* Tracking

Tail Docking. *See* Body Modification

Taxidermy

Taxidermy, deriving from two Greek words meaning arrangement of skins, is the craft of preparing and mounting animal skins to appear "lifelike" (Patchett 2015). Often considered creepy and gruesome, the lifelikeness of taxidermy animals is a source of discomfort for many. Moreover, with actual skin and bone reflecting the killing practices that went into their making, the reasons for ownership of taxidermy specimens and displays have been called into question. As a result, many personal collections have been relegated to back rooms while those on institutional display are not to be replaced. Yet it is precisely their provocative presence that has inspired a new wave of interest in, and reuse of, taxidermy specimens and displays. Whether it is in museums and galleries, designer boutiques, or homes, taxidermy animals are once again making their presence felt. They have even found a home in academia, where they are being utilized as important resources for telling complex histories of human-animal relations.

The craft of taxidermy emerged in response to one of the major technical challenges confronting 18th-century European naturalists: how to preserve animal specimens for taxonomic study (the description, identification, and naming of species). Enormous quantities of animal skins were being sent back to Europe from Africa, Asia, and the New World by naturalist-explorers. However, the skins were often in poor condition due to the crude preservation techniques administered. In 1748, French naturalist René-Antoine Réaumur (1683–1757) published a small pamphlet describing all known methods for preserving animal skins, which included stuffing. Stuffing, a rudimentary form of taxidermy, consisted of drying or tanning animal skins and then literally stuffing them with cotton or wood wool (shavings of wood). However, this technique on its own was inadequate for maintaining permanent study collections as it failed to tackle the problem of insect attack. This was remedied when French apothecary Jean-Baptiste Bécœur (1718–1777) devised his "arsenical soap" skin treatment, and versions of this formula have been used by museums as an insecticide until relatively recently.

Taxidermy teacher Kyle Tubbs helps student Paige Kasper with the placement of a rabbit's pelt during a 2016 class at Croswell-Lexington High School in Croswell, Michigan. While often seen as something only hunters do, taxidermy has a long history in education as a tool of natural sciences, and today is enjoying an expansion of interest as people look for ways to have hands-on engagement with animals. (AP Photo/Carlos Osorio)

With adequate preservation of skins ensured, the keepers of natural history collections turned their attentions to putting them on display. This required sculpting a body, often using the skeletal structure as a base and then binding wood wool around it to create the bodily form, before arranging the skin on top. Dynamic displays using this technique were showcased at the Great Exhibition of London (the first international display of manufactured items) in 1851. Hugely popular with the general public, they increased demand for more engaging trophy and decorative taxidermy.

Trophy and decorative taxidermy had emerged alongside scientific taxidermy, when wealthy sportsmen-naturalists sought to amass their own personal "natural history" collections. However, instead of reflecting scientific knowledge, these collections were intended to showcase their owner's colonial conquests and hunting prowess. By the late 19th century, large trophy taxidermy firms were competing to secure the lucrative hauls of these big-game sportsmen. For example, to outdo their rivals, the company Rowland Ward's of London developed a technique of modeling the flesh and muscles of animals out of clay, which enabled them to meet their clients' demands for dynamic poses and animated (e.g., snarling) expressions. They even developed a range of animal furniture, such as zebra-hoof inkwells and elephant-foot umbrella stands. Although abhorrent today, at the time they were merely a way to make something, and money, out of the "waste products" of big-game taxidermy because clients really only wanted the heads.

Museum taxidermy was also evolving in response to displays at the Great Exhibition, shifting from taxonomic to diorama display toward the end of the 19th century. Diorama display took realism to the next level by presenting animal mounts in a re-creation of their natural habitat. The taxidermist Carl Akeley (1864–1926) is considered to have created some of the finest examples of habitat dioramas as part of his African Hall of Mammals at the American Museum of Natural History in New York City. Although designed and built by teams of people, including groundwork and scenic artists, the dioramas were brought together by Akeley's ruling artistic vision: to produce scenes of "nature in perfection" (Haraway 1989, 42). To ensure this, Akeley had led several large-scale big-game hunting trips to Africa to procure and preserve the best pelts to work with. However, in creating pristine scenes of African wildlife back at the museum, Akeley, according to social theorist Donna Haraway's famous critique of the hall, had cleaned up the "violence against nature" that went into their making (Haraway 1989, 42).

Thus, today, an increasingly conservation-conscious museum public has questioned the legitimacy of having "death on display" (Alberti 2011). However, a new wave of curators and academics are harnessing the provocative presence and difficult histories of taxidermy animals to engage with modern audiences (Patchett and Foster 2008). Outside of museums and galleries, taxidermy remains a major side-industry of the hunting and trapping economies, with auctions and sales being held regularly in the United States and online. In addition, taxidermy has even become part of the urban hipster culture with "rogue" taxidermy—the creation of fantasy style animals such as rats with wings (Palet 2014). What is clear from this short survey is that taxidermy animals, from the earliest trade skins to their lifelike and even tacky manifestations, offer rich resources for exploring the complexity of human-animal relations over time and place. So, the next time you meet the glass or hollow-eyed stare of a taxidermy animal, critically consider its journey from life to death and back again.

Merle Patchett

See also: Ethics; Hunting; Trophy Hunting; Zoology

Further Reading

Alberti, S., ed. 2011. *The Afterlives of Animals: A Museum Menagerie.* Charlottesville, VA: University of Virginia Press.

Haraway, D. J. 1989. *Primate Visions: Gender, Race, and Nature in the World of Modern Science.* New York: Routledge.

Palet, L. 2014. "Art from Death: Taxidermy as a Creative Hobby." Accessed January 25, 2016. http://www.npr.org/2014/08/09/338940500/art-from-death-taxidermy-as-a-creative-hobby

Patchett, M. 2015. "The Taxidermist's Apprentice: Stitching Together the Past and Present of a Craft Practice." *Cultural Geographies* 23(3): 401–419.

Patchett, M., and Foster, K. 2008. "Repair Work: Surfacing the Geographies of Dead Animals." *Museum and Society* 6(2): 98–122.

Poliquin, R. 2012. *The Breathless Zoo: On Taxidermy and the Cultures of Longing*. University Park, PA: Pennsylvania State University Press.

Wonders, K. 1993. *Habitat Dioramas: Illusions of Nature in Museums of Natural History*. Uppsala, Sweden: Almsqvist and Wiksell.

Taxonomy

The word "taxonomy" has a number of different definitions and usages but overall relates to classification and ordering. The word's origins come from the Greek *taxis*, meaning "arrangement," and *nomia*, meaning "method." With regard to human-animal relations, the term's most important usage is in the context of the biological sciences, where the word "taxonomy" generally refers to either the branch of this scientific field that identifies, describes, orders, and classifies organisms or to the classification system itself. A taxonomist uses morphological (physical appearance) characteristics and, increasingly today, genetic information about organisms in order to classify them.

In Western society, evidence of early taxonomic systems appears in ancient Greece in the work of the philosophers Aristotle (384–322 BCE) and Theophrastus (370–285 BCE) and ancient Rome with Pliny the Elder (23–79 CE). Classification based on these systems persisted until the Swedish botanist, physician, and zoologist Carl Linnaeus (1707–1778) published his *Systema Naturae* (1735 and 1758) and *Species Plantarum* (1753). In seeking to describe and classify the entire natural world, Linnaeus (referred to as the "father of modern taxonomy") developed a hierarchical system based on morphology within three broad "kingdoms"—Animalia, Mineralia (although not living organisms), and Vegetabilia. Although used before him, Linnaeus was the first to regularly use the Latin binomial nomenclature (two-part names comprised of genus and species) that is now the international scientific standard. An example of this nomenclature is the biological name for humans, *Homo sapiens*, given by Linnaeus in his recognition of humans as a species of animal. Linnaeus's taxonomic system is still the one used in Western science today, although it has been modified for new biological discoveries and knowledge.

It is generally believed that all human societies have practiced some form of taxonomy to categorize and organize the world. This is and would have been helpful in understanding, remembering, and communicating the functions of things in the natural world with respect to, at a minimum, usefulness (e.g., edible or medicinal plants) and danger (e.g., insects that sting), and would not have to be as comprehensive or detailed as the Linnaean system in order to be useful for daily life. The Linnaean system is, and has been, useful in biology because it has provided a systematic format for classification that has been broadly generalizable, proving useful in categorizing a spectrum of life forms that includes fungi, microbes, and humans, and species that are found both throughout the globe and only in one location. It has also been useful as a standard scientific format that can be used by both professional and amateur scientists and students from differing backgrounds and cultures.

To categorize only living organisms, the Linnaean system was modified to include only the Animalia and Plantae kingdoms as the most comprehensive levels of classification. Additional kingdoms (and other names) were proposed after Linnaeus to reflect knowledge gained about the extent and diversity of microorganisms. Although still a subject of debate, the two-kingdom model was modified relatively recently. The new system (proposed in 1990 by American biologist Carl Woese [1928–2012] and now generally accepted throughout the biological sciences) has as the highest levels three *domains*—Archaea, Bacteria, and Eukaryote. The first two domains contain only single-celled organisms with no nuclei, and the second contains all the multicellular organisms, which includes all animals and plants.

This new division of life reflects that the animal and plant life forms that were, for centuries, thought to comprise most of the living organisms on Earth actually are only a small part. The categories in this new system from most general to most specific are now domain, kingdom, phylum, class, order, family, genus, and species. There can also be subordinate categories as well, such as subphylum and subspecies. In this system, a dog, for example, would be classified as follows: domain—*Eukaryote*; kingdom—*Animalia*; phylum—*Chordata* (animals with a spinal column); class—*Mammalia*; order—*Carnivora*; family—*Canidae* (includes coyotes, foxes, and wolves); genus—*Canis* (includes coyotes and wolves); species—*Canis familiaris* (The dog's two-part species name reflects the standard binomial nomenclature).

Although early taxonomic classification was done primarily through identification of similar physical characteristics, today these are only one facet. For example, biologists may encounter an animal (currently living or thought to be extinct) that they believe may be an undocumented species, and initially they will compare the morphology of this animal with earlier records—perhaps descriptions, drawings, and/or specimens from centuries ago. Further analysis may involve dissection, and frequently DNA comparisons with existing species are performed. Based on all the findings, the scientists will determine if the animal is truly a new species. If so, it will be classified based primarily on its genetic relationship to other species previously identified, and the species description will include how it differs from these others. Finally, the new species will be given a name with the Latin binomial structure.

Despite its status as the accepted standard within the biological sciences, the Linnaean system has been challenged and critiqued. Critics assert that this system, rather than being based on objective, natural categories, represents a worldview arising from a particular historical and cultural context. For example, the historian of science and feminist scholar Londa Schiebinger (1952–) has pointed out that Linnaeus's class of *mammalia*, based on the ability for females to secrete milk, was challenged by other taxonomists who proposed classifications based on skin covering (e.g., hair or feathers) and heart structure.

Connie L. Johnston

See also: Animals; Biodiversity; Biogeography; Endangered Species; Humans; Hybrid; Microbes; Species; Zoology

Further Reading

Hickman, C., Keen, S., Larson, A., Eisenhour, D., I'Anson, H., and Roberts, L. 2013. *Integrated Principles of Zoology,* 16th ed. New York: McGraw Hill.

Linnaeus, C. 1735. *Systema Naturae.* Stockholm: Laurentius Salvius.

Linnaeus, C. 1753. *Species Plantarum.* Stockholm: Laurentius Salvius.

Schiebinger, L. 2008. *Nature's Body: Gender in the Making of Modern Science.* New Brunswick, NJ: Rutgers University Press.

Woese, C. R., Kandler, O., and Wheelis, M. L. 1990. "Towards a Natural System of Organisms: Proposal for the Domains Archaea, Bacteria, and Eucarya." *PNAS* 87: 4576–4579.

Tigers

Cats evolved around 30 million years ago, and the tiger, *Panthera tigris*, emerged in Asia about 2 million years ago. At the beginning of the 19th century, over 100,000 tigers ranged throughout Asia. Since then, human pressures have forced tiger subspecies into extinction and a fourth is extinct in the wild. Fewer than 3,200 wild tigers now live in 13 tiger range countries: Bangladesh, Bhutan, Cambodia, China, India, Indonesia, Lao PDR, Malaysia, Myanmar, Nepal, Russia, Thailand, and Vietnam. The remaining subspecies (Bengal, Malayan, Sumatran, Indochinese, and Amur) are found in isolated pockets of just 7 percent of their former range. Tigers have coexisted with humans for over 55,000 years. The tiger's deeply embedded cultural significance and ecological importance, contrasting with human pressures that compromise their survival, make tigers a key topic for Human-Animal Studies.

Habitat destruction has contributed to the decline of tigers, but hunting is the primary driver for rapid tiger population decline. The British ruling class extensively hunted tigers during Britain's occupation of India (1858–1947). After India gained independence (1947), hunting was no longer reserved for royalty, and hunting of tigers and their prey dramatically increased. Lack of wild prey has forced tigers (usually elderly, sick, or nursing individuals) to enter villages to kill livestock. Others have adapted to hunt humans. Human-tiger conflict is a great concern in conservation, as conflict can result in retaliatory killing of tigers or intolerance for coexistence.

Tigers are the largest cat species. Amur tigers, commonly known as Siberian tigers, can grow to 700 pounds. The Bengals of Eastern India and Bangladesh are the smallest tigers, with an average weight of 150 pounds. Tigers are apex predators (top-level feeders on which no other animal preys) with exceptional physical ability. They see as well as humans in daylight and six times better at night. They are expert swimmers, they can climb trees, and they can jump 6 meters (nearly 20 feet) in a single leap. Even a massive Amur tiger can run more than 50 miles per hour. They primarily hunt pigs and deer but have been known to eat fish, birds, bears, and even crocodiles. Tigers are a keystone species (one that provides a critical ecosystem function), so their survival in the wild has tremendous ecological importance.

The tiger has great cultural significance, and contemporary culture is filled with tiger representations in everything from plush toys to tattoos. Human representations of the tiger date from 4000–3500 BCE. Celebrated in poetry, art, song, dance, and story, tigers have been depicted flying, opposing dragons, in the stars, and with gods on their backs. They are honored as forest guardians in all tiger range areas. Many cultures bestow the tiger with supernatural powers and identify the tiger as a god or a direct ancestor. Some Indian paintings depict the tiger living peacefully among humans in villages, but the tiger is also feared. People living near tiger territory leave shrines to tiger gods in exchange for protection from tigers when entering the forest.

The tiger is the third sign of the Chinese zodiac and is woven into Chinese culture as a symbol of valor, virility, and majesty. Tigers are believed to protect against fire, theft, evil, and disaster. The stripes on each tiger are unique, but every tiger displays the Chinese *wang* symbol for King (王) on its forehead, similar to the way domestic tabby cats sport a forehead letter "M." The tiger is a foundational figure in the Chinese practice and philosophy of *Feng shui*, which seeks to bring people and their environments into harmony by balancing female and male energy. The white tiger represents the female energy *yin* that balances the male *yang*. According to *Fung shui*, the most powerful *chi* (life force) cannot be attained without the tiger.

Today, the tiger's mystical life force is its greatest threat because the tiger is being hunted into extinction for its vital qualities. Traditional Chinese medicine (TCM) philosophy holds that one can enhance one's own characteristics by consuming those of an animal. Chinese medicine practitioners use tiger parts to improve sexual vitality, bring fertility, calm fright, cure ulcers, relieve cramps, prevent infection, prevent demonic possession, and provide relief from countless other ailments. Throughout modern Asia, beliefs persist that the tiger can bring rain, stop children's nightmares, provide protection, combat evil, heal the sick, and bring light and peace to Earth. The World Federation of Chinese Medicine Societies decries the use of tiger parts, but the demand for tiger parts for Chinese medicine is the primary driver for poaching (illegal killing) in every tiger range country.

The UN Convention on International Trade of Endangered Flora and Fauna (CITES) has banned trading of tigers and tiger parts since 1975, but poaching and trafficking continues. Killing of tigers continues throughout the world because tiger protection is regulated differently at national and regional levels. More tigers exist in captivity than in the wild. In China, tigers are farmed for their body parts, and in some U.S. states tigers can legally be owned as pets or used in "canned hunts" (hunts that confine tigers, making it easier for hunters to kill them).

Tiger conservation is supported worldwide, and in the last decade efforts to protect them have improved. In 2010, the 13 remaining tiger range countries signed the St. Petersburg Declaration on Tiger Conservation, a pledge to work collaboratively to restore wild tiger populations.

Anita Hagy Ferguson

See also: Agency; Black Market Animal Trade; Canned Hunting; Cats; Deforestation; Ecotourism; Endangered Species; Extinction; Fur/Fur Farming; Human-Wildlife Conflict; Keystone Species; Poaching; Taxidermy; Traditional Chinese Medicine (TCM); Trophy Hunting; Wildlife; Wildlife Management; Zoos

Further Reading

Boomgaard, P. 2001. *Frontiers of Fear: Tigers and People in the Malay World, 1600–1950*. New Haven, CT: Yale University Press.

Ellis, R. 2005. *Tiger Bone and Rhino Horn: The Destruction of Wildlife for Traditional Chinese Medicine*. Washington, DC: Island Press.

Green, S. 2006. *Tiger*. London: Reaktion Books.

Jalais, A. 2010. *Forest of Tigers: People, Politics and Environment in the Sundarbans*. New Delhi: Routledge.

Meacham, C. 1997. *How the Tiger Lost Its Stripes: An Exploration into the Endangerment of a Species*. New York: Harcourt Brace & Company.

Saunders, N. J. 1991. *The Cult of the Cat*. London: Thames and Hudson.

Tilson, R., and Nyhus, P. J., eds. 2010. *Tigers of the World: The Science, Politics and Conservation of "Panthera Tigris,"* 2nd ed. London: Elsevier.

Totemism. *See* Indigenous Religions, Animals in

Tracking

Tracking refers to practices through which people learn about the location, movements, migratory patterns, habits, and behaviors of wild animals, from a distance. People use tracking to hunt prey animals or study elusive or easily disturbed species

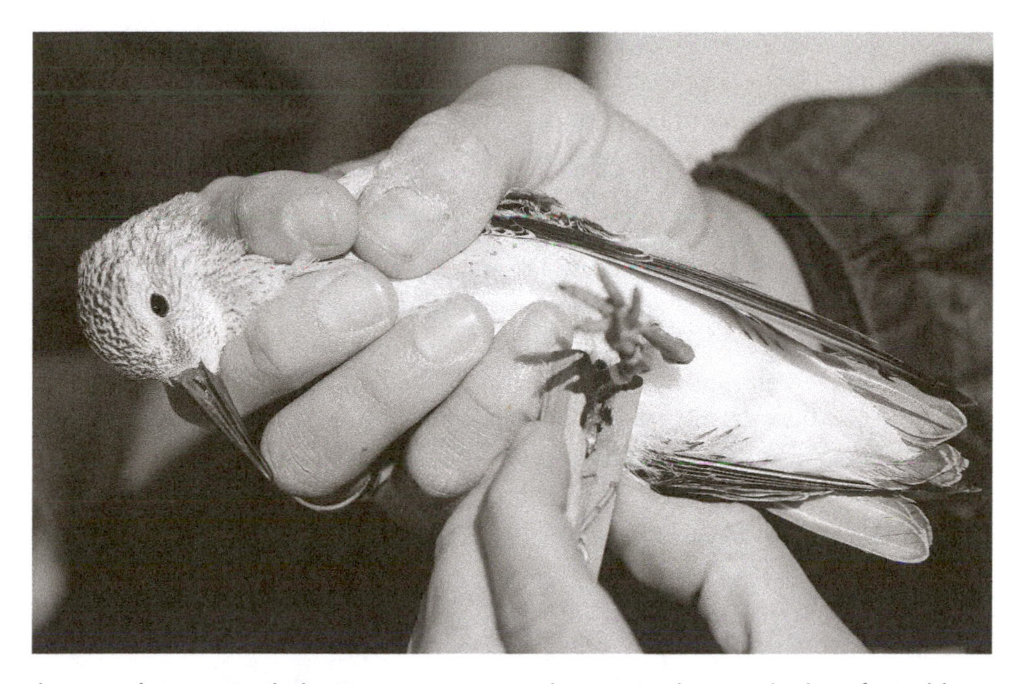

A researcher uses a clothespin to secure a geolocator in place on the leg of a red knot shorebird in Eastham, Massachusetts, in 2013. Red knots are one of the longest-distance migrants in the animal kingdom, traveling as much as 9,300 miles from the Arctic south to the Caribbean and South America on a mere 20-inch wingspan. Tracking technologies help hunters locate and catch animals, but technologies are also essential to field-based studies of animal behavior such as this one. These studies can help us understand the conservation needs of different species. (AP Photo/Stephan Savoia)

when continuous or close observation is difficult or impossible. They utilize various methods, including the collection of evidence onsite, using other animals such as hunting dogs, and the application of technologies such as radio collars. Because these methods and tools evolve over time, the costs versus benefits of tracking practices for both humans and nonhumans must be continuously reassessed.

Since ancient times, traditional peoples and skilled naturalists have practiced the art of tracking to find the animals they depended on for subsistence or hunted for pleasure. They look for footprints, broken branches, hair or fur samples, territorial markings (e.g., tree scratches), and impressions left on the ground. This is because certain species—such as nocturnal solitary creatures who inhabit dense forest, subterranean burrowing animals, long-range migrants, or deep-diving fish—are harder to find and monitor than other species. To help hunters find or retrieve prey animals in the field, certain dog types, such as retrievers, fox hounds, and bloodhounds, were bred specifically for their tracking abilities.

For conservation, scientists collect information on animals using the most sophisticated tracking technologies. Since the mid-1900s, conservation biologists have utilized high-tech equipment for animal tracking, often repurposed from the military, including radar, sonar, radio collars, animal-borne video cameras, light-sensitive geolocators, sensors inserted below the skin, and GPS transmitters (glued on or worn in a harness), which communicate with satellites in real-time. Such devices have been frequently applied to beloved, charismatic species such as elephants, grizzly bears, whales, condors, wolves, and jaguars. As a result, tracking has solved many mysteries, such as where migratory species go, while producing new questions, such as whether subspecies populations with vastly different migration routes should still be considered the same species. Improvements in tracking devices—such as reduced size, weight, and cost; multipurpose functionality; increased battery life; and more and finer details (such as solar-powered tracking collars that also monitor body heat, heart rate, and temperature)—have allowed more species and different kinds of animals to be successfully tracked. For example, it is now possible to place tracking devices on a monarch butterfly or a hummingbird, as well as deep-diving, highly migratory fish such as tuna.

One example is the case of the red knot (*Calidris canutus rufa*), a migratory shorebird with one of the longest migrations in the world, almost 20,000 miles a year, a hemisphere-wide migration route impossible for researchers to survey in full. Only in the past few years have tracking devices become light and small enough to be worn on the birds' backs. First results from these tracked journeys have changed the way researchers think about and protect the *rufa* subspecies, as they discovered populations that either fail to migrate the full distance or travel to destinations previously unknown to researchers.

The purpose, sustainability, and methods of animal tracking are controversial. Critics charge that hunting and tracking practices are inhumane and unethical, and many wonder if the benefits of tracking are worth the costs for the animal and for society. With better, cheaper, and more widely available technology, the impacts of tracking for different purposes are amplified. For instance, the proliferation of increasingly inexpensive fish-finder devices on fishing boats leaves fish with no place to hide, leading to overfishing and the collapse of global fisheries. Similarly, the increased use of helicopters and drones to hunt big game, such as wolves and elephants, leaves wide-roaming animals who range across open habitats similarly exposed. Tracking for research causes stress to the animal through the use of tranquilizer guns, darts, and/or anesthetic drugs. Sometimes animals are hurt, maimed, or killed in the process of netting, tagging, or recapturing.

Invasive methods of tracking, such as the placement of computer chips under the skin of animal research subjects (without the individual animal's consent), could be considered a type of injury caused in the name of science. Scientists don't know for sure the negative effects on behaviors and fitness borne by animals who wear tags; for example, they can only guess the effects of drag when a diving bird's tracking tag hits the water, or if a harness worn on an animal compromises its ability to attract a mate or clean itself fully. Finally, park managers wonder if seeing a radio collar or identification tag on park wildlife ruins visitors' experience of nature, raising the question of whether an animal whose movements are monitored at all times remains a wild animal. Despite these concerns, tracking continues to be a powerful way for people to monitor wildlife in their habitats (in situ), providing greater proximity to nature, advantages for hunting and fishing, and a fuller scientific understanding of the daily activities of free-ranging animals.

Jenny R. Isaacs

See also: Bluefin Tuna; Hunting; Wildlife Forensics; Wildlife Management

Further Reading

Benson, E. 2010. *Wired Wilderness: Technologies of Tracking and the Making of Modern Wildlife*. Baltimore: JHU Press.

Brown, T. 1986. *Tom Brown's Field Guide to Nature Observation and Tracking*. New York: Berkley Books.

Cagnacci, F., Boitani, L., Powell, R. A., and Boyce, M. S. 2010. "Animal Ecology Meets GPS-Based Radiotelemetry: A Perfect Storm of Opportunities and Challenges." *Philosophical Transactions of the Royal Society B: Biological Sciences* 365(1550): 2157–2162.

Moll, R. J., Millspaugh, J. J., Beringer, J., Sartwell, J., and He, Z. 2007. "A New 'View' of Ecology and Conservation through Animal-Borne Video Systems." *Trends in Ecology & Evolution* 22(12): 660–668.

Niles, L. J., Burger, J., Porter, R. R., Dey, A. D., Minton, C. D. T., González, P. M., Baker, A. J., Fox, J. W., and Gordon, C. 2010. "First Results Using Light Level Geolocators to Track Red Knots in the Western Hemisphere Show Rapid and Long Intercontinental Flights and New Details of Migration Pathways." *Wader Study Group Bulletin* 117(2): 123–130.

Wall, J., Wittemyer, G., Klinkenberg, B., and Douglas-Hamilton, I. 2014. "Novel Opportunities for Wildlife Conservation and Research with Real-Time Monitoring." *Ecological Applications* 24(4): 593–601.

Weidensaul, S. 2012. "Unlocking Migration's Secrets." *Audubon Magazine*. Accessed July 9, 2015. http://www.audubon.org/magazine/march-april-2012/unlocking-migrations-secrets?page=5

Traditional Chinese Medicine (TCM)

Traditional Chinese medicine (TCM) is a general term in Western society that refers to the diagnosis, treatment, and prevention of illness based on ideas derived from the Chinese medical system. TCM has also had an influence on the development of the medical traditions of Korean, Japanese, Vietnamese, and Tibetan cultures. We can identify three main aspects of TCM, comprehensively, which can appear either separately or intermixed as 1) a philosophical framework in which Chinese medical knowledge can interact with Confucianism, Taoism, and Buddhism;

2) a medical practice based on the doctor-patient relationship; and 3) daily health care regimens. Animals play a variety of roles in TCM, and the relationship between TCM and animals varies under different contexts.

The earliest record of medical practice in China dates back to around the 12th century BCE and was deciphered from an oracle bone script (text written on turtle shells and other animal bones). Sometime during the 1st century BCE, the first systematic and comprehensive TCM text—*The Yellow Emperor's Inner Canon* (author unknown)—was compiled; it conceptually emphasizes a holistic and balanced philosophy, which forms the basis of this whole system of medicine. The earliest manual of TCM medical materials, *Shennong Emperor's Classic of Materia Medica* (author unknown), was compiled between about the 2nd century BCE and the 2nd century CE.

One role that animals play in TCM is as symbols or metaphors. Many cases of animal images have been found in *The Yellow Emperor's Inner Canon*. This fundamental medical text demonstrates how a human body corresponds to the universe or natural world by using abundant landscape and animal imagery as metaphors to support the use of clinical techniques. For example, with arterial pulse evaluations (touching the pulse to gather physical information), the imagery of a fish swimming and rising to the surface is used to describe a kind of pulse movement.

Animals have inspired people to create a range of Chinese exercise forms, such as *Dao yin*, which is a series of body-stretching exercises with a special breathing technique and thought to have been a common form of health regimen in early China. It is well known for its animal-based poses, such as Bear-hanging and Ape-call, which mimic the physical movements of these animals. *Dao yin* is also associated with the developments of *Qigong* (a practice that combines breathing, movement, and mindfulness) and Chinese martial arts. All of these exercise forms share the same basic TCM view of the human body.

Animals are also used as ingredients in traditional Chinese medicinal cuisine, either as health foods or as food therapy. TCM dietary theory determines the kind of food that is good or bad for each individual specifically by taking into consideration an individual's constitution, physical status, and the seasons, as well as which foods can go together. Accordingly, a person who is considered to have a cold/frail body constitution is advised that eating foods believed to have hot or warm energy (such as Chinese herb soup with mutton [sheep meat]) is appropriate.

In the TCM pharmacognosy (the study of drugs derived from natural sources), animals (in addition to plants and minerals) are exploited as the source of medicinal ingredients. Humans obtain the necessary parts through both lethal and non-lethal approaches; an example of the latter would be gathering the ecdyses (shedded skins or casings) of snakes and some insects. The practices of killing or directly harming animals for medicines include hunting wildlife (e.g., for tiger bones as drugs for rheumatic disorders) and farming animals to periodically obtain body parts (e.g., deer antlers as drugs for several acute infectious diseases) or specific substances (e.g., bear gall bladder bile to treat human liver/gall bladder disorders).

It is controversial to use body parts from animals, particularly from tigers, rhinos, and bears, and advocates have launched campaigns at both international and local levels. The grounds for opposition are 1) unreliable therapeutic properties, 2) availability of more effective alternatives, 3) conservation of threatened animal species, and 4) animal cruelty. During the 1990s, in response to pressure from the United States and several European countries, Chinese and Taiwanese authorities banned the use of tiger and rhino parts from their respective markets. However, the illegal trade engaging in the TCM medicine and food supplement market has not collapsed. Also, bear parts are still used legally in China, which is one of the crucial ethical issues for many animal advocates. Bears are usually kept confined in small cages while being continuously attached to catheters that remove the bile from their gall bladders. Bears may be kept alive in these conditions for years, and many people see it as unnecessary cruelty as there is no evidence that bear bile is an effective medicine and plenty of alternatives exist.

Animals have also taken part in TCM as patients. In ancient China, such TCM treatments were primarily for working animals. The earliest text involving official TCM practitioners who specifically dealt with the health care of animals appeared in *The Zhouli* (Bureaucracy and Government of the Zhou Dynasty; author unknown) around the mid-second century BCE. Moreover, several ancient medical books were dedicated to domesticated animals. Today, companion animals are the foremost beneficiaries of these treatments. At present, some veterinarians use therapies based on TCM theories, and this practice is called traditional Chinese veterinary medicine (TCVM), a specialized field emerging in the West. Common treatments in TCVM include practices such as acupuncture, Chinese massage, food therapy, and traditional TCM medicines.

Yu-ling Kung

See also: Black Market Animal Trade; Eastern Religions, Animals in; Ethics; Non-Food Animal Products; Social Construction

Further Reading

Animals Asia Foundation. 2015. "End Bear Bile Farming." Accessed June 28, 2015. https://www.animalsasia.org/intl/our-work/end-bear-bile-farming

Hinrichs, T., and Barnes, L., eds. 2013. *Chinese Medicine and Healing: An Illustrated History*. Cambridge, MA: Harvard University Press.

National Geographic. 2016. *Treating Animals with Acupuncture*. Accessed May 8, 2016. http://video.nationalgeographic.com/video/news/160331-news-animal-acupuncture-vin

Xiong, E., Yuanzhong, C., and Qiao, T. 2011. *Moon Bear*. Accessed June 28, 2015. https://www.youtube.com/watch?v=3-eOlbWXJCY

Trafficking, Animal. *See* Black Market Animal Trade; Endangered Species

Trophy Hunting

Trophy hunting is a form of hunting in which the hunter takes as a "trophy" part of the hunted animal—usually the skin, antlers, horns, or head. The morality of trophy hunting is questioned because, in contrast to subsistence hunting (hunting for food and/or needed supplies), it is unnecessary for survival. Some animal protection organizations encourage lawmakers to ban the practice, whereas other groups, the most famous of which is Safari Club International (SCI), create the hunting experience for tourists. Trophy hunting may have positive effects on species conservation funding but negative effects on local animal populations.

Because hunters display trophies at home or in hunting lodges, many pursue big "game" (the term for animals who are hunted) such as large cats and bears, or the largest and "best" of other species. In that they are kept objects taken from nature, trophies are similar to totems (sacred or religious natural objects) used by certain indigenous tribes around the world. However, totems differ from trophies in the spiritual power ascribed to them by their possessors, while modern Western trophies display the hunter's power over the killed animal.

Trophy hunting may occur, apart from in the wild, at ranches in the United States and some African nations. Ranchers stock their land with game, and some offer "canned hunts" for animals who have been made easier to kill: raised by humans and kept in enclosures from which they can't escape. This practice is controversial even among hunters, as it falls outside of the concept of "fair chase," an important piece of the hunters' code of ethics. "Fair chase" describes the situation in which a hunter, on foot, has a chance to kill a wild animal, but the animal will generally escape (Posewitz 1994).

The continent of Africa is a popular site for trophy hunts, partially due to the presence of the "Big Five": lions, leopards, elephants, rhinoceroses, and water buffalo—the favorite African game. SCI hosts contests there, and in North America, for the "North American Twenty-Nine": all species of bear, bison, sheep, moose, caribou, and deer on the continent. The organization keeps a record book of kills to encourage competition. Greenland is a third popular hunting site, for reindeer and musk oxen.

There are three categories of concerns about trophy hunting. One, expressed by (largely Western) animal protection organizations includes questioning the rationale for killing individual animals if they are not needed for survival. This includes questioning the importance of both sport and trophy and also the intentions of hunting organizations that profit from the hunts. In 2015, the illegal killing of Zimbabwe's "Cecil the Lion" by an American hunter for $50,000 drew strong international criticism—Cecil had been both a tourist favorite and part of a scientific study. The backlash strengthened political measures against trophy hunting.

A second concern about trophy hunting, in places like Africa, is that it appears to mimic social structures between black Africans and white non-Africans from the colonial period. During this period, which lasted from the 19th to approximately mid-20th centuries, European powers ruled and exploited much of the continent for natural resources. The majority of trophy hunters are wealthy and white and may pay a great sum of money for the hunting experience, whereas many local (poorer, black) residents are legally forbidden from hunting for wildlife conservation reasons. In addition, upper-class landowners with ranches benefit the most from trophy hunting—and they are also mostly white males.

The third body of concern over trophy hunting surrounds its negative impact on conservation and acknowledges that some of the species targeted are endangered. Regulating bodies such as the Convention on International Trade in Endangered Species of Wild Fauna and Flora (CITES) permit and regulate the trade and hunting quotas for endangered species within participating countries. For example, the hunt of an endangered black rhino, of which only 4,000–5,000 exist, sold for $350,000 in 2015. In the United States, the Endangered Species Act (ESA) permits the sale of hunting permits for endangered species if the generated funds will support the species' survival. Some animals move from threatened lists to game lists and back as their populations wax and wane in response to hunting: Gray wolves in Michigan and black bears in Florida represent two such species in the early 2010s.

Trophy hunting may decrease the numbers of large adult males and affect the social structures and/or the genetic traits of the species that are hunted, as well as decrease local populations and affect other species that feed on the leftover carcasses. Quotas imposed by regulating bodies such as CITES limit the numbers that are taken to minimize potential negative effects. Other alternatives, such as wildlife viewing and photography, may bring in funds without the same negative effects.

Trophy hunting also has positive effects on conservation. Wildlife conservation efforts require large amounts of funding that may be (and are, in many countries) generated through fees associated with hunting, in general, and trophy hunting especially. Fees from wildlife viewing alone have not always been enough. Trophy hunters may inhibit poaching (hunting that is illegal due to location, method, species hunted, individual hunted, or other reasons) and help protect wildlife. There continues to be a debate as to whether or not these potential positive effects outweigh the negative ones.

Heather Pospisil

See also: Canned Hunting; Elephants; Endangered Species; Human-Wildlife Conflict; Hunting; Poaching

Further Reading

Crosmary, W-G., Cote, S. D., and Fritz, H. 2015. "The Assessment of the Role of Trophy Hunting in Wildlife Conservation." *Animal Conservation* 18: 136–137.

The League Against Cruel Sports. 2004. "The Myth of Trophy Hunting as Conservation." Accessed June 30, 2015. http://www.league.org.uk/~/media/Files/LACS/Reference-material/League-Against-Cruel-Sports-2004-The-Myth-of-Trophy-Hunting-as-Conservation.pdf

Lindsey, P. A., Roulet, P. A., and Romanach, S. S. 2007. "Economic and Conservation Significance of the Trophy Hunting Industry in Sub-Saharan Africa." *Biological Conservation* 134: 455–469.

Posewitz, J. 1994. *Beyond Fair Chase: The Ethic and Tradition of Hunting*. Guilford, CT and Helena, MT: Morris Book Publishing, LLC.

U.S. Fish & Wildlife Service. 1973. *The Endangered Species Act.* Washington, DC: Department of the Interior. Accessed June 30, 2015. http://www.fws.gov/endangered/esa-library/pdf/ESAall.pdf

V

Veal Crates. *See* Factory Farming

Veganism

Veganism is a form of dietary choice in which one abstains from consuming all meat and animal products, including milk and eggs. For almost all vegans, this abstinence extends to the use of all kinds of animal-derived products, including fur, wool, leather, and animal fats that are used for other consumer goods, like soap. Concerns over animal rights and animal welfare are the driving force behind veganism, also known as ethical veganism. Specifically, vegans aim to exclude all forms of exploitation of, and cruelty to, animals, even where no animal life is taken for human use. In terms of setting an ethical relationship between humans and animals, then, vegans are exemplary in their desire to not harm animals as much as practically possible. The extent of such a commitment has led many to view veganism as a lifestyle choice underpinned by a strong moral philosophy.

While the adherents of selected ancient religions, like Jainism (a small Indian religion with about 6 million present-day followers), are commonly vegans, veganism has had a shorter history than vegetarianism in the Western world. It was only as recent as 1943 that a member of the British Vegetarian Society (the oldest organization established to promote vegetarianism) argued against vegetarians consuming milk. By late 1944, a breakaway faction of the Vegetarian Society began publishing a competing newsletter called "Vegan News" and establishing a rival organization called the "Vegan Society." The society, led by Donald Watson (1910–2005), who also coined the term "vegan," advocated that humans' diet should be derived from fruits, nuts, vegetables, and grains. It also encouraged the use of alternatives to all animal-derived products.

Vegans argue that it is unethical to consume animal products (like eggs and milk), even though the animal is not killed directly in the production process, because production causes animal suffering and, indirectly, killing. In the dairy industry, for example, the most unresolvable welfare issue is that cows need to be constantly impregnated in order to lactate (produce milk). This and the desire for high milk production severely compromise the life span of dairy cows, with many living only, at most, 5–6 years (compared to the 20 years that an average cow might live). This is due to a number of reasons. First, the typical dairy cow has been bred to produce

Animal rights activists promote a vegan diet in Boston in 2007. The vegan diet, relying entirely on plants, is noted by many experts to be more environmentally friendly and sustainable than a diet that includes meat. This assertion is based on the fact that meat production is one of the largest producers of greenhouse gases worldwide. (AP Photo/Lisa Poole)

as much as 7–8 times more milk a day than they would need to feed their own calves. This additional weight that the cows have to carry means that lameness is a common ailment. Moreover, high milk production also results in a higher incidence of the painful udder infection mastitis. Last but not least, the cows' ability to become pregnant and lactate begins to decline after only a few years and therefore they become economically unproductive. Additionally, most male calves are economically unproductive for the dairy industry and are slaughtered or placed into veal production soon after birth.

Despite strong moral reasons to become vegan, it is a lifestyle choice that comes with considerable challenges. In the first place, omitting the use of animals in any form from our daily lives is wrought with practical difficulties. At the broadest scale, veganism is arguably not workable in particular places and among particular peoples. Nomadic tribes and other people who live in harsh environments (e.g., the Inuit of Canada) have to consume/use animal products (e.g., fur, meat, and fats) for survival as the weather conditions do not allow for most kinds of permanent agriculture. This suggests that a vegan lifestyle for them is impractical, if not outright impossible. Even those who are able to commit to a vegan lifestyle are faced with other practical obstacles, one of which is the traceability of the food we consume and the products we use. Vegans must rely on the rigor of food and product labeling to ensure that no animal products or byproducts have been used. Yet, there are cases where animal-derived products have slipped into the production process unknown. These might include the inclusion of animal-based gelatin in confectionery (e.g., marshmallows) or edible food dyes that are derived from insects and used in a range of products like lipstick, candy, and ice cream. Even with perfect traceability, studies have also shown how vegans sometimes feel socially excluded and in extreme cases discriminated against due to their beliefs and dietary choices. With the gradual increase in restaurants and supermarkets that offer vegan options, such feelings of exclusion may be lessened, however.

Finally, a more fundamental debate over veganism revolves around the possibility of consuming animal products that are produced in ways that do not harm the animal. The case of honey is interesting because the crux of the issue is whether bees are *necessarily* "exploited" when we harvest honey. The consensus is that honey is nonvegan because in addition to a concern for animal exploitation, cruelty, and lives, implicit in the belief of many vegans is the notion that the animal kingdom does not exist for the sake of humans. Rather, animals such as bees have their own natural right to flourish on Earth. Hence, to "use" animals for food or other things, even without harm, cruelty, or necessarily causing the deaths of the animals, is a form of exploitation. This implies that the rationale for veganism goes beyond the simple notion of prevention of harm. Hence, although some ethicists argue that vegans can consume food animals like oysters that presumably cannot feel pain or fear and therefore can experience no meaningful harm, it is not an argument that is accepted by many vegans.

Harvey Neo

See also: Bees; Ethics; Livestock; Rights; Vegetarianism; Welfare

Further Reading

Cherry, E. 2006. "Veganism as a Cultural Movement: A Relational Approach." *Social Movement Studies* 5(2): 155–170.

Leneman, L. 1999. "No Animal Food: The Road to Veganism in Britain, 1909–1944." *Society and Animals* 7(3): 219–228.

McPherson, T. 2014. "A Case for Ethical Veganism." *Journal of Moral Philosophy* 11(6): 677–703.

Vegetarianism

While most people adopt a diverse diet comprised of meat and nonmeat (e.g., fruits, vegetables, and legumes), a significant minority of the global population abstain from the consumption of meat for a variety of reasons. There also exist many advocacy groups that aim to persuade more people to not consume meat.

In a 2010 study based on 29 countries (representing 54 percent of the world's population), Leahy et al. extrapolated that there are 75 million vegetarians by choice in the world and almost 1.5 billion vegetarians by necessity. The latter are defined as people who are too poor to afford meat but would likely consume meat if/once their income level improves. Vegetarians by choice are people who make the conscious decision not to consume meat. Vegetarianism is a general term that encompasses different degrees of meat abstinence. The most common understanding of vegetarianism is lacto-ovo vegetarianism. This is a diet that abstains from meat consumption but includes milk and eggs. Ovo-vegetarians are those who would consume eggs but not milk. Veganism is a diet that abstains from consuming all meat and other animal byproducts. Besides these, there are other forms of diets that are considered "semi-vegetarianism" by those who practice them. For example, pescetarians consume seafood but abstain from meat. In recent years, flexitarianism has become popular. It is a form of diet akin to semi-vegetarianism with someone consuming mostly plants, dairy, and eggs but occasionally eating meat and fish. In most cases, flexitarians will often consume meat only if the animals are raised ethically.

Studies have shown that there are three somewhat interrelated reasons driving consumers to become vegetarians. They are concerns over animal welfare, the environment, and health. Ethical vegetarians are people who avoid eating meat because they recognize the dismal animal welfare implications of the modern meat-producing sector and the suffering that food animals experience. In some cases, such ethical vegetarians are also driven by the dictates of their religion to adopt a meat-free diet (for example, followers of Jainism—an Indian religion founded in the sixth century BCE with about 6 million present-day followers—are vegetarians, as are many Hindus and some Buddhists). Environmental vegetarians are those who are particularly concerned with the environmental impacts of meat production. A 2006 UN Food and Agriculture Organization report concluded that the livestock sector is one of the most significant contributors to the most serious environmental problems, at every scale from local to global. Finally, the last group of vegetarians are motivated by health issues to turn to a nonmeat diet, believing that omitting meat will lead to better health. These three reasons are often present with any vegetarian, albeit in different degrees.

Ethically reared food animals to some extent will moderate both the suffering of the animals as well as the negative environmental impacts in the meat production system. However, for many ethical vegetarians, it is not enough to minimize the suffering of food animals because they view the killing of animals for human consumption as fundamentally wrong. In any case, public education by nongovernmental organizations (NGOs) is imperative in persuading people to adopt a vegetarian diet. In most of such campaigns, NGOs appeal to any combination of the three reasons outlined above.

Although the number of vegetarians is increasing, meat consumption has increased steadily as well because those who continue to consume meat are consuming it at much higher quantities. Vegetarian societies have a long history, with the first known advocacy group for vegetarianism, "The British and Foreign Society for the Promotion of Humanity and Abstinence from Animal Food" (the precursor for the United Kingdom's "Vegetarian Society"), established in 1843 in England. Today, animal rights groups like People for the Ethical Treatment of Animals (PETA) are active in publicizing the cruelty of the modern meat industry in an attempt to convince more people to stop eating meat.

Of late, there have been large scale events to promote vegetarianism. For example, "Meatout," a result of grassroots social activism, is held on the first day of spring in the United States to educate people about reducing meat consumption. First organized in 1985, the nationwide event aims also to persuade people that a vegetarian diet is more wholesome. There have also been city-level efforts to promote vegetarianism through the symbolic declaration of a meat-free day each week. The earliest city to adopt this was the Belgian city of Ghent, where each Thursday is designated as a "Veggie Day." Other cities that have adopted a meat-free day each week include Cape Town in South Africa, Bremen in Germany, and São Paulo in Brazil. While such a gesture is impossible to be legally enforced, it is to the credit of vegetarianism advocates to have successfully launched such high-profile sociopolitical campaigns amid skepticism and resistance from various quarters (for example, the restaurant industry). Indeed, there are signs that "Meatout" initiatives are increasingly being replicated in other places like universities and private companies.

In the final instance, due to the political and economic clout of global meat-producing companies, coupled with the ever-declining costs of producing meat, the demand for meat is projected to increase in the immediate future. Despite the realities of the harmful effects of the modern meat-production system in terms of animal welfare and environmental well-being, there exists a cognitive dissonance within the average consumer where simple awareness of such realities is not sufficient to prompt them to make a drastic dietary change. Nonetheless, there remains much scope to persuade them to consume less (as opposed to no) meat.

Harvey Neo

See also: Ethics; Livestock; Veganism; Welfare

Further Reading

Emel, J., and Neo, H., eds. 2015. *The Political Ecologies of Meat.* London: Routledge.

Leahy, E., Lyons, S., and Tol, R. 2010. "An Estimate of the Number of Vegetarians in the World." *ESRI Working Paper,* No. 340.

Neo, H. 2016. "Ethical Consumption, Meaningful Substitution and the Challenges of Vegetarianism Advocacy." *Geographical Journal* 182(2): 201–212.

Steinfeld, H., Gerber, P., Wassenaar, T., Castel, V., Rosales, M., and De Haan, C. 2006 Rome: *Livestock's Long Shadow: Environmental Issues and Options*. Rome: Food and Agriculture Organization of the United Nations.

Vermin. *See* Nuisance Species

Veterinary Medicine

Veterinary medicine's first known description of an animal doctor is seen in the tomb of King Ur-Ningursu of Lagash (ca. 2200–2000 BCE), whose empire stretched from the Persian Gulf to Syria. Its formal recognition as a profession did not occur until France opened the world's first veterinary school in 1762. Iowa State University opened the first American veterinary medical school in 1879; there are now hundreds of veterinary schools worldwide.

American and Canadian veterinary programs are generally 8 years—4 years of undergraduate and 4 years of post-baccalaureate study. The first 2-1/2 to 3 years of the professional program are lecture-based; the last 1 to 1-1/2 years are clinical. During the didactic portions of the program, students learn the basic biological, chemical, and physical sciences of medicine. Students also learn how to recognize normal and abnormal functions and presentations of seven core species: dogs, cats, cattle, horses, pigs, goats, and birds. During the clinical year, students learn to write medical notes, obtain accurate histories, perform thorough physical exams, assess patients, create diagnostic plans, diagnose, and create treatment plans for patients under the direct guidance of licensed veterinarians. Students are responsible for all aspects of patient care, including the technical skills, such as placing catheters, drawing blood, and safe animal handling and restraint.

Clinical rotations vary depending upon the program, but generally provide experience in large and small animal internal medicine, emergency medicine, surgery (including orthopedic, soft tissue, and general), radiology, anesthesia, community practice, and clinical and gross pathology. Elective rotations covering topics such as exotic animal medicine, oncology, shelter medicine, neurology, cardiology, dentistry, and theriogenology (reproduction) are also usually offered.

Each rotation has different requirements. Anesthesia, for example, consists of determining appropriate sedatives (pre-medications) and anesthetics for each patient based on age, species, breed, surgical procedure, and overall health. Students must then calculate appropriate doses, administer medications, and monitor each patient from injection of pre-medications to recovery. During an oncology rotation, however, students diagnose and identify specific types of cancer, create and present appropriate treatment options (e.g., surgery, radiation therapy, chemotherapy) to clients, and assist the clinical professor in administering any therapy the client has decided to pursue.

Veterinary medical credentials vary globally, with only the United States and Canada awarding doctoral degrees, although bachelor's degrees from the accredited

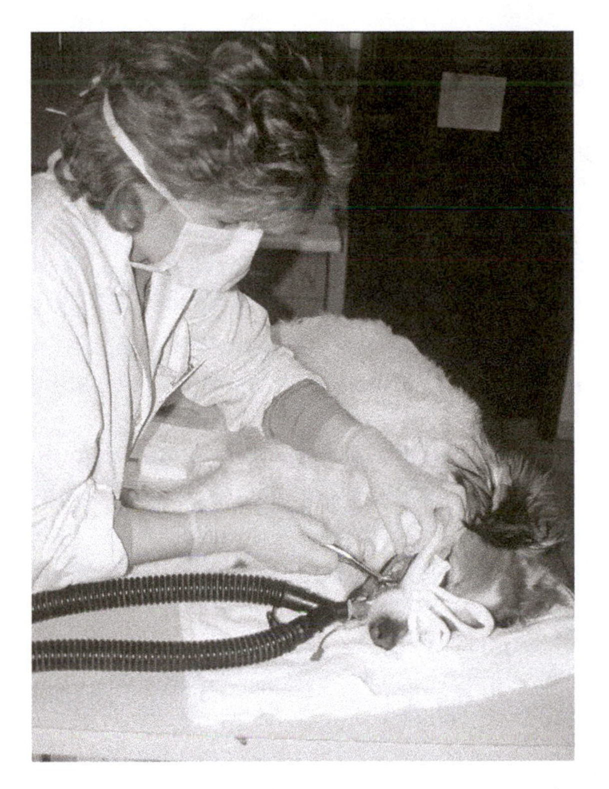

A veterinarian operates on a dog. In addition to companion animals, veterinarians treat agricultural animals, horses, exotic animals, and wildlife. In the United States, there is high demand for veterinarians in practices that treat companion animals. (Corel)

foreign veterinary programs are considered equivalent. Accredited programs are overwhelmingly located in highly industrialized regions (North America, Western Europe, and Australia), where higher disposable incomes allow for advanced medicine and surgical techniques, primarily focused on companion animals. In regions with lower incomes, such as Africa and Asia, veterinarians work with older technology and medicines and are more focused on livestock.

All accredited programs have met the minimum standards set forth by the American Veterinary Medical Association (AVMA), the organization charged with establishing guidelines that ensure every graduate has obtained the entry-level skills and competencies needed to pass the North American Veterinary Licensing Exam (NAVLE). Veterinary students must pass the NAVLE to be eligible to apply for state licensure. Each state sets forth its own licensing requirements, and veterinarians must obtain licenses to practice in each state in which they wish to work.

Students graduating from non-accredited foreign veterinary programs who would like to practice in the United States must successfully complete the Educational Commission for Foreign Veterinary Graduates (ECFVG) certification program to become eligible for a license to practice. Applicants must pass a series of exams covering English language proficiency, knowledge of basic and clinical veterinary science, and clinical veterinary medical skills.

Accreditation of foreign veterinary medical schools is controversial. Domestic critics believe the AVMA should focus solely on new U.S. veterinary programs, and accrediting foreign schools catering to Americans is causing a saturation of the workforce. Global critics question the idea that the United States is viewed as the "gold standard" for veterinary education, and they propose that animal health needs differ from country to country, making a "one size fits all" approach inappropriate. Supporters believe that workforce issues should not be considered and that accreditation elevates the standards for graduating veterinarians in the United States and around the world.

As the most expensive of all health-sciences programs, significant debt is driving the majority of veterinarians into clinical practice, where salaries are highest. Few enter research and academia, leaving the profession at the back edge of research, teaching, and inter-professional collaborative efforts. Yet, the United States continues to increase the number of graduating veterinarians through expanding class sizes and opening new veterinary schools.

Some argue the changing demographic of American veterinary practitioners has also driven this shortage of non-clinical veterinarians. The profession officially became predominantly female in 2007, a trend that shows no signs of abating. Currently about 75 percent of veterinary students are female, although there is still a distinct lack of racial and ethnic diversity within the profession.

Veterinarians treat food animals, horses, companion animals, wildlife, and exotic animals. Many pursue additional training in one of 40 different specialties, each eligible for board certification through one of 21 recognized veterinary specialty organizations. To become a board-certified specialist, veterinarians must successfully complete an internship and residency and pass a credential review and rigorous exam administered through the given specialty organization. This additional training ranges from 2–6 years.

Not all veterinarians go into clinical practice. A small number go into research, teaching, government (including politics, food inspection, and the military), industry, pharmaceuticals, epidemiology, and public health.

Katherine Fogelberg

See also: Cats; Dogs; Exotic Pets; Horses; Husbandry; Livestock; Pets; Shelters and Sanctuaries; Spay and Neuter

Further Reading

Dunlop, R. H., and Williams, D. J. 1996. *Veterinary Medicine: An Illustrated History*. St. Louis: Mosby.

Fiala, J. 2013. "Overseas Opinion Dulls Repute of AVMA Accreditation as 'Gold Standard.'" Accessed May 27, 2015. http://news.vin.com/VINNews.aspx?articleId=28587

Fiala, J. 2014. "Bid to End Foreign Veterinary Accreditation Dies at AVMA Meeting." Accessed May 27, 2015. http://news.vin.com/VINNews.aspx?articleId=30444

McPheron, T. 2007. "2007 Is DVM Year of the Woman." Accessed May 27, 2015. https://www.avma.org/News/JAVMANews/Pages/070615d.aspx

National Research Council. 2011. "Workforce Needs in Veterinary Medicine." Accessed May 27, 2015. http://www.nap.edu/catalog.php?record_id=13413

Vivisection

From the Latin *vivus* (living) and *secare* (cut), the word "vivisection" means "to cut the living." The term mirrors the word "dissection" (used by anatomists), which referred to the cutting of tissue from dead animals. Vivisection was used by 19th-century physiologists (scientists studying the normal functions in living organisms) to refer to the act of performing scientific experiments on living animals. The term publicly came to connote all kinds of painful experiments on animals and was skillfully used by the European antivivisection movement of the latter part of the 19th century in their campaigns. The word "vivisection" therefore became a political term by the late 19th century and represented the moral dangers to society of uncontrolled scientific advancement. All use of animals for scientific purposes came to be called vivisection.

As a research method, animal experimentation is as old as medical science itself. During antiquity, experiments on, and dissections of, living animals were performed in places like Alexandria, Egypt, where anatomical and physiological sciences flowered. During the Middle Ages the practice became less frequent; however, experiments with poisons were performed on both humans and animals. With the Renaissance and the Scientific Revolution of the 16th and 17th centuries, the method became fashionable again. With the discovery by William Harvey (1578–1657) of the circulatory system at the beginning of the 17th century, vivisection as a method was widely legitimized and the number of vivisections increased. Harvey studied the circulatory system in vivisected animals by observing the heart pump and showing how blood only flowed in one direction.

Vivisection was criticized early on scientifically as well as morally. Scientifically, the arguments against vivisection said that the differences between humans and animals were significant enough that the practice would give little or no valuable information about the human body. Furthermore, it was argued that the vivisected animal was under such extreme conditions of pain and agony that its bodily functions could not be expected to be normal. The philosophical teachings of René Descartes (1596–1650) about animals being senseless automata (machines without feelings), however, were used by some scientists to justify the practice. At the same time, vivisection as a method was popularized and public displays of the practice were sometimes performed. This evoked disgust and horror in those not convinced by Descartes' teachings.

The first organized opposition against vivisection as a method of science started in the 1860s with a group of British women led by the feminist and animal activist Frances Power Cobbe (1822–1904). The debate about vivisection really heated up in the 1870s. The cause of this debate was a textbook in physiological methods that was published in 1873. The book sparked opposition because it did not discuss sedation or methods of pain relief for the animals. One of the authors of the book replied to the critics that pain relief was only necessary to relieve the scientist from being scratched or bitten by the animals. The debate following the comments on the book became very animated and ultimately resulted in the first law regulating experiments on animals, the Cruelty to Animals Act, in 1876 in England. The law stated the terms under which painful experiments on animals were allowed and stipulated that vivisection could only be used in medical science specifically aimed at saving human lives. In addition, dogs, cats, donkeys, and horses were not to be used in experiments. The English law on vivisection became a prototype of laws in several European countries, such as Germany and Sweden, during the following decades.

The movement against the use of animals for vivisection became relatively weak during the first half of the 20th century due to the impact of the two world wars. In the early 1970s opposition resurfaced alongside a rising animal advocacy movement. The psychologist Richard Ryder invented the term "speciesism" (as a parallel to racism and sexism) to explain the mechanisms of unfair treatment of nonhuman animals. The moral philosophies of the Australian philosopher Peter Singer (1946–), the British philosopher Mary Midgley (1919–), and the American philosopher Tom Regan (1938–) fueled the arguments against animal experimentation. The term "vivisection" thereby once more became fashionable for animal activists and was now used to represent all practices involving animals in science. As the term became a call to arms for the animal advocacy community, scientists began to shy away from using it themselves to avoid controversy over the use of animals in research.

The anti-vivisection movement has been partly successful even though large-scale experimentation (not only vivisection) is still performed all over the world on approximately 115 million animals each year. The largest victory has been won at the level of public opinion, which has led to increased regulations in many countries. Today, demanding an ethical perspective on all experiments involving living animals is not considered to be a suspect or fanatical idea, as was the case in the 19th century. The word "vivisection" is rarely used in the debate at the present time, but animal experimentation continues to be an issue of conflict in society. A wide variety of animals are exploited in a broad range of uses, which include medical testing, genetic research, chemical testing, and military purposes.

Karin Dirke

See also: Animal Law; Animal Liberation Front (ALF); Dissection; Ethics; People for the Ethical Treatment of Animals (PETA); Research and Experimentation; Speciesism

Further Reading

Birke, L., and Hubbard, R. 1995. *Reinventing Biology: Respect for Life and the Creation of Knowledge.* Bloomington, IN: Indiana University Press.

Hamilton, S. 2006. *Frances Power Cobbe and Victorian Feminism.* Basingstoke, UK: Palgrave Macmillan.

Linzey, A., and Linzey, C., eds. 2015. *Normalising the Unthinkable: The Ethics of Using Animals in Research. A Report by the Working Group of the Oxford Centre for Animal Ethics.* Accessed June 1, 2015. http://www.oxfordanimalethics.com/wpcms/wp-content/uploads/Normalising-the-Unthinkable-Report.pdf

Midgley, M. 1978. *Beast and Man: The Roots of Human Nature.* Ithaca, NY: Cornell University Press.

Regan, T. 1983. *The Case for Animal Rights.* London: Routledge & Kegan Paul.

Rupke, N., ed. 1987. *Vivisection in Historical Perspective.* London: Croom Helm.

Ryder, R. D. 2000. *Animal Revolution: Changing Attitudes Toward Speciesism,* revised and updated ed. New York and Oxford: Bloomsbury Academic.

Singer, P. 2002 [1975]. *Animal Liberation.* New York: Ecco, Harper Collins Publishers.

War. *See* Military Use of Animals

Welfare

Animal welfare (or animal well-being) is a concept that has developed substantially since the mid-20th century. Although many societies now utilize this concept and have related laws, ideas about what constitutes good or poor welfare can vary considerably, even within the same society. In Western society, animal welfare is frequently discussed as synonymous with animal rights; however, there are important differences. While focused on animals' well-being, animal welfare is a human-centered concept in that its purpose is to define and/or measure something that many humans have decided is important.

The concept of welfare relates to physical and/or emotional states of being, including actions that have an impact on such states. Human actions that are thought to contribute to good animal welfare are usually associated with being humane, for example, not only attending to animals' physical needs and minimizing suffering but also showing kindness. Promoting good welfare, therefore, is different from preventing cruelty, in that good welfare is seen as a *positive* state, whereas the goal of preventing cruelty is the absence of acts or conditions with *negative* effects. Basic needs for any species include adequate food, water, and shelter, but needs beyond these vary by species. For example, a polar bear's needs for shelter will be different from an iguana's, and social species such as humans, chimpanzees, and dogs need companionship. Although not always the case, in the present day most definitions of welfare assume that an animal must be "sentient" (having conscious awareness of experience).

Welfare is also technically distinct from rights. Advocates for animal rights are concerned with more than promoting humane treatment and good welfare. The rights position holds that (at least sentient) animals have a moral *right*, as individual beings, not to be exploited by humans. Although rights advocates are concerned with the treatment of animals on farms, in zoos, in laboratories, and so forth, their ultimate goal is not to simply ensure their good welfare, but for them to not be used by humans at all. Therefore, although in practice the distinction between welfare and rights advocacy may be difficult to see, animal welfare organizations primarily work within, rather than challenge, conventional ethical perspectives and legal institutions, which hold that nonhumans can be owned and used by humans.

Major questions related to animal welfare are how to know what factors contribute to good welfare and how to measure these factors. These questions have concerned scientists and animal advocates in locations such as North America and Europe throughout much of the 20th century and continue today. For example, in the United States, after animal farming became industrialized in the mid-20th century, it became common to focus on productivity as a measure of good welfare. That is, if a pig was growing quickly or a cow producing the expected amount of milk, their welfare was considered to be good. The animals' emotional states were not considered important factors. These views are changing, although somewhat slowly, due in large part to pressure from animal advocates and consumers. In farming and other captive environments (e.g., zoos), animals' welfare may also be measured in terms of stress, with high stress levels indicating poor welfare. Stress has been measured by the level of the cortisol (a stress hormone) in blood samples. Although this measurement is still used/useful in the present day, animals' behavior is also observed. For example, animals on farms, while productive, and in zoos have frequently exhibited stereotypic (repetitive, nonfunctional) behaviors, which indicate extreme boredom or frustration and, therefore, poor welfare. In all contexts where animal welfare is a concern, research is being done on the main factors, how to measure them, and how to improve the situation.

Laws are one way that a society as a whole expresses concern for animal welfare, and these laws vary widely. For example, a country may have few national, but relatively more local, laws. Such is the case with the United States, which nationally has only the broad (and many say too limited) Animal Welfare Act and two pieces of legislation covering farmed animal slaughter and transportation, but many welfare laws at the state and local levels. The European Union overall, and in particular Austria, Germany, and the United Kingdom, have some of the most protective animal welfare laws. At the other end of the spectrum, many African and Middle Eastern countries currently have no national animal welfare legislation.

Worldwide, many nongovernmental organizations (NGOs) are dedicated to animal welfare, working on such things as educating the public, lobbying for legal change, and promoting welfare certification/labeling on animal products. Their concerns include rescuing stray and injured animals, changing factory farming practices, promoting reductions in drug and products testing on animals, and protecting wildlife. Two longstanding national NGOs are the American and Royal Societies for the Prevention of Cruelty to Animals in the United States (ASPCA) and United Kingdom (RSPCA), respectively. International NGOs include Compassion in World Farming and World Animal Protection. World Animal Protection has taken the lead on the Universal Declaration on Animal Welfare (UDAW). Originally drafted in the early 2000s, UDAW's goal is to have nonbinding intergovernmental agreement on welfare standards and animal sentience. Although still in process, UDAW has achieved broad support, including from the Food and Agriculture Organization of the

United Nations (FAO), a number of governments (e.g., the United Kingdom and Costa Rica) and national veterinary associations (e.g., Chile and New Zealand), and the World Organisation for Animal Health (OIE).

Connie L. Johnston

See also: Advocacy; Animal Geography; Animal Law; Cruelty; Emotions, Animal; Empathy; Ethics; Ethology; Factory Farming; Humane Education; Humane Farming; Institutional Animal Care and Use Committees (IACUCs); Rights; Sentience; Stereotypic Behaviors; World Organisation for Animal Health (OIE); Zoos

Further Reading

Appleby, M. C., Hughes, B. O., Mench, J. A., and Olsson, A. 2011. *Animal Welfare,* 2nd ed. Wallingford, UK: CABI.

Bekoff, M. 2007. *The Emotional Lives of Animals: A Leading Scientist Explores Animal Joy, Sorrow, and Empathy—and Why They Matter.* Novato, CA: New World Library.

Broom, D. M. 2014. *Sentience and Animal Welfare*. Wallingford, UK: CABI.

Global Animal Law Project. Accessed February 1, 2016. https://www.globalanimallaw.org/index.html

Regan, T. 1983. *The Case for Animal Rights*. Berkeley: University of California Press.

Rollin, B. E. 1995. *Farm Animal Welfare: Social, Bioethical, and Research Issues.* Ames, IA: Iowa State University Press.

World Animal Protection. Accessed February 1, 2016. http://www.worldanimalprotection.org/take-action/back-universal-declaration-animal-welfare

Western Religions, Animals in

Judaism, Christianity, and Islam—often called the Western, or Abrahamic (traced back to the biblical figure Abraham), religions—have a mixed record with regard to human-animal interactions. All three traditions regard animals as subordinate to human beings because of a special connection between God and humanity. Adherents of all three traditions kill and eat animals, though the dietary restrictions of Judaism and Islam are concerned with quick and respectful slaughter. Each tradition also holds strong religious impulses toward compassion and kindness. Accordingly, many Jews, Christians, and Muslims provide religious reasons for becoming vegans, vegetarians, and animal activists. This brief entry describes mainstream teachings and practices in each tradition and does not claim to represent the many sects and dissenting positions.

Judaism exalts human beings as the most important of Earth's creatures yet also sets human beings into moral and spiritual community with nonhuman animals. One of the creation stories from the Hebrew Scriptures attributes the following words to God: "Let us make humankind in our image, according to our likeness; and let them have dominion over the fish of the sea, and over the birds of the air, and over the cattle, and over all the wild animals of the earth" (Genesis 1:26). This text asserts a special connection between human beings and God that all other animals lack. On this basis, the lives of animals are subordinated to human interests. However, numerous teachings in both the Torah (the heart of the Hebrew Scriptures) and rabbinic texts (e.g., Talmud and Midrash) prohibit cruelty and mandate attentive care for agricultural animals. For example, working animals participate in Shabbat—the weekly day of rest (Exodus 23:12)—and likewise, must be permitted to benefit from their labor ("Do not muzzle the ox while it is treading out the grain" [Deuteronomy 25:4]). This compassion reflects the conviction that God tends to the needs of all nonhuman animals, so that unnecessary cruelty stands in opposition to God's mercy. The dietary regulations of Judaism—Kashrut, or Kosher laws—distinguish between animals that may or may not be eaten and require that a special blessing be recited before food animals are slaughtered by cutting the throat with a knife. Leather and fur garments are not worn during the annual celebration of Yom Kippur (Day of Atonement) because it is unseemly to pray for divine compassion while wearing the skins of animals who were not shown similar compassion.

Originally a sect within ancient Judaism, Christianity shares many of Judaism's texts and teachings; the Hebrew Scriptures are also the Christian Old Testament. Notably, Christian teachings justify a rigid hierarchy subordinating nonhuman animals to human interests through the ideas of the image of God and human dominion mentioned above. The teaching that Jesus of Nazareth is the incarnation of God often intensifies this hierarchy, insofar as it represents another exclusive link between humanity and God. Popular Christian teaching about salvation focuses on the resurrection of human beings to a heavenly existence, offering little reference to other creatures even though the Christian New Testament provides a much more expansive view of the redemption of the whole cosmos. This narrow view of salvation restricts the importance of animal life to this world, denying animals a spiritual and eternal existence. Christians generally eat meat without restriction and have no prohibition against wearing fur and leather. Nevertheless, some mainstream Christian teachings support compassionate interaction with animals. Even dominion—the idea that God has placed creation under human authority—is frequently used to condemn cruelty to animals insofar as cruelty represents a distortion of prudent stewardship. Similarly, in Christian hagiography (fantastic stories of the saints) cooperative and communicative relationships with animals signify strong moral and spiritual character. Francis of Assisi (1181–1286), for example, is known for preaching to birds and communicating kindly with a wolf.

The Arabic root of the word "Islam" means submission, and submission to the will of God (*Allah*, in Arabic) represents the heart of the Islamic religious tradition. In Islamic teaching, animals and the whole natural world live in submission to God, who subsequently places them under human authority—a similar arrangement to that in Christianity and Judaism. Similar to the Kosher laws of Judaism,

Islamic dietary restrictions (*halal*, Arabic for permissible) forbid the eating of pork and require that animals are killed by cutting the throat with a knife after pronouncing the name of God. While both Christianity and Judaism have histories of ritual animal sacrifice, Muslims continue the practice in the present. The holiday Eid al-Adha commemorates the story of Abraham's near-sacrifice of his son Ishmael. The rituals and prayers of Eid al-Adha include the sacrifice of a valuable domestic animal (e.g., a cow) whose meat is shared with family, friends, neighbors, and the poor. Islam also contains numerous impulses toward compassion. There are many *hadith*—stories about the Prophet Muhammad that provide guidelines for proper behavior—telling of the Prophet's kindness toward animals. Likewise, in the Qur'an (the Islamic scriptures) God tests the people of Thamud by sending a camel to them. The people's failure to show compassion to the camel causes divine judgment to fall on them in the form of an earthquake. Finally, many Sufi texts (Sufism is a branch of Islam emphasizing mystical prayer and personal intimacy with God) celebrate empathy for animals—feeding stray dogs, and unburdening heavily loaded camels, for example.

Eric Daryl Meyer

See also: Eastern Religions, Animals in; Indigenous Religions, Animals in; Slaughter

Further Reading

Gross, A. 2015. *The Question of the Animal and Religion: Theoretical Stakes, Practical Implications*. New York: Columbia University Press.

Perlo, K. W. 2009. *Kinship and Killing: The Animal in World Religions*. New York: Columbia University Press.

Waldau, P., and Patton, K., eds. 2006. *A Communion of Subjects: Animals in Religion, Science, and Ethics*. New York: Columbia University Press.

Whaling

Whaling is the pursuit, capture, and killing by humans of whales and other cetaceans (a category that also includes dolphins and porpoises). This ancient method of food and resource acquisition reached its peak in the early 20th century and still occurs today, despite an international moratorium on commercial whaling. Whaling remains a controversial activity, primarily in countries where it was once, but is no longer, practiced. Much variety exists in the methods used for whaling and in the legal structures that govern whaling.

Whaling is historically and geographically widespread. It originated independently in antiquity in various locations, primarily throughout the Arctic, but also in Europe and Japan (Reeves and Smith 2006). From these and other points of origin, the techniques of whaling spread to additional locations. It was, and is, primarily a method of subsistence—the production of food and other necessities. The Basques, a people native to the region near the border of present-day France and Spain, were the first to commercialize whaling during the 11th century. When commercial whaling became widespread beginning in the 17th century and increasing into the 20th, whale populations began to decline severely, in some cases experiencing as much as an 80 percent reduction (Pfister and Demaster 2006). Today whaling is still widespread, occurring in all of the world's oceans by people of many nations, employing diverse techniques, and producing a variety of food and other products (Robards and Reeves 2011).

The main products of whaling are meat and blubber. Both are primarily used as food for human consumption. During the era of widespread commercial whaling before the 1986 moratorium, oil—used primarily for lighting, manufacturing, and lubrication—was the most valuable product from the whaling industry but has now been largely replaced by petroleum or other natural or synthetic oils. Other formerly valuable products include baleen (the bristled plates in the mouths of filter-feeding whales), used for clothing and umbrella supports; teeth, used in the art form known as *scrimshaw*; skin and organs used as food (skin was also sometimes tanned as leather); bones, made into tools and used as structural support for buildings; and ambergris (a biological waste product formed in the digestive tracts of sperm whales), used as an ingredient in perfumes. In nearly all cases, these products have been replaced by synthetics or have otherwise become obsolete.

The practice of whaling can take a variety of forms. The primary implement associated with whaling is the harpoon, which can be either thrown by hand or fired from a gun. The harpoon itself can be the instrument of death—if fitted with an exploding head—or can simply serve to attach the whale to the boat or to a floating buoy. If not killed by the harpoon, the whale's death usually comes from a lance or a gunshot. In some cases, the whale is simply allowed to die of exhaustion. The length of the process can range from a few seconds to several hours.

Whaling may also be conducted without a harpoon. Inuit (the Arctic aboriginal people of North America and Greenland) and other Arctic whalers typically shoot whales with rifles when they rise to breathe. Another method involves using multiple boats to drive whales onto the shore or into a net. Other whaling methods have made use of weirs (systems of posts placed in the shallow water to entrap whales or fish) or gates, behind which whales would become trapped during the ebbing tide.

Whaling vessels range in size from the tiny kayaks of the Inuit to the enormous factory ships employed by the Japanese. After killing a whale, whalers butcher the carcass, removing the meat, blubber, and any other usable parts in a process known as *flensing*. The meat and blubber are prepared according to local culinary traditions. Both can be eaten raw or cooked and can also be preserved by drying, salting, or freezing.

Throughout most of its history, whaling was not controversial. The "Save the Whales" movement began in the 1960s. Herman Melville's *Moby Dick*, published in 1851, anticipated some of the controversy that would later arise regarding issues of animal cruelty and overhunting of whales by wondering whether whales "can long endure so wide a chase, and so remorseless a havoc," or whether they "must not at last be exterminated from the waters" (1851, 432). People oppose whaling for a variety of reasons but most cite the intelligence of whales, the suffering inflicted upon whales as they are killed, or the risk of extinction of certain whale species. Proponents of whaling often cite cultural tradition and food security as the reasons whaling should continue.

The legality of whaling depends largely on the nationality of the whalers and the species of whale. The International Whaling Commission (IWC) is the global organization that regulates whaling. In 1986 the IWC enacted a moratorium on commercial whaling, meaning that its member countries agreed not to hunt whales commercially. Despite this moratorium, several ways exist by which whaling still occurs.

Countries that have not signed the IWC cannot be bound by the moratorium. Individuals in IWC member countries may hunt whales under two exceptions to the moratorium: the scientific exception and the aboriginal subsistence exception. Scientific whaling is permitted by the IWC if a member country shows evidence for the scientific value of data obtained through lethal research methods. Aboriginal subsistence whaling is permitted when an IWC member country shows that its population, or a subset thereof, has a nutritional or cultural need that must be met through whaling. Whaling operations that target cetaceans of species other than those that the IWC regulates—the 12 species of baleen whale and the sperm whale—are not forbidden by the moratorium. In these cases, only national policies—which vary from one country to another—regulate whaling. Finally, some "pirate whaling" does occur that is in open objection to, or in clandestine violation of, the IWC moratorium or national policy.

Russell Fielding

See also: Advocacy; Animal Law; Cruelty; Endangered Species; Ethics; Extinction; Indigenous Rights; Literature, Animals in

Further Reading

International Whaling Commission. 2015. Accessed March 20, 2015. https://iwc.int/home

Melville, H. 1851. *Moby-Dick, or, the Whale*. New York: Harper.

Pfister, B., and DeMaster, D. P. 2006. "Changes in Marine Mammal Biomass in the Bering Sea/Aleutian Islands Region before and after the Period of Commercial Whaling." In Estes, J. A., DeMaster, D. P., Doak, D. F., Williams, T. M., and Brownell, Jr., R. L., eds., *Whales, Whaling, and Ocean Ecosystems*, 116–133. Berkeley: University of California Press.

Reeves, R. R., and Smith, T. D. 2006. "A Taxonomy of World Whaling: Operations and Eras." In Estes, J. A., DeMaster, D. P., Doak, D. F., Williams, T. M., and Brownell, Jr., R. L., eds., *Whales, Whaling, and Ocean Ecosystems*, 82–101. Berkeley: University of California Press.

Robards, M. D., and Reeves, R. R. 2011. "The Global Extent and Character of Marine Mammal Consumption by Humans: 1970–2009." *Biological Conservation* 144: 2770–2786.

Starbuck, A. 1878. *History of the American Whale Fishery from its Earliest Inception to the Year 1876*. Waltham, MA: published by author.

Wildlife

Most simply, wildlife refers to non-domesticated animals. Domesticated animals are tame animals that people have developed through selective breeding. Yet the word "wildlife" and the animals to which the term refers are far more complicated than one might think. Wildlife is a term people have created to apply to certain animals, much like the words "pets" or "pests." Complicating matters further, the animals that societies have classified as wildlife have changed over time.

The word "wildlife" is a recently coined term. It came into common use during the 1930s as the field of wildlife management emerged in the United States as a way to study and conserve non-domesticated animals. At the time, wild animals were more commonly called "game," a revealing term that shows how hunters valued wild animals for sport. Wildlife managers saw their reason for existence as managing and perpetuating wildlife largely for the benefit of recreational hunters from cities. Prior to then, the two words "wild life" often "signified a way of life or even a lifestyle" and it could be applied to the actions of people as well as animals (Benson 2011, 419). By the late 19th and early 20th centuries, influential conservationists such as William Hornaday (1854–1937) began referring to certain groups of animals as wild life. These were animals in decline or threatened by extinction during the time as cities, industries, and agriculture expanded and took a toll on animal habitat.

Although most people in industrial countries have little contact with wildlife, they remain a subject of intense fascination. National parks, wildlife refuges, and other protected areas where the public can view wild animals remain popular. To take one example, bird watching, perhaps the most common form of wildlife viewing, is a popular activity enjoyed by millions of North Americans. The primary way people view wildlife, however, is not in the wild or in a zoo but through television, films, and online media. Wildlife films have been a staple of documentary filmmaking for over 70 years and wildlife programming is common on cable television. Because most peoples' direct interactions with wild animals are so limited, these cinematic depictions of wildlife play a key role in sculpting viewers' understanding of wildlife, for better or worse.

Some ostensibly domesticated animals can, under certain circumstances, be considered wildlife. Horses were domesticated by people thousands of years ago, and they are used today for, among other things, recreation and for racing. Yet in parts of the American West, some horses run free. Ranchers and environmentalists contend that the horses compete with cattle and some species of wildlife for grazing land. They see the horses not as wildlife but as feral (domesticated animals who have returned to a wild state) creatures and perhaps even pests. Others argue that despite once being domesticated, these horses are now wildlife and deserve the same sort of protection afforded other species such as deer, elk, and pronghorn antelope.

Some species of fish also complicate the distinction between wild and domestic. Rainbow trout are by the far the most popular fish sought by fishermen in North America. Yet many of the rainbow trout they see were bred and hatched in government fish hatcheries before being released into the wild. In many ways trout are raised in the same sort of industrial fashion as the cattle and chicken people eat for food.

Although wildlife generally refers to animals for which people have a favorable view, some wildlife can become pests, or nuisance species. Two types of wild animals—white-tailed deer and Canada geese—are considered pests in some parts of North America. Throughout much of the Northeast United States, white-tailed deer were extirpated (reduced to zero in a particular area) primarily by overhunting during the 19th century, but with enforced game laws, the regrowth of forests on abandoned farmland, and the decline in hunting during recent decades, the number of deer has exploded. White-tailed deer are now a regular sight in the backyards and streets of the region's suburbs, devouring gardens in the process, much to the dismay of homeowners. In the early 20th century, some feared Canada geese might go extinct due to hunting and habitat loss, but the animals have returned with a vengeance, finding the ponds and turf in suburban parks and golf courses as suitable for habitat as remote marshes and grasslands. People in cities find the copious droppings from the geese to be a real nuisance. The situation of these two animals points to how space is absolutely crucial to how we perceive animals, especially wildlife. For many Americans, wildlife ceases to be wildlife when they are no longer in wilderness areas and the countryside but instead venture into cities and suburbs.

Wildlife are also cultural icons. For instance, in the mid-19th century, the American bison, which once numbered in the tens of millions, nearly went extinct. For elite, white Americans, the destruction of the bison was a symbol of how modern industrial society ravaged wildlife and the natural world in general. For some Native Americans at the time, such as the Paiute and other tribes of the interior western United States, it was believed that the bison might return if Native peoples carried out a sacred dance. More recently, the polar bear has become, perhaps, the principal icon of climate change. Scientists and environmentalists fear that global warming will diminish the Arctic sea ice habitat polar bears depend on for survival and lead to their extinction. That an animal so few people have seen in captivity and fewer people still have seen in the wild has aroused such concern is testament to how Americans remained fascinated by, and concerned about, wild animals.

Robert M. Wilson

See also: Biodiversity; Ethics; Nuisance Species; Social Construction; Wildlife Management

Further Reading

Alagona, P. S. 2013. *After the Grizzly: Endangered Species and the Politics of Place in California*. Berkeley: University of California Press.

Benson, E. 2011. "From Wild Lives to Wildlife and Back." *Environmental History* 16: 418–422.

Lorimer, J. 2015. *Wildlife in the Anthropocene: Conservation after Nature*. Minneapolis: University of Minnesota Press.

Mitman, G. 2009. *Reel Nature: America's Romance with Wildlife on Film*. Seattle: University of Washington Press.

Warren, L. S. 2011. "Animal Visions: Rethinking the History of the Human Future." *Environmental History* 16: 413–417.

Wilson, R. M. 2010. *Seeking Refuge: Birds and Landscapes of the Pacific Flyway*. Seattle: University of Washington Press.

Wildlife Crime. *See* Wildlife Forensics

Wildlife Forensics

When crimes against wild animals occur, wildlife forensics is used to solve them and prosecute violators in order to prevent future abuses. Wildlife forensic scientists work in labs analyzing evidence, but they also visit crime scenes or testify in court on cases of illegal trafficking in protected wildlife, fishing and timber harvesting violations, poaching, animal cruelty, bioterrorism, oil spills, and other ecological disasters. Many challenges confront practitioners, including too many species and not enough experts, too many cases and too few certified labs, high costs of collecting and analyzing evidence, poor understandings of differences between species and subspecies, few reference samples available for many rare species, lack of professional standards, and the challenge of working between many agencies and countries. Wildlife forensics is an important tool of law enforcement, but it is limited in its capacity to handle the growing number of cases and causes of wildlife crime worldwide.

Demand for wildlife forensics grew with increasing public concern over biodiversity loss as well as the link between organized and wildlife crime operations. Researchers and practitioners drew together a network of scientists to help solve wildlife crimes, specifically within organizations such as the TRACE Wildlife Forensics Network (est. 2006) and the Society for Wildlife Forensic Science (est. 2009), which now offers professional certification to wildlife forensic scientists. In labs such as the National Oceanic and Atmospheric Administration Marine Forensics Unit and the U.S. National Fish and Wildlife Forensic Laboratory, these scientists examine, identify, and compare evidence from wildlife crime scenes through scientific processes such as toxicology, chemical analysis, and physical examination in order to link suspect and victim. When possible, they establish the identity of the animal, cause of death, and where it came from, as well as the time of the crime.

Because there are many kinds of wildlife crimes and the list of species encountered in wildlife forensic science is extensive, a huge variety of evidence types comes into the handful of wildlife forensic laboratories for identification. Analysis is requested for intact or processed parts of dead or live animals or plants and often must be done on the basis of fragmentary, modified, or badly damaged evidence. Methods such as chemical, fingerprint, or DNA analysis are often used. For example, because the capture of endangered Atlantic blue marlin brings criminal penalties, but the import and sale of blue marlin from the Pacific or Indian oceans is legal in the United States, forensic sampling is necessary; in this case, forensic tests determine those specimens sold in U.S. seafood markets which were taken illegally from the Atlantic Ocean.

There are several challenges in wildlife forensics. Because the field is new, there are only a few experts working with a limited database of genetic reference material to consult when identifying species or individuals; as a result, identification techniques are often improvised, based on experience with closely related species. Additionally, the transnational geography of wildlife crime is challenging, as trade and trafficking of endangered species frequently crosses international borders, making identification a global process. For instance, wildlife products are often harvested in one country and sent to foreign markets or sold on the Internet within the global economy—such instances are commonly prosecuted in the United States under the Endangered Species Act and the Convention on the International Trade in Endangered Species of Wild Fauna and Flora (CITES). These problems are compounded by the fact that certified wildlife forensics laboratories are few in number and expensive to operate, funding to fight these crimes is low, fines and punishments are weak compared to potential profits for smugglers, and costs are high for enforcement agents who engage in surveillance, which is resource intensive and time consuming. Effective enforcement requires not just forensics but a coordinated network of police and enforcement agencies, customs, quarantine, border control agencies, and nongovernmental organizations. Finally, the potential for wildlife forensics to disrupt wildlife crime networks is limited; while its applications might positively identify blood of an endangered species detected on a suspected poacher's clothing, resulting in conviction, in many instances that poacher was only one poorly paid link within an international smuggling ring. For these reasons, there is much room for growth and improvement within wildlife forensics.

Jenny R. Isaacs

See also: Elephants; Endangered Species; Ivory Trade; Poaching; Tracking; Traditional Chinese Medicine (TCM)

Further Reading

Burnham-Curtis, M. K., Trail, P. W., Kagan, R., and Moore, M. K. 2015. "Wildlife Forensics: An Overview and Update for the Prosecutor." *United States Attorneys Bulletin*. May: 68–69.

Cooper, J. E., and Cooper, M. E. 2013. *Wildlife Forensic Investigation: Principles and Practice*. Boca Raton, FL: CRC Press/Taylor & Francis.

Huffman, J. E., and Wallace, J. R. 2012. *Wildlife Forensics: Methods and Applications*. West Sussex, UK: John Wiley & Sons.

Iyengar, A. 2014. "Forensic DNA Analysis for Animal Protection and Biodiversity Conservation: A Review." *Journal for Nature Conservation* 22(3): 195–205.

Neme, L. A. 2009. *Animal Investigators: How the World's First Wildlife Forensics Lab Is Solving Crimes and Saving Endangered Species*. New York: Simon and Schuster.

Ogden, R. 2010. "Forensic Science, Genetics and Wildlife Biology: Getting the Right Mix for a Wildlife DNA Forensics Lab." *Forensic Science, Medicine, and Pathology* 6(3): 172–179.

Wildlife Management

While people have sought to manage wild animals for centuries, wildlife management today refers mainly to the ways governments have sought to regulate the harvesting of species and to modify animal habitats to decrease, sustain, or increase the numbers of wild animals. Wildlife management decisions are informed by conservation science and influenced by the public, who often demand that wildlife be managed, and advocacy groups, who want wild animals protected outright.

Wildlife management is as much about managing people as it is about managing wildlife, perhaps even more so. This was clearly the case with the development of hunting regulations, the first and most common way government agencies sought to manage wildlife. Governments have done this by setting bag limits (the number of wild animals that hunters can legally take), designating hunting seasons, and requiring hunters to obtain hunting licenses. Also, countries must determine who owns the wildlife. In the United States for instance, individual states own most species of wild animals. Individuals can own wildlife after they kill them through hunting, provided they have hunted the animals legally and have the necessary licenses.

Government wildlife management agencies developed hunting regulations to manage wildlife in the 19th century, and they began to employ other methods to manage wild animals in the 20th century. Scientists and conservationists recognized that they needed to protect wildlife habitat—the places where animals fed, bred, and raised their young—as well as regulate the harvesting of species. Early national parks in the United States, Canada, and other countries often functioned as wildlife refuges, although most were primarily established to preserve spectacular natural scenery rather than wildlife or their habitat.

Refuges for birds were the first protected areas established expressly to protect wildlife in the United States. In 1903, U.S. president Theodore Roosevelt (1858–1919) created the Pelican Bird Reservation (reservation being the commonly used term for refuges at the time), and he established dozens of other refuges during his presidency. Most of these refuges were only a few dozen acres and designed to protect the rookeries (nesting areas) of marine or plume-bearing birds that had been overhunted for their feathers, which were used in women's hats. Wardens were assigned to patrol these sanctuaries and, on occasion, apprehend poachers (illegal hunters) as needed.

Wildlife management in the United States has sometimes met with stiff resistance, especially by rural residents and Native peoples. Rural people have often relied on the harvesting of wild animals and then selling the meat, fur, or feathers. But many late 19th-century rural Americans, whether they were Native Americans or not, hunted animals for subsistence rather than monetary gain. Some of the people most affected by the government imposing these regulations and creating protected areas resisted these new regulations, sometimes fiercely and violently. Rural residents often regarded these efforts to curtail hunting and protect wildlife as benefiting urban middle-class or wealthy hunters at the expense of those who harvested deer, elk, bison, and other creatures to feed their families This sentiment endures in some rural areas.

The conflicts over managing wildlife remain particularly contentious in places such as Africa, where wildlife agencies seek to manage wild animals in areas still intensively used by rural residents and where poaching (illegal hunting) remains common. In Africa's oldest national park, Virunga National Park in the Democratic Republic of the Congo (DRC), poachers illegally hunt animals such as elephants and mountain gorillas. The tusks (or ivory) from elephants fetch a hefty price on the illegal wildlife-parts market. Some of these poachers, in turn, funnel this money to support militia groups trying to overthrow the DRC's government. Such violence and political turmoil amid protected areas demonstrate that wildlife management is complicated and connects to matters that at first glance would seem unrelated to conserving wildlife.

Wildlife managers have also sought to focus on individual species and boost their populations by manipulating the environment to improve the animals' habitats. Such measures might include constructing ponds and marshes for migratory birds or thinning forests and creating meadows for the benefit of deer. This narrow focus on particular species is not without its downsides. Improving habitat for one animal often comes at the expense of other animals. Also, wildlife managers have learned that the natural world is not like a machine where tinkering with different parts can increase the output of wild animals for people to hunt or view, but a complicated and interdependent system.

Endangered species law, in the United States and elsewhere, has provided wildlife managers with a powerful tool to regulate land and water use for the benefit of wild animals. Because this sort of legislation typically covers all types of animals and plants, not just charismatic megafauna (animals people easily relate to) such as elk, wolves, and eagles, it has forced wildlife managers to broaden their horizons and consider the welfare of the other species when creating protected areas and managing habitat. As with earlier eras of wildlife management, trying to protect endangered wildlife is extremely contentious because saving endangered wildlife entails placing real limits on the way people use the environment, both within protected areas as well as on private land.

Robert M. Wilson

See also: Black Market Animal Trade; Ethics; Hunting; Wildlife

Further Reading

Alagona, P. S. 2013. *After the Grizzly: Endangered Species and the Politics of Place in California*. Berkeley: University of California Press.

Coleman, J. 2004. *Vicious: Wolves and Men in America*. New Haven, CT: Yale University Press.

Haggerty, J., and Travis, W. 2006. "Out of Administrative Control: Absentee Owners, Resident Elk and the Shifting Nature of Wildlife Management in Southwestern Montana." *Geoforum* 37(5): 816–830.

Jacoby, K. 2001. *Crimes Against Nature: Squatters, Poachers, Thieves, and the Hidden History of American Conservation*. Berkeley: University of California Press.

Lorimer, J. 2015. *Wildlife in the Anthropocene: Conservation after Nature*. Minneapolis: University of Minnesota Press.

Warren, L. S. 1997. *The Hunter's Game: Poachers and Conservationists in Twentieth-Century America*. New Haven, CT: Yale University Press.

Wilson, R. M. 2010. *Seeking Refuge: Birds and Landscapes of the Pacific Flyway*. Seattle: University of Washington Press.

Wildlife Rehabilitation and Rescue

Since its inception in the United States in 1982, the organized practice of wildlife rehabilitation and rescue (WRR) has grown in popularity, reaching around the world. Wildlife hospitals now exist in Turkey, Mexico, Belize, China, Thailand, India, and throughout Africa. Post-release studies of rehabilitated wildlife have shown the practice to be successful at returning healthy and functioning individuals, often damaged by human causes, to their breeding groups and/or original territories. Because of this, it offers insights into repairing human relationships with other animal species and the natural world. WRR has informed other fields' knowledge about wild populations, ecosystems, avian medicine, cognition, reproduction, native species, conservation of endangered species, and emerging wildlife and zoonotic diseases. Wildlife rehabilitators and rescuers act as advocates for nonhuman animals on a larger scale as well—many work to support nonhuman animals through public education.

A worker of the Sumatran Orangutan Conservation Programme carries a tranquilized Sumatran orangutan as it is being prepared to be released into the wild at a rehabilitation center in North Sumatra, Indonesia, in 2015. Wildlife rescue work is an essential aspect of conservation not only because it assists individual animals, but also because it often raises public awareness of and action on behalf of animals in trouble. (AP Photo/Binsar Bakkara)

Wildlife rehabilitation is the practice of medically caring for injured, orphaned, sick, or otherwise-disabled wildlife with the goal of releasing them back to the place where they were found. One of the first rehabilitation centers was the Lindsey Wildlife Museum in California, which began accepting wildlife patients in the late 1960s. According to Jay Holcomb (1951–2014), a pioneer in the professional wildlife rehabilitation field, the practice gained popularity in the Western world in the 1970s and 1980s in response to a series of oil spill mortality events in California (Dmytryk 2012). Americans across the country were alarmed when the spills harmed thousands of birds, and local communities joined together in an attempt to rescue them.

Wildlife rescue and care are more ancient practices in non-Western areas of the world. In ethnographic (detailed studies of social groups) literature, for example, there are many cases of indigenous hunter-gatherers "adopting" and caring for wildlife babies. Jainists, as well, have a long history of helping wildlife. Those who follow the religion of Jainism believe that humans have a duty to care for and protect all animals, and the present-day Jain Bird Hospital in Delhi, India, treats all bird species. WRR centers in different global locations may focus on different human-wildlife conflicts. While many centers in the United States receive patients with injuries from cars, domestic cats, and guns, ARCAS Wildlife Rehabilitation Center in Guatemala receives 200–700 animals each year who were confiscated from poachers in the black market pet trade (Collard 2013).

Wildlife patients are brought to rehabilitators by members of the public or by rescue organizations. Individual rehabilitators are licensed by government agencies, organized through national agencies such as the National Wildlife Rehabilitation Association (NWRA) in the United States, and must work underneath the supervision of veterinarians. Both individual rehabilitators and rehabilitation organizations are paid through grants and by donations, and individuals sometimes volunteer their time; they are not funded by state or federal governments.

Animal-related and medical fields may focus on either individual animals or species as a whole. Wildlife rehabilitation, so far, is the only field that focuses on wild animal individuals, whereas wildlife biologists and conservationists work with wild animal species, and veterinarians work with domestic animal individuals. Many wildlife rehabilitators believe that human responsibilities to individuals of other species include reparations for anthropogenic (human-caused) injuries to them, placing more, or equal, importance on them as to their species populations and environments.

Critics of WRR have three primary concerns. The first has been that the practice could prevent nature from "taking its course," and thus obstruct what they claim are normal evolutionary and ecosystemic processes. However, rehabilitators argue that injured wildlife are often in need of help because of humans—and not natural causes. A second concern is that WRR efforts could have no effect upon species conservation and thus waste resources that could be allocated elsewhere. However, rehabilitators argue that, in cases where each individual of the species is important to its survival, WRR plays a significantly helpful role.

A third concern about WRR is that it may have a detrimental effect by helping members of invasive, pest, or nuisance species (the latter being animals like raccoons, opossums, and skunks, who have adapted successfully enough to human presence to create obstacles to human goals). Recently, the U.S. state of Alabama acted upon this worry. Licensed wildlife rehabilitators received notice from the state government that they would no longer be able to rehabilitate fur-bearing animals and instead must euthanize those individuals—including babies—upon intake. The reasons given for this order included overpopulation and zoonotic disease (diseases, like rabies, that can pass from animals to humans) concerns. However, due to the ethical implications of euthanizing healthy individuals, Alabama rehabilitators successfully petitioned the state to retract these policies.

As a secondary goal to rescuing and treating injured wildlife, many wildlife rehabilitation centers also provide public education about wildlife with the use of education, or ambassador, animals: wild animal individuals who are kept in captivity because they are unable to survive in the wild due to injury or imprinting (in which the animal identifies with humans). Education animals can teach members of the public about natural history, ecological niches, and human-wildlife conflict. In addition, the feeling of close connection formed during the face-to-face interaction can inspire protective concerns for those species. While these benefits are great, due to both the artificial setting in which the presentation occurs and to the dominant position of the human educator, WRR educators are challenged to avoid misrepresenting the animal as a pet or entertainment.

Heather Pospisil

See also: Advocacy; Black Market Animal Trade; Endangered Species; Humane Education; Human-Wildlife Conflict; Invasive Species; Nuisance Species; Pets; Poaching; Wildlife Management; Zoonotic Diseases

Further Reading

Bickett, D. 2014. "Alabama Department of Conservation and Natural Resources: Stop the Euthanization of Orphaned Wildlife in Alabama." Accessed October 30, 2015. http://www.change.org/p/alabama-department-of-conservation-and-natural-resources-stop-the-euthanization-of-orphaned-wildlife-in-alabama

Collard, R-C. 2013. "Putting Animals Back Together, Taking Commodities Apart." *Annals of the Association of American Geographers* 104(1): 151–165.

Dmytryk, R. 2012. "History of Wildlife Rehabilitation." Accessed October 30, 2015. http://wildlifeemergencyservices.blogspot.com/2012/10/history-of-wildlife-rehabilitation-part.hml

Duke, G. E., Frink, L., and Thrune, E. 1998. "Why Wildlife Rehabilitation Is Significant." *NWRA Quarterly Journal* 16(4): 14–16.

Friedman, E. 2004. "Rehabilitation through the Eyes of the Animal." *Wildlife Rehabilitation Bulletin* 22(1).

O'Connor, T. 2013. *The Past and Present of Commensal Animals*. East Lansing, MI: The Michigan State University Press.

Wolves

Wolves have been idealized for their strength, intelligence, and strong family bonds but also criminalized for their vigor, persistence, and cunning. In early human history, wolves ranged in vast, unpopulated wild lands. As human societies grew, wolf habitat shrank and conflict arose over resources and space. Present global distribution of wolves includes North America, Europe, Asia, and northeast Africa.

Early canids (the biological family of dog-like carnivorous mammals) evolved between 4.5 and 9 million years ago and diverged approximately 2 million years ago into coyotes, wolves, jackals, and other species. The wolf's closest relative today is the domestic dog. Wolf origin is unclear, but it is widely accepted that wolves spread from Eurasia to North America in several waves, evolving into four distinct species: gray wolf (*Canis lupus*), red wolf (*Canis rufus*), eastern wolf (*Canis lycaon*), and the Dire wolf (*Canis diris*). These species coexisted in North America until the Dire wolf went extinct in the late Pleistocene, roughly 10,000 years ago. The genetically different Ethiopian wolf (*Canis simensis*) is believed to have evolved separately.

Most wolves are gray wolves (also known as timber wolves or western wolves), but there are many gray wolf subspecies. Red wolves and eastern wolves (recognized as a separate species in 2012) are currently thought to have no subspecies. Classification of wolves as unique species/subspecies influences the degree of legal protection and conservation management. Gray wolves, with greater global distribution and numbers, do not have the high-level protections of the endangered red and eastern wolves or of Mexican gray wolves (a subspecies) and are now legally hunted in parts of the United States and Canada.

Wolves live in groups called packs and have highly developed social structures. They mate for life and maintain strong pack bonds. They are intelligent animals able to anticipate the actions of others, solve new problems based on experience, and pass along information generationally, including established migration routes and denning sites (places wolves raise pups or shelter in inclement weather). Wolves communicate using complex vocal expressions and body language.

These positive qualities observed in wolves led to their personification in many indigenous cultures as a "totem animal" (sharing kinship with and acting as a guide or protector for a person or clan). Recognition of human-wolf kinship has engendered a tolerance for coexistence with wolves in many indigenous cultures, including the Arikara, Ojibwe, Nez Perce, Hopi, and Tanaina of North America, the Dukha of Mongolia, and the Chechen of Eastern Europe.

In Western society, wolves are often depicted as tricksters, liars, or thieves. Stories such as *Little Red Riding Hood* (Perralt 1697) and *The Three Little Pigs* (Halliwell-Phillipps 1886) have introduced many children to the wolf in a negative context. In film, wolves are commonly depicted as vicious killers or as werewolves. Negative depictions of wolves have been shown to contribute to fear of wolves.

Contrary to popular portrayals, humans are not a prey species for wolves, and attacks are rare, although they do still occur, mostly in Russia, India, and Europe. Wolves prey primarily on ungulates (hooved animals such as deer, elk, and sheep) and generally avoid humans. The rabies virus causes wolves to behave aggressively and is responsible for many attacks on humans over the last 400 years. However, human activity is widely considered to be directly or indirectly responsible for most attacks on humans, with major contributing factors including human provocation, habituation (wolves losing their fear of humans through recurrent exposure/contact), significant modification of wolf habitat, and prey scarcity. In continental

Europe, many attacks occurred during the early modern period (1500–1900), a time of heavy landscape modification in which humans depleted wolf prey, domestic dogs bred with wolves, and unaccompanied children were shepherding livestock. Most victims of documented predatory attacks over the last 400 years globally have been children (Linnell et al. 2002).

Lacking the physical ability to ambush large prey in the manner of big cats, wolves must hunt in packs to avoid being killed by their prey, and they must get close to and test their prey for weakness by encircling, staring, and charging prior to attack. They attack progressively by biting the legs and then underside of the prey, often bringing a slow, painful death. Wolves commonly do not consume an entire kill, often abandoning partially consumed prey. Wolves often return to a kill, although other carnivores including bears, raptors, and nonpredatory scavengers that rely on carrion (dead animals) may also consume the remains.

When natural prey became scarce, wolves began to hunt livestock, generating significant conflict. Subsequently, wolves were systematically eradicated (commonly through shooting, trapping, and poisoning) from most of Europe and North America by the 17th and mid-20th centuries, respectively. Where eradicated, no other predator assumed the wolf's ecological role as a top predator. The result was an overpopulation of ungulates that overgrazed on new trees, thereby preventing healthy forest growth.

In the late 20th century, wildlife agencies, recognizing the wolf's critical ecological role, began reintroducing and protecting wolves in many U.S. and European forests. Reintroductions are frequently opposed by ranchers and rural communities accustomed to living in landscapes without wolves, and the return of the wolf has magnified tensions between rural communities and government agencies that previously supported elimination of wolves as pests. Wolves still prey on livestock but in very low numbers. Despite programs to reimburse for livestock losses, rural communities remain largely intolerant of wolves. Although wolf conservation is generally supported on a national scale in most wolf-range countries, the future of wolves remains uncertain.

Anita Hagy Ferguson

See also: Agency; Deforestation; Endangered Species; Extinction; Human-Wildlife Conflict; Literature, Animals in; Poaching; Trophy Hunting; Wildlife; Wildlife Management

Further Reading

Coleman, J. T. 2004. *Vicious: Wolves and Men in America.* New Haven, CT and London: Yale University Press.

Emel, J. 1998. "Are You Man Enough, Big and Bad Enough? Wolf Eradication in the U.S." In Wolch, J., and Emel, J., eds., *Animal Geographies: Place, Politics, and Identity in the Nature-Culture Borderlands,* 91–116. New York: Verso.

Halliwell-Phillipps, J. O. 1886. "The Three Little Pigs." In *Nursery Rhymes of England.* London: F. Wayne and Co.

Linnell, J. D. C., Andersen, R., Andersone, Z., Balciauskas, L., Blanco, J. C., Boitani, L., Brainerd, S., Breitenmoser, U., Kojola, I., Liberg, O., Løe, J., Okarma, H., Pedersen, H. C., Promberger, C., Sand, H., Solberg, E. J., Valdmann, H., and Wabakkenet, P. 2002. *The Fear of Wolves: A Review of Wolf Attacks on Humans*. Trondheimm, Norway: Norsk Institutt for Naturforskning (NINA).

Marvin, G. 2012. *Wolf*. London: Reaktion Books.

Mech, D., and Boitani, L., eds. 2003. *Wolves: Behavior, Ecology, and Conservation*. Chicago and London: The University of Chicago Press.

Perralt, C. 2010. "Little Red Riding Hood." In *Charles Perralt: The Complete Fairy Tales*. Translated by Christopher Betts. Oxford: Oxford University Press.

Wool. *See* Non-Food Animal Products

Working Animals

Working animals are one of the main categories of human-animal relations in the domestic sphere alongside pets/companions and livestock and farmed aquatic species. They are important to understand as one of the many ways animals' special abilities are used to benefit human society.

Working animals can be understood as having three often intersecting roles. The first is service, where animals are used to carry out specific tasks that help humans. The second is education, where animals are used to help further human understandings of science and animal behaviors. The third role is entertainment, where animals are used to fulfill human desires for storytelling and/or feats of accomplishment. The history of working animals goes back to the earliest days of animal domestication—especially of dogs, who were domesticated thousands of years ago, becoming partners with humans in hunting and used as guard animals.

The role of service animals includes a wide range of species and practices. Perhaps the biggest role, both historically and geographically, has been as draft animals, valued for their strength and endurance. Horses, oxen, donkeys, camels, goats, and mules (a cross between a horse and donkey) have been used to plough fields, transport materials and humans, and pump water. Horses have been used to pull logs, and elephants both push down trees and haul logs. Pictorial evidence from ancient Egypt (around 4000 BCE) shows oxen pulling ploughs. It is thought, however, that animals as diverse as llamas in South America and water buffalos and elephants in South Asia may have been trained for service anywhere from 8000 to 4000 BCE. Draft animals continue to be used extensively across Africa and South and East Asia, and even South America today, but in the western cultures of North America and Europe draft animals are used by only a small segment of traditional farmers, such as the Amish in the United States.

Dogs have been used extensively as service animals to assist in herding livestock and as hunting dogs to locate and retrieve prey. Breeds such as German Shepherds

Horses have been used as workers in a variety of ways since their domestication thousands of years ago. Police horses are especially useful for crowd control, officer safety, and community relations. (Charlotte Leaper/Dreamstime.com)

and Belgian Malinois have been part of police and military operations. They have been trained to serve as guard animals, to sniff out bombs, drugs, and track criminals, conduct search and rescue missions, and even to parachute with human soldiers to conduct operations. Horses and elephants were used for centuries in warfare before motorized equipment. Horses continue to be used by police forces worldwide for crowd control and public relations.

Service animals have been used in a much different way as support animals for people with mental and/or physical disabilities. The first seeing-eye dogs were trained in the early 20th century to assist blind persons, and today animals such as dogs, birds, ponies, and even monkeys have been trained to perform tasks such as opening doors, retrieving objects, assisting with movement, or monitoring for possible seizures.

Working animals in the education role include those species and breeds used for veterinary and human medical research, product testing, and basic science research/instruction. The places where these animals are found include research laboratories, corporate research and development departments, classrooms, militaries, and zoos. The earliest known documented example of animal research comes from the

Greek philosopher Aristotle (384–322 BCE), who dissected animals to study the differences between them. The use of animals for research is a highly controversial topic, with many animal advocates believing that causing any type of harm to other species is unethical. The controversy has even extended to the classroom, where many schools today offer students the option to use alternative anatomy models rather than having to dissect an animal.

While we have evidence for zoos—as collections of animals—going back to ancient Egypt, today's modern zoo began in Europe in the 18th century. As Europeans were colonizing (taking control over) different parts of the world, there was a dramatic increase in the desire to understand the animals and people they were finding. Animals were brought back so they could be studied by both the public and scientists and were placed in zoos in barren cages. Today our notion of zoos—especially in North America and Europe—continues to see these spaces as locations of both education and research, but we try and focus more on the welfare of the animals than before.

Zoos are also an example of the third role of working animals—human entertainment. Many people are simply entertained by seeing novel animals, especially during feeding times. Circuses, racing, and fighting animals all have long histories around the world. Some of the earliest circuses using animals date to Roman times (31 BCE–476 CE) and horse racing probably developed as the horse was being domesticated around 5,000 years ago in central Asia. More modern forms of entertainment animals include rodeos and animals used in television and film. Yet many forms of animal entertainment have their detractors, who believe the unique abilities of other species should not be used purely for human enjoyment. Dogfighting—which was once a very popular and public sport in the early 20th century in the United States—has now become illegal. Countries such as Costa Rica and Sweden have banned circuses that feature animals, and there is currently a huge push to free orcas (killer whales) from marine mammal parks such as SeaWorld.

Julie Urbanik

See also: Bullfighting; Circuses; Domestication; Elephants; Ethics; Fox Hunting; Greyhound Racing; Horse Racing; Horses; Marine Mammal Parks; Military Use of Animals; OncoMouse; Research and Experimentation; Service Animals; Vivisection; Xenotransplantation; Zoos

Further Reading

Bumiller, E. 2011. "The Dogs of War: Beloved Comrades in Afghanistan," May 2. Accessed December 2, 2015. http://www.nytimes.com/2011/05/12/world/middleeast/12dog.html?_r=1&ref=unitedstatesspecialoperationscommand

Hancocks, D. 2001. *A Different Nature: The Paradoxical World of Zoos and Their Uncertain Future.* Berkeley: University of California Press.

Urbanik, J. 2012. *Placing Animals: An Introduction to the Geography of Human-Animal Relations.* Lanham, MD: Rowman and Littlefield.

World Organisation for Animal Health (OIE)

The World Organisation for Animal Health is the English-language name for the Organisation Mundiale de la Santé Animale, based in Paris. Started in 1924, it is an intergovernmental organization whose mission is to understand, control the spread of, and eradicate animal and zoonotic diseases worldwide. The organization is concerned with both animal and human health and well-being. As of early 2016, there were 180 member countries.

The World Organisation for Animal Health was originally named the Office International des Epizooties (OIE), which translates into English as the International Agency of Animal Disease. (Because it is based in Europe, the British spelling, "organisation," is used.) The organization began as a response to a cattle disease, known as rinderpest, discovered in Belgium in the early 1920s that came from Indian zebu (also called Brahman) cattle being transported through the port of Antwerp. It was felt by many at the time that an intergovernmental organization would be able to respond more efficiently than a multitude of individual governments to animal disease threats. Over the ensuing decades, the organization has worked closely with many important global entities, such as the League of Nations (which became the United Nations [UN] in 1945), the World Health Organization (WHO), the European Community (known as the European Union [EU] since 1993), and the World Bank. In 2003, the OIE officially changed its name to the World Organisation for Animal Health but continues to use OIE as its official acronym and in its URL.

The OIE is run by what is called a World Assembly of Delegates and a director general, elected by the delegates. The delegates are individuals appointed by the OIE member country governments. The OIE is funded by both mandatory and voluntary annual contributions from member countries. The 180 member countries (out of approximately 195 independent countries worldwide), from all world regions, range from less to more financially well-off (e.g., Afghanistan and Rwanda to Germany and the United States) and from monarchies (e.g., Saudi Arabia) to communist dictatorships (e.g., North Korea) to democracies (e.g., Argentina).

The OIE's mission has six parts. The first is to promote transparency in global animal and zoonotic (transmissible between species, including to humans) disease situations. Member countries report disease occurrences within their borders to the OIE, which then transmits the information globally. Second, the OIE is a clearinghouse for scientific developments in animal disease, providing member countries with the latest information to assist in their veterinary efforts. Third, the OIE seeks to unite all member countries in controlling and eradicating animal/zoonotic diseases. One of the main efforts here is to provide assistance and expertise as needed to poorer countries. The fourth part of the OIE's mission is to develop animal and animal products health standards in order to protect international trade. For example, the OIE published (first in 1996 and in its 18th edition in 2015) the *Aquatic*

Animal Health Code to set standards in identifying, reporting, and controlling diseases in aquatic animals and products before their entry into international trade circuits. The fifth part of the mission is to promote veterinary services worldwide, especially in countries that are less economically well-off. Finally, the OIE is concerned with food safety and (primarily health-based) animal welfare.

From the overall mission, the OIE engages in a variety of activities and provides services to its members. One of its main functions is creating centers of scientific research and expertise. An outcome from this function that is of primary importance is publications providing technical guidance, for example on diagnosing a particular animal disease. The scientific research also contributes to setting international standards (such as those related to aquatic animal health referenced above). The OIE also maintains the World Animal Health Information System (WAHIS), a database that is continually updated as new information is obtained. WAHIS contains information on specific diseases, areas (including maps), and control measures. In response to member concerns, in 2004 the OIE became more explicitly engaged with animal welfare issues, publishing, in 2005, its first international animal welfare standards. The OIE currently uses what are known as the Five Freedoms (Farm Animal Welfare Council 1979) in its definition of good welfare. These freedoms are *from* hunger and thirst; fear and distress; physical discomfort; pain, injury, and disease; and *to* express normal behaviors.

There are a number of specific issues that the OIE works on at any given time. For example, issues in early 2016 included antimicrobial resistance and rabies eradication. Antimicrobial resistance refers to the increasing resistance of microbes, particularly bacteria, to treatment drugs. This resistance has occurred in large part through the overuse of antibiotics in industrial animal agriculture, and it directly threatens both human and animal health. The OIE has worked with other international organizations such as the WHO to develop standards for the use of antimicrobial agents in veterinary practice, assemble a network of experts, and hold a 2013 international conference. With regard to rabies eradication, the OIE notes that it is one of the deadliest zoonotic diseases and indicates that global human deaths—mostly of children in poorer nations, many in Africa and Asia—approach 70,000 annually. Because almost all of the cases arise from dog bites, the OIE's campaign focuses on vaccinating dogs, with a goal of a 70 percent vaccination rate where rabies exists. They state that this rate would substantially eradicate the incidence of rabies in humans, with the cost of the vaccinations being almost 10 times less than treatment after a bite.

Connie L. Johnston

See also: Aquaculture; Biodiversity; Factory Farming; Fisheries; Husbandry; Livestock; Mad Cow Disease; Meat Eating; Microbes; Slaughter; Veterinary Medicine; Welfare; Zoonotic Diseases

Further Reading

Farm Animal Welfare Council. 1979. Press Statement. Accessed February 21, 2016. http://webarchive.nationalarchives.gov.uk/20121007104210/http:/www.fawc.org.uk/freedoms.htm

OIE. 2015. *Aquatic Animal Health Code*. Accessed February 2, 2016. http://www.oie.int/en/international-standard-setting/aquatic-code/access-online

OIE. 2015. *Terrestrial Animal Health Code.* Accessed February 2, 2016. http://www.oie.int/en/international-standard-setting/terrestrial-code/access-online

World Organisation for Animal Health. Accessed February 2, 2016. http://www.oie.int/en

Xenotransplantation

Xenotransplantation (*xeno* from the Greek for strange/foreign) is the use of nonhuman animal parts in human medical treatments. While this practice has a long history, it is a key part of life-saving surgery options for patients today because there is such a shortage of human organ donations. Xenotransplantation is a key topic for human-animal relations because it shows simultaneously how closely some species are linked biologically with humans and yet also calls into question the ethics of creating animals just to harvest their organs.

Xenotransplantion is part of the larger industry of animal biotechnology, which targets genetic material and tissue for human health, animal agriculture, basic science research, and commercial purposes. Providing animal material for human organ/tissue purposes is a growing field because of the disparity between people who need organs and what is available. The U.S. Department of Health and Human Services (USDHHS) reports that every 10 minutes someone is added to the transplant waiting list, and an average of 22 people die each day for lack of a needed organ (USDHHS 2016). Additionally, in the United States about 14,000 organs were donated (from living and deceased donors), yet more than 121,000 people needed organs in 2013. Medical science has turned to animals as a solution to this problem.

While xenotransplantation might seem to be a modern invention because of the technology involved, it has a history that goes back to the 1600s in Europe, when a French doctor transfused the blood of a lamb into a teenage boy (Deschamps et al. 2005). The transfusion was successful in the sense that the blood circulated; however, the boy did not live. And in 1682 a Russian received a piece of skull from a dog (Warmflash 2015). It wasn't until the 20th century, however, that xenotransplantation became a true focus of human medical research. In the 1960s and 1970s transplants of chimpanzee and baboon hearts and kidneys into humans were seen as major breakthroughs—even though none of the humans survived longer than nine months (most died within days or hours). In 1984, "Baby Fae" made international headlines as the recipient of a baboon heart. She passed away after 20 days because her body rejected the organ. Although organ transplants do not have a strong track record of success, xenotransplantation of other types of tissue is now quite routine. For example, heart valves from cows and pigs are placed in humans, whose bodies seem much better able to adapt to nonhuman tissue parts rather than whole organs (although there is always the risk of rejection). In this instance, the choice of whether to use an animal or artificial valve is now down to personal preference.

Rejection of what the body sees as foreign material is what makes organ and tissue transplants so difficult. Our immune systems are tasked with protecting the integrity of the body and actively fight any perceived "invaders." This is why it is so important to get close matches for human-to-human organ donations and why an individual must take heavy doses of immunosuppressive drugs. These drugs try and stop the immune system from fighting the foreign object until it has time to become integrated.

Xenotransplantation research is heading in two directions. The first includes continued work on using animal material. Pigs are seen as the most promising species because of their physiological similarity to humans. For xenotransplants to work, however, the pigs have to be "humanized" to have a better chance of producing materials that won't be rejected. This has been accomplished by using animal biotechnologies to breed miniature pigs that have more comparable organ sizes and by using genetic technologies that remove markers that identify pig tissue as pig tissue. Harvard University has recently announced that it has created the "most extensively gene-edited organism ever, a line of pigs that have had more than 60 genes modified" (Templeton 2015). The second direction is to use animals as hosts on, or in, which to grow human material using a person's own stem cells. Stem cells are cells that are able to morph into any type of bodily tissue. The use of stem cells is very controversial because they are easiest to obtain from human fetuses, and some people object to using humans as research objects.

Many have raised concerns about the use of animals as "spare parts" for humans. These concerns range from animal rights groups opposed to the use of animals for human research purposes at all, to those concerned about the health consequences of spreading diseases from animals to humans. Proponents argue that breeding animals for medical purposes is really no different from breeding them for food and that government oversights for safety are in place globally. The U.S. Food and Drug Administration requires lifelong monitoring for anyone who receives a transplant. The United States banned all xenotransplants between humans and primates in 1999 because of concerns over disease transmission. Author Brenda Peterson, in her 2004 novel *Animal Heart*, raises more philosophical and experiential questions about what it means to have other animals' organs inside humans through a character who gets a baboon heart and then begins to dream of African savannahs and "being" baboon. In spite of the controversies surrounding xenotransplantation, it is considered the most viable life-saving option for humans until more are willing to become organ donors themselves.

Julie Urbanik

See also: Animals; Biotechnology; Ethics; Humans; Research and Experimentation; Species

Further Reading

Deschamps, J. F. R., Sai, P., and Gouin, E. 2005. "History of Xenotransplantation." *Xenotransplantation* 12: 91–109.

Institute of Medicine (U.S.) Committee on Xenograft Transplantation: Ethical Issues and Public Policy. 1996. *Xenotransplantation: Science, Ethics, and Public Policy*. Washington, DC: National Academies Press.

Peterson, B. 2004. *Animal Heart*. San Francisco: Sierra Club Books.

Simmonds, F. 2001. "Organ Farm." Accessed January 3, 2016. http://www.pbs.org/wgbh/pages/frontline/shows/organfarm

Templeton, G. 2015. "With the Most-Edited Genomes of All Time, Harvard's Pigs Could Spark a Transplant Revolution." October 9. Accessed January 5, 2016. http://www.extremetech.com/extreme/215767-with-the-most-edited-genomes-of-all-time-harvards-pigs-could-spark-a-transplant-revolution.

USDHHS. 2016. "The Need Is Real." Accessed January 2, 2016. http://www.organdonor.gov/about/data.html

Warmflash, D. 2015. "Coming Age of Xenotransplantation: Would You Accept an Organ from a Pig to Save Your Life?" Accessed January 2, 2016. https://www.geneticliteracyproject.org/2015/02/12/coming-age-of-xenotransplantation-would-you-accept-an-organ-from-a-pig-to-save-your-life

Z

Zoogeography. *See* Animal Geography; Biogeography

Zoogeomorphology

Zoogeomorphology is the study of the effects of animals as agents of erosion, transportation, and depositing of rocks, sediment, and soil, both on and beneath Earth's surface; as such, zoogeomorphology encompasses principles from the subdisciplines of geomorphology (the study of landforms and landform-shaping processes) and biogeography. These studies allow scientists to understand how animals interact with their environment and how those interactions vary across different locations. The size of animals that have such impacts ranges from extremely small invertebrates such as ants and termites to extremely large vertebrates such as elephants and grizzly bears.

Animals alter Earth's surface and landforms in a variety of ways, including burrow excavation, digging for food, mounding of sediments, trampling and chiseling the landscape, wallowing, eating of soil and rocks, and by construction (in the case of beavers) of dams that impound water, create ponds, and accumulate sediments. Among the earliest scientists who examined animals as geomorphic agents was the British naturalist Charles Darwin (1809–1882), whose final published work detailed the burrowing and depositional activities of earthworms.

Animals excavate burrows into the ground for a variety of reasons, including shelter from harsh elements and predators, giving birth, access to underground food sources, and hibernating. During the process of excavation, sediments and rocks are deposited on the surface in impressive amounts. Surface mounds may reach several meters (yards) in diameter. Burrow excavation as well as digging for food also mix sediments and soils, a process known as biopedoturbation or bioturbation. Excavated burrow systems and associated surface mounds created by animals living in colonies, such as prairie dogs, may cover many acres and substantially alter the surface and subsurface water systems of the local area.

Wallowing is when animals roll on the surface in dust or mud, in the process coating their bodies with dry or moist sediment. An adult elephant may remove as much as 3.5 cubic feet of sediment from the surface; that sediment is subsequently transported elsewhere on its body. Wallowing in the same general location over an extended period of time may lead to the creation of an uneven, large surface depression. Such "wallow holes" may accumulate and temporarily pond water after episodes

An elephant wallow in Kruger National Park, South Africa. Note the several large elephant footprints in the sediment in the floor of the wallow site. (David R. Butler)

of rain or snowmelt, because the wallowing action causes the sediments in the base of the wallow hole to become compacted. This sediment compaction prevents water from draining down into the soil. Over longer periods of time, large wallow holes may become semipermanent or seasonal waterholes on a landscape. The Great Plains of North America were at one time covered with hundreds of thousands of so-called "buffalo wallows" associated with the wallowing activities of seasonally migrating bison.

Among the most impressive animals in terms of their ability to shape the landscape are the two living species of beavers, the North American beaver *Castor canadensis* and the European beaver *Castor fiber*. Both species construct dams on smaller streams in order to create ponds in which the beavers can live, find food sources, and have protection from predators. Beaver dams may be as much as 2 meters high, and reports of dams of several hundred meters in length are not uncommon in the scientific literature. Beaver dams and their associated ponds substantially alter a landscape's surface hydrology and how much sediment is being carried downstream. Stream velocities (the rates at which stream waters move) are dramatically slowed down by the dams and their impounded waters, and the lower stream velocities reduce stream erosion both above and below the dam. The reduced velocity of water flowing into the pond results in the deposition of fine-grained sediments in the water at the base of the pond. Sediments accumulated at the base of beaver ponds can reach well over a meter in depth. Eventually, most beaver ponds undergo sufficient siltation (accumulation of mud) on pond bottoms that the pond becomes too shallow for the beavers to survive during winter when the pond could freeze solid, and the beavers move on to create another dam and pond, leaving behind a rich, boggy meadow environment of fine-grained, silt-laden soils. Some beaver dams undergo failure during times of high rainfall and may lead to outburst flooding downstream from the failed beaver dam.

Scientists who study the geomorphic effects of animals are known as zoogeomorphologists, and zoogeomorphology is a branch of the broader subdiscipline of biogeomorphology that examines the landform-related effects of both plants and animals. Studies to date have examined a diversity of animals, including the effects of domesticated animals such as cattle; feral (domesticated but free-roaming) animals such as burros and pigs; and populations of wild, free-ranging animals in nature reserves such as grizzly bears, elephants, and hippopotami. Recent attention has been given to how climate change is influencing food and water resources available to geomorphologically significant animals, and how those shifting food resources may alter the spatial patterns and intensities of activities such as digging for food, burrowing for shelter, and dam building by beavers.

David R. Butler

See also: Biogeography; Climate Change; Earthworms; Feral Animals; Wildlife

Further Reading

Butler, D. R. 1992. "The Grizzly Bear as an Erosional Agent in Mountainous Terrain." *Zeitschrift für Geomorphologie* 36(2): 179–189.

Butler, D. R. 1995. *Zoogeomorphology: Animals as Geomorphic Agents.* Cambridge, UK and New York: Cambridge University Press.

Darwin, C. 1881. *The Formation of Vegetable Mould, through the Action of Worms, with Observations on Their Habits.* London: John Murray.

Gurnell, A. M. 1998. "The Hydrogeomorphological Effects of Beaver Dam-Building Activity." *Progress in Physical Geography* 22: 167–189.

Jones, C. G. 2012. "Ecosystem Engineers and Geomorphological Signatures in Landscapes." *Geomorphology* 157–158: 75–87.

Naylor, L. A. 2005. "The Contributions of Biogeomorphology to the Emerging Field of Geobiology." *Palaeogeography, Palaeoclimatology, Palaeoecology* 219: 35–51.

Wilkinson, M. T., Richards, P. J., and Humphreys, G. S. 2009. "Breaking Ground: Pedological, Geological, and Ecological Implications of Soil Bioturbation." *Earth-Science Reviews* 97: 257–272.

Zoology

From the Greek *zoon* ("animal") and *logos* ("argument from reason"), zoology refers to a field of study concerning characteristics of nonhuman organisms and populations as well as their social and ecological interactions with other organisms and with the abiotic (that is, nonliving chemical and physical) environment. Having developed from natural history, early zoology was grounded in observations within natural settings and emphasized organism-level documentation of morphology (form or structure) and physiology (function) to understand the formation and functioning of different species. Facilitated by evolutionary theory, zoology more recently questions how morphology and physiology are integrated and shaped through broader processes of biological and ecosystem change. Expanding from the original focus on organisms, contemporary zoology considers a broad spectrum of biological levels of organization—from genes to ecosystems—and incorporates methods and concepts from related fields, notably physiology, genetics, animal behavior, and ecology.

Since at least the Neolithic era (10,000–2,000 BCE), knowledge about domestic and wild animals has been critical for livelihoods of people worldwide. For example, animal herder and hunter-gatherer societies frequently cultivate deep understanding of the form, function, and behavior of the animals they depend on and encounter on a daily basis. Yet the origin of the Western scientific practice of zoology is commonly attributed to curiosity-driven knowledge, particularly the ancient Greek study of natural history.

A practice of describing individual organisms in their environment, natural history originated in 4th-century-BCE Greece with Aristotle's (384–322 BCE) *Historia Animalium.* Aristotle's concepts were further developed in the Middle East with the *Kitab al-Hayawan* (Book of Animals) by Muslim scholar Al-Jahiz (776–868), whose understanding of interspecies competition foreshadowed the concept of natural selection (a key evolutionary mechanism whereby organisms better adapted to their environment are more likely to survive, reproduce, and pass on beneficial traits). In Renaissance Europe (14th–17th centuries), a resurgent importance of empiricism (conviction that knowledge comes primarily through sensory experience) and extensive global exploration (which led to encounters with exponentially more species) spurred a renewed interest in natural history. Much in line with Aristotle's approach 2,000 years earlier, 16th-century European natural history emphasized descriptive study of animal morphology, physiology, and behavior through direct observation in order to analyze and catalog diversity into taxonomic groups (or groups defined on the basis of shared, mostly morphological, characteristics).

Along these lines, "bestiaries" such as Conrad Gessner's (1516–1565) *Historiae Animalium* (1551–1558) developed as an early zoological genre, merging fantastical accounts of animals with detailed description and artistic depiction based on specimens collected throughout the world and brought back to Europe. Toward the end of the 17th century, protocols for naming and classifying species were managed through national entities such as England's Royal Society of London and the French Académie des Sciences, raising informal natural history practice to a professional and often elite level. Working with these groups, early 19th-century naturalist-explorers like the Prussian Alexander von Humboldt (1769–1859) adapted natural history analysis to systematically and quantitatively study the relationship between animals and environment, paving the way for zoological specializations of behavioral ecology (which considers how behaviors evolve according to ecological factors) and biogeography (which documents how species, populations, and ecosystems vary across space and time).

Following Humboldt's methodology, Charles Darwin (1809–1882) amassed evidence to support his theory of how species come into being through natural selection, which he documents in his manuscript *On the Origin of Species by Means of Natural Selection* (1859). Darwin's research paralleled work by fellow naturalist Alfred Russel Wallace (1823–1913), who independently developed a theory of species origin. Alongside then-recent paleontological evidence of large-scale species extinction and social concerns about resource scarcity, these theories captivated the scientific community as well as the general public almost immediately. In putting forth hypotheses of verifiable mechanisms by which organisms change in relation to "nonliving" physical and chemical properties, the theories of Wallace and Darwin helped elevate zoology to a science that emphasized connections between morphology, physiology, and environment.

Advances in microscope technology moved zoological science increasingly into laboratory settings by the 20th century. Contemporary zoology considers animal form and function through both experimental and field-based approaches. Evidencing its deep ties to ecology, zoology has recently been recast as integrative biology in a number of university programs. Zoologists commonly specialize in subfields such as animal physiology, behavioral ecology, biogeography, and primatology. Zoologists and those with training in the field often work within universities, zoos, and laboratory or field research sites. While methods and findings of zoological research are often published in academic journals (such as *Science, Nature*, or *Zoology*), many insights are also adapted for more general audiences through popular magazines (for example, *National Geographic*). Zoos, wildlife sanctuaries, and natural history museums also strive to provide up-to-date information regarding species on exhibit.

Although zoological research rarely explicitly addresses interactions between humans and nonhumans, the practices and concepts of zoology are windows to such interaction. Attention to the production of zoological knowledge affords a unique opportunity to understand how science is shaped by and shapes human-nonhuman relations. For example, the geographer Sarah Whatmore has written about how zoological scientific practice has been shaped by both individual humans and nonhumans using the example of dynamic interactions within zoos and national parks that have shaped our understanding of elephants. Similarly, scholars advocate attention to non-Western zoological thought, as in historian/geographer Diana Davis's exploration of indigenous zoological knowledge through her accounts of veterinary practices among North African pastoralists (nomadic herders).

Nathan Clay

See also: Animals; Biodiversity; Biogeography; Ethology; Evolution; Research and Experimentation; Pastoralism; Species; Taxonomy; Zoos

Further Reading

Darwin, C. 1964. *On the Origin of Species*. Facsimile of the 1st ed. Cambridge, MA: Harvard University Press.

Davis, D. 1996. "Gender, Indigenous Knowledge and Pastoral Resource Use in Morocco." *The Geographical Review* 86(2): 284–288.

Hickman, C., Keen, S., Larson, A., and Roberts, L. 2013. *Integrated Principles of Zoology*, 16th ed. New York: McGraw-Hill Education.

Russell, E. 2011. *Evolutionary History: Uniting History and Biology to Understand Life on Earth.* New York: Cambridge University Press.

Whatmore, S. 2002. *Hybrid Geographies: Natures Cultures Spaces.* London: Sage.

Zoonotic Diseases

Zoonotic diseases, or Zoonoses, are diseases that are transmissible from nonhuman animal hosts to humans. Zoonoses is a broad descriptive term, and such diseases may be viral, bacterial, fungal, or parasitic. Zoonotic diseases are an important aspect of human-animal relations because they can affect the way humans interact with other animal species due to real and/or perceived health threats, and also because of the potential for widespread outbreaks due to global trade and travel.

Zoonoses range from mild to lethal, with symptoms ranging from rashes to severe hemorrhaging (internal bleeding). They are spread primarily by insects (e.g., fleas, mosquitoes, and ticks) but also by livestock, pets, and wild animals. Zoonotic diseases are quite common and infrequently make news. One exception is the Zika virus in South America, whose emergence has been linked to an increase in mosquitos due to global warming. Zika is spread by the *Aedes aegypti* mosquito, which also transmits other zoonoses and is associated with the birth defect microcephaly, in which a baby's head is abnormally small. Another exception is the Ebola virus, which causes viral hemorrhagic fever and made global news in 2014 after outbreaks that caused more than 11,000 deaths, primarily in the West African nations of Guinea, Liberia, and Sierra Leone (CDC 2016). Ebola is carried by fruit bats and nonhuman primates (e.g., monkeys and apes), who are hosts to the virus. Some Africans hunt fruit bats and/or nonhuman primates for food (bushmeat) and then acquire the virus from them. It also is thought that the HIV virus, another zoonotic disease, was originally transmitted to humans by bushmeat from chimpanzees.

Many zoonotic diseases can be deadly. One example is malaria, which was diagnosed in almost 200 million cases, causing 500,000 deaths worldwide in 2013 (CDC 2015). Malaria causes flu-like symptoms; complications, including death, can occur if not treated. It is spread by female mosquitos of the *Anopheles* genus and is prevalent in tropical and subtropical locations, such as Sub-Saharan Africa. Another example is rabies, which affects the central nervous system and is usually fatal unless treated within a certain period of time. It is transmitted between mammals, frequently dogs to humans or wild animals to dogs, typically through saliva or blood. In areas with low dog vaccination rates, such as much of Asia and Africa, annual human rabies deaths number in the tens of thousands (WHO 2015).

A large number of human zoonotic infections are influenza strains originating in wild animals, such as migratory geese, ducks, and bats, who carry the viruses and infect farmed animals such as chickens and swine. In these species, the virus mutates and causes outbreaks of influenza such as those designated H1N1, or swine flu, and avian influenza/bird flu, to which humans are highly susceptible. The famous influenza global pandemic of 1918–1919, which killed an estimated 2 percent of humans, was an avian influenza virus.

Most of the cases of avian influenza occur in humans who have had direct contact with live birds, primarily on farms and in live poultry markets. China maintains the largest amount of domestic poultry production in the world, and consequently most cases of avian influenza arise there. Where avian influenza has been discovered, public health measures are enacted. These consist of quarantining and treating those humans infected, culling (killing) of animals, and imposing travel restrictions/monitoring travelers. The economic cost of zoonoses, therefore, can be considerable for producers who may be forced to cull animals, consumers who will pay higher retail prices, and governments who muster public health resources or reimburse farmers.

Arguments have been made that large, industrial-scale animal farms ("factory farms"), often called concentrated animal feeding operations (CAFOs), promote conditions that allow for rapid transmission of zoonotic diseases because of birds and swine who are raised in large numbers in crowded conditions. Counterarguments maintain that because the birds and swine are confined, they are less likely to come into contact with wild animals who might serve as vectors for such viruses. Regardless of these two opposing perspectives, data are increasingly available that show that the extensive use of antibiotics in animal agricultural production are contributing to antibiotic resistance of disease-causing bacteria that can infect both humans and animals.

Concerns about bioterror, weapons that use zoonotic viral or bacterial agents, are increasingly worrisome, and bioweapon research focuses on methods of spreading, and protecting from, zoonotic diseases, such as plague, influenza, and anthrax. In 2003, the U.S. Department of Transportation's Volpe National Transportation Systems Center commissioned a study, *The Economic Impacts of Bioterrorist Attacks on Freight Transport Systems in an Age of Seaport Vulnerability*, which estimated the cost of a bioterror attack upon a metropolitan seaport in the trillions of dollars.

Zoonotic diseases are not *caused* by the animal hosts, who only serve as vectors (sources of transmission). Humans' domestication of, and coexistence with, animals have a twofold effect in that they provide both the abundance of animal products, which benefit humans, and the corresponding diseases by which humans experience harms. These issues remain a matter of great concern not only in the context of animal agriculture but also with respect to climate change, which is changing the life cycles and geographic distributions of many insect vectors, making them hardier and more widespread.

John T. Maher

See also: Climate Change; Concentrated Animal Feeding Operation (CAFO); Dogs; Factory Farming; Microbes; Mosquitoes; Nuisance Species; Species

Further Reading

Abt, C. C., Rhodes, W., Casagrande, R., and Gaumer, G. 2003. "Executive Summary." *The Economic Impacts of Bioterrorist Attacks on Freight Transport Systems in an Age of Seaport Vulnerability.* Cambridge, MA: Abt Associates, Inc.

Centers for Disease Control and Prevention (CDC). 2015. "Malaria Facts." Accessed February 29, 2016. http://www.cdc.gov/malaria/about/facts.html

Centers for Disease Control and Prevention (CDC). 2016. "2014 Ebola Outbreak in West Africa." Accessed February 29, 2016. http://www.cdc.gov/vhf/ebola/outbreaks/2014-west-africa/index.html

Ling, F., Chen, E., Liu, Q., Miao, Z., and Gong, Z. 2015. "Hypothesis on the Source, Transmission and Characteristics of Infection of Avian Influenza A (H7N9) Virus–Based on Analysis of Field Epidemiological Investigation and Gene Sequence Analysis." *Zoonoses Public Health* 62(1): 29–37.

Olival, K. J., and Hayman, D. T. S. 2014. "Filoviruses in Bats: Current Knowledge and Future Directions." *Viruses* 6(4): 1759–1788.

Pantin-Jackwood, M. J., Miller, P. J., Spackman, E., Swayne, D. E., Susta, L., Costa-Hurtado, M., and Suarez, D. L. 2014. "Role of Poultry in the Spread of Novel H7N9 Influenza Virus in China." *Journal of Virology* 88(10): 5381–5390.

Phillips, C. J. C. 2015. *The Animal Trade.* Wallingford, UK: CABI.

World Health Organization (WHO). 2015. "Rabies." Accessed February 29, 2016. http://www.who.int/mediacentre/factsheets/fs099/en

Zoophilia

The term "zoophilia" was coined in 1886 by Austro-German psychiatrist and sexologist Richard von Kraft-Ebbing (1840–1902). Zoophilia has since been defined according to psychiatric and medical institutions as a sexual fixation on or attraction to nonhuman animals. Zoophilia is therefore distinct from bestiality, defined as human-animal sexual *acts*. Numerous debates surround zoophilia and bestiality, often invoking questions of morality, proper or improper sexuality, animal rights and welfare, and consent within human-animal relationships. To understand zoophilia, however, it is necessary to look back at its history as a category that emerged within psychological and medical institutions but that has more recently become connected to other social categories.

A number of terms exist to name different sexual practices and erotic attractions between humans and animals. For instance, many scientific scholars and self-identified zoophiles distinguish zoophilia from zoosadism, which refers to the deriving of pleasure from inflicting pain on animals and may or may not involve sexual acts. The American Psychiatric Association's fifth edition of its *Diagnostic and Statistical Manual of Mental Disorders* (2013)—often considered the leading authority on psychiatric knowledge—classifies zoophilia as a "specific paraphilic disorder"—paraphilia meaning a sexual "perversion" or "deviation" from typical human sexual behavior. Even more specifically, researcher Anil Aggrawal, in the *Journal of Forensic and Legal Medicine* (2011), classifies almost a dozen different distinct types of zoophilia, including human-animal sexual role-play and zoophilic fantasies, among others.

Since the late 20th century, an increasing number of zoophiles have found community support and information exchange on the Internet, producing cyberspaces where they can spark public debate around their sexual desires and practices. These moves to publicize zoophilia have coincided contentiously with increasing efforts to criminalize bestiality as an act of animal cruelty. For instance, in response to a proposed anti-bestiality law in Denmark, zoophiles have sought to redefine zoophilia as an innate sexual orientation and sexual identity similar to lesbian, straight, or bisexual identity and therefore deserving of legitimacy, rights, social acceptance, and legal protections.

While these moves to associate bestiality with homosexuality have taken place within popular culture, several scholars of queer theory, which seeks to understand and often challenge culturally dominant ideas and values around sexuality, have also taken up zoophilia as a subject of study. For instance, many thinkers have considered the ways the stigma against zoophilia and bestiality has reinforced culturally dominant norms about sexuality and human-animal relations. For instance, on the one hand, owning and loving pets in an "appropriate" or "proper" way has been a hallmark in popular culture representations of happy, heterosexual nuclear family life, as in the 1954–1973 American television show *Lassie*. Yet, on the other hand, people who love their pets "inappropriately" or too much—and this may or may not involve what we think of as sex—are frequently seen as weird, strange, and/or queer. For instance, one urbandictionary.com definition of "crazy cat ladies" is women who love cats when they should have a husband and children. Drawing on queer theory, which warns against all-encompassing assumptions about or definitions of human sexuality, feminist philosopher Kathy Rudy argues that if we stop assuming "a coherent and agreed upon definition of sex . . . the line between 'animal lover' and zoophile is not only thin, it is nonexistent" (Rudy 2012, 611). Instead of worrying about separating "good" from "bad" human-animal love, Rudy insists that we focus on the ways that a love for animals can create possibilities for ethical relationships when we do not assume a human-animal hierarchy that denies animals' their agency and unique personhood in these relationships. As one example, Tim Treadwell, whose life and death became the subject of the documentary *Grizzly Man* (2005), loved grizzly bears so much that he rejected humans' superiority over other animals, expressed a desire to become a bear himself, and spent much of his life living among them.

These debates over zoophilia reflect the tension between psychiatric and popular understandings of sexuality and desire. They also reflect ethical debates over human sexuality and the proper treatment of animals. Many zoophiles argue that they develop mutually beneficial relationships with their animals. On the other hand, scientific research suggests that *both* abuse and care occur within zoophilic relationships (Beetz and Podberscek 2005). As popular attitudes around sexuality and human-animal relationships shift, these debates are likely to continue. By knowing the history of zoophilia and its creation as a social category, we can better understand these conversations.

William L. McKeithen

See also: Agency; Bestiality; Popular Media, Animals in

Further Reading

Aggrawal Anil. 2011. "A New Classification of Zoophilia." *Journal of Forensic and Legal Medicine* 18(2): 73–78.

American Psychiatric Association. 2013. *Diagnostic and Statistical Manual of Mental Disorders*, 5th ed., Arlington, VA: American Psychiatric Publishing.

Beetz, A. M., and Podberscek, A. L. 2005. *Bestiality and Zoophilia: Sexual Relations with Animals*. West Lafayette, IN: Purdue University Press.

Carman, C. 2012. "Grizzly Love: The Queer Ecology of Timothy Treadwell." *GLQ: A Journal of Lesbian and Gay Studies* 18(4): 507–528.

Rudy, K. 2012. "LGBTQ . . . Z?" *Hypatia* 27(3): 601–615.

Tsoulis-Reay, A. "What It's Like to Date a Horse." *Science of Us*. Accessed March 18, 2015. http://sousocial.nymag.com/scienceofus/2014/11/what-its-like-to-date-a-horse.html

Zoos

Zoos are highly controversial, of historical cultural importance, and a significant medium through which people are exposed to a variety of exotic animals. As such, zoos offer a lens through which we can view human-animal relations and see how debates about the continued existence of zoos and their changing image and roles have been driven by changes in perceptions of animal rights and welfare in society.

Zoos are places where people can learn about and be entertained by captive animals, but they are also places where an increasingly controversial separation of humans and animals occurs. (AP Photo/Mel Evans)

When was the first zoo established? This is actually a difficult question to answer as the term "zoo" is far more modern than the practice on which zoos are based: namely, the collecting and displaying of live wild animals for human enjoyment. Such practices can be traced back over 4,500 years. Menageries, the forerunner of modern zoos, were the creation of European aristocrats, primarily during the Renaissance period (1300–1700), for their private entertainment.

In contrast, zoological gardens, originating in the 1800s, were supposed to offer a location for the scientific study of animals. Not initially open to the public, this situation soon changed, as in the case of London Zoo, which was created in 1828 and opened to the public in 1847. The term zoo is simply a shortening of zoological gardens. Since the era of the zoological garden, we have seen the emergence of aquaria and oceanariums, focused on aquatic animals, and wildlife parks, safaris, and bioparks (or biological parks). Arguably, all of these, to varying extents, are based on the same principle as the menagerie: human entertainment. It is as a result of this that all of them can be grouped together under the heading zoo.

Schönbrunn Zoo in Vienna, Austria, claims to be the oldest zoo in the world, having opened in 1752. It is worth noting that it began life as a menagerie and only opened to the public in 1779. Other notable zoos (opening dates in parentheses) around the world include London Zoo (1828), San Diego Zoo (1916), Singapore Zoo (1973), and Durrell Wildlife Park (located on the island of Jersey, UK) (1959). The World Association of Zoos and Aquariums (WAZA) now consists of more than 1,200 establishments, attracting over 600 million visitors each year. Yet these establishments represent only a fraction of the total number of zoos in the world, as there is no requirement for a zoo to be a member of WAZA. WAZA is a voluntary organization whose stated goal is to support and encourage zoos in the areas of animal welfare, conservation, and environmental education.

Zoos are at least partially focused on providing visitors with a leisure experience that is situated around the viewing of wild and exotic animals. This is because of the financial reality that makes almost all of them dependent on income generated from visitors and because of a continuing demand for such experiences from members of the public. However, societal acceptance of zoos has shifted significantly with the rise of the animal rights movement and consequent concerns about the appropriateness of animal enclosures and the utilization of animals as objects of human entertainment. Consequently, we have seen zoos rebranding themselves as sites where endangered animals may be protected from extinction, with the idea that they can then at some unspecified future date be utilized to repopulate the "wild." Zoos have also branded themselves as educational centers, where researchers can learn more about wild animals and members of the public can be educated about the importance and value of wildlife conservation.

Have zoos been successful in rebranding themselves? The answer to this from a financial viability perspective would certainly be yes. London Zoo is an excellent example of this, having almost had to close in 1991 due to declining visitor numbers but now being one of London's significant leisure/tourism attractions. However, the debate about whether zoos such as London's have or can actually successfully undertake a meaningful education, conservation, and/or research role is ongoing. Zoos and their supporters point to their successes in animal conservation and educational theories about their potential to change societal attitudes about conservation. Anti-zoo protagonists highlight the limited successes of zoos in preserving endangered animals, the lack of evidence of their ability to influence public opinion about conservation, and the welfare costs borne by animals housed in zoos. Much of the debate is arguably intensified by feelings and opinions related to animal rights that may cloud the judgement of those both for and against zoos.

The reality is that there are some excellent zoos in the world today (e.g., Durrell Wildlife Park) that are doing all they can to ensure the welfare of their animals, to aid wildlife and natural landscape conservation, and educate the general public at the same time as having to remain economically viable. A visit to the website of the Durrell Wildlife Conversation Trust, of which the Wildlife Park is an integral part, highlights the fact that entertaining visitors is only one small part of the work they do. Ensuring the survival of threatened species and their native habitat is the clear underlying ethos of this zoo and its parent organization, an ethos matched by deeds. These excellent zoos are juxtaposed against ones that are clearly more interested in the provision of entertainment and the maximization of profit than animal well-being. Visiting the websites of zoos and seeing the emphasis placed on human entertainment, conservation, research, and well-being is one way to begin to judge the nature of a zoo. The challenge is not necessarily to see the closure of all zoos but to see that the leading zoos, in which animal welfare is prioritized, are encouraged while others are required to either move in a similar direction or shut down.

Neil Carr

See also: Advocacy; Aquariums; Biodiversity; Endangered Species; Ethics; Extinction; Stereotypic Behaviors; Wildlife; Wildlife Rehabilitation and Rescue

Further Reading

Carr, N., and Cohen, S. 2011. "The Public Face of Zoos: Balancing Entertainment, Education, and Conservation." *Anthrozoos*. 24(2): 175–189

Durrell, G. 1976. *The Stationary Ark*. London: Collins.

Durrell Wildlife Conservation Trust. 2015. "Introduction." Accessed June 18, 2015. http://www.durrell.org

Jamieson, D. 1985. "Against Zoos." In Singer, P., ed., *In Defence of Animals*, 108–117. Oxford: Basil Blackwell.

Kisling, Jr., V., ed. 2001. *Zoo and Aquarium History: Ancient Animal Collections to Zoological Gardens*. London: CRC Press.

PRIMARY DOCUMENTS

DOCUMENT 1: Animal Damage Control Act (1931)

This act was passed into law by the Congress of the United States in 1931. The name was changed to Wildlife Services (WS) in 1997. It gives the U.S. Department of Agriculture (USDA), through the Animal Health and Plant Inspection Service (APHIS), the authority to use lethal means to remove wild animals considered to be pests and/or conflicting with economic interests such as ranching. Since its inception, 80 million animals have been killed, including prairie dogs, coyotes, wolves, foxes, bears, and eagles. The law is controversial for some biologists who argue that removing these animals may damage local ecosystems by removing predators and increasing prey populations who overeat plant species. Additionally, wildlife advocates argue that it is unethical and hypocritical that one part of the government, WS, focuses on killing animals while another (U.S. Fish and Wildlife Service) is charged with protecting it. The USDA has been implementing less lethal approaches to human-wildlife conflict and increasing farmer/rancher education programs to combat the program's negative image. This excerpt shows how the statute has been amended from its inception to the present day.

Section 426 of the 1931 Animal Damage Control Act: The Secretary is authorized to conduct investigations, experiments, and tests to determine the best methods of eradication, suppression, or bringing under control mountain lions, wolves, coyotes, bobcats, prairie dogs, gophers, ground squirrel, jack rabbits, and other animals injurious to agriculture, horticulture, forestry, animal husbandry, wild game animals, fur-bearing animals and birds. Another purpose of these investigations is to protect stock and other domestic animals through the suppression of rabies and tularemia in predatory or other wild animals. The Secretary is also directed to conduct campaigns for the destruction or control of these animals. In carrying out the Act, the Secretary may cooperate with states, individuals, agencies and organizations. . . . Section 426 of the current (2011) code. Predatory and Other Wild Animals: The Secretary of Agriculture may conduct a program of wildlife services with respect to injurious animal species and take any action the Secretary considers necessary in conducting the program. The Secretary shall administer the program in a manner consistent with all of the wildlife services authorities in effect on the day before October 28, 2000. 426c. Control of nuisance mammals and birds and those constituting reservoirs of zoonotic diseases; exception: On and after

December 22, 1987, the Secretary of Agriculture is authorized, except for urban rodent control, to conduct activities and to enter into agreements with States, local jurisdictions, individuals, and public and private agencies, organizations, and institutions in the control of nuisance mammals and birds and those mammal and bird species that are reservoirs for zoonotic diseases, and to deposit any money collected under any such agreement into the appropriation accounts that incur the costs to be available immediately and to remain available until expended for Animal Damage Control activities. Prevention of Introduction of Brown Tree Snakes to Hawaii from Guam: (a) In general. The Secretary of Agriculture shall, take such action as may be necessary to prevent the inadvertent introduction of brown tree snakes into other areas of the United States from Guam. (b) Introduction into Hawaii. The Secretary shall initiate a program to prevent, the introduction of the brown tree snake into Hawaii from Guam. In carrying out this section, the Secretary shall consider the use of sniffer or tracking dogs, snake traps, and other preventative processes or devices at aircraft and vessel loading facilities on Guam, Hawaii, or intermediate sites serving as transportation points that could result in the introduction of brown tree snakes into Hawaii. (c) Authority. The Secretary shall use the authority provided under the Federal Plant Pest Act (7 U.S.C. 150aa et seq.) [section 150aa et seq. of this title] to carry out subsections (a) and (b).

Source: Animal Damage Control Act of March 2, 1931 (46 Stat. 1468). 7 USC 426–426d.

DOCUMENT 2: Animal Enterprise Terrorism Act (2006)

The Animal Enterprise Terrorism Act (AETA) was passed into law by the U.S. Congress under the George W. Bush administration in 2006, amending the Animal Enterprise Protection Act (AEPA) of 1992. Both Acts were strongly supported by biomedical and animal agricultural interests, who claimed that federal law was needed to protect against violent acts against businesses using animals. AEPA provided a legal foundation to limit protest and investigative activities of animal protection advocates. AETA not only significantly expanded these limitations but also legally raised these acts to the level of terrorism. This expansion effectively eliminated constitutional First Amendment (free speech) protections for a number of activities, making certain acts formerly considered lawful, nonviolent civil disobedience into illegal terrorist acts. Under AETA, acts do not have to cause physical harm to persons or damage property to be violations but may simply cause profit loss. Because of the breadth and vagueness of its language, challengers of the Act have called its effects "chilling" on the rights of animal advocates to inform the public about animal treatment. This excerpt shows the Act's language that makes these activities terrorism, two relevant definitions, and the minimal levels of offenses that are now subject to penalty.

An Act

To provide the Department of Justice the necessary authority to apprehend, prosecute, and convict individuals committing animal enterprise terror.
[. . .]

Sec. 2. Inclusion of Economic Damage to Animal Enterprises and Threats of Death and Serious Bodily Injury to Associated Persons.

[. . .]
"(a) OFFENSE.—Whoever travels in interstate or foreign commerce, or uses or causes to be used the mail or any facility of interstate or foreign commerce—

"(1) for the purpose of damaging or interfering with the operations of an animal enterprise; and

"(2) in connection with such purpose—

"(A) intentionally damages or causes the loss of any real or personal property (including animals or records) used by an animal enterprise, or any real or personal property of a person or entity having a connection to, relationship with, or transactions with an animal enterprise;

"(B) intentionally places a person in reasonable fear of the death of, or serious bodily injury to that person, a member of the immediate family (as defined in section 115) of that person, or a spouse or intimate partner of that person by a course of conduct involving threats, acts of vandalism, property damage, criminal trespass, harassment, or intimidation; or

"(C) conspires or attempts to do so;

shall be punished as provided for in subsection (b).

"(b) PENALTIES.—The punishment for a violation of section (a) or an attempt or conspiracy to violate subsection (a) shall be—

"(1) a fine under this title or imprisonment not more than 1 year, or both, if the offense does not instill in another the reasonable fear of serious bodily injury or death and—

"(A) the offense results in no economic damage or bodily injury; or

"(B) the offense results in economic damage that does not exceed $10,000;

"(2) a fine under this title or imprisonment for not more than 5 years, or both, if no bodily injury occurs and—

"(A) the offense results in economic damage exceeding $10,000 but not exceeding $100,000; or

"(B) the offense instills in another the reasonable fear of serious bodily injury or death;

[. . .]

"(d) DEFINITIONS.—As used in this section—

"(1) the term 'animal enterprise' means—

"(A) a commercial or academic enterprise that uses or sells animals or animal products for profit, food or fiber production, agriculture, education, research, or testing;

"(B) a zoo, aquarium, animal shelter, pet store, breeder, furrier, circus, or rodeo, or other lawful competitive animal event; or

"(C) any fair or similar event intended to advance agricultural arts and sciences;

[. . .]

"(3) the term 'economic damage'—

"(A) means the replacement costs of lost or damaged property or records, the costs of repeating an interrupted or invalidated experiment, the loss of profits, or increased costs, including losses and increased costs resulting from threats, acts or vandalism, property damage, trespass, harassment, or intimidation taken against a person or entity on account of that person's or entity's connection to, relationship with, or transactions with the animal enterprise; but "(B) does not include any lawful economic disruption (including a lawful boycott) that results from lawful public, governmental, or business reaction to the disclosure of information about an animal enterprise. . . .

Source: The Animal Enterprise Terrorism Act (AETA) of November 27, 2006 (18 U.S.C. § 43).

DOCUMENT 3: *Animal Liberation* by Peter Singer (1975)

Peter Singer (1946–) is a philosopher who focuses on ethics. Although Singer's philosophy is not based on the concept of rights, his 1975 publication, Animal Liberation, *is considered one of the founding texts of the modern animal rights movement. His documentation of the inhumane treatment of animals on farms, in laboratories, and in other areas of life were shocking and disturbing, as most people were unaware of these practices. As a philosopher, he advocated for animals' liberation from human domination, and the extension of the utilitarian concept of the "greatest good for the greatest number," previously only applied to humans, to all other animal species. He argued that we should give other animals ethical consideration as we do humans because animals can and do experience suffering and, therefore, we need to act to reduce suffering to the extent possible. In the following excerpt, from the preface, he outlines how he sees the connection between human prejudice against animals and prejudices that exist between humans in order to demonstrate 1) the difficulty in recognizing bias when you are the beneficiary and 2) how deeply society must go into its moral framework to liberate animals from systemic prejudice and abuse.*

The title of this book has a serious point behind it. A liberation movement is a demand for an end to prejudice and discrimination based on an arbitrary characteristic like race or sex. The classic instance is the Black Liberation movement. The immediate appeal of this movement, and its initial, if limited, success, made it a model for other oppressed groups. We soon became familiar with Gay Liberation and movements on behalf of American Indians and Spanish-speaking Americans. When a majority group—women—began their campaign some thought we had come to the end of the road. Discrimination on the basis of sex, it was said, was the last form of discrimination to be universally accepted and practiced without secrecy or pretense, even in those liberal circles that have long prided themselves on their freedom from prejudice against racial minorities.

We should always be wary of talking of "the last remaining form of discrimination." If we have learned anything from the liberation movements we should have learned how difficult it is to be aware of latent prejudices in our attitudes to particular groups until these prejudices are forcefully pointed out to us.

A liberation movement demands an expansion of our moral horizons. Practices that were previously regarded as natural and inevitable come to be seen as the result of an unjustifiable prejudice. Who can say with any confidence that none of his or her attitudes and practices can legitimately be questioned? If we wish to avoid being numbered among the oppressors, we must be prepared to rethink all our attitudes to other groups, including the most fundamental of them. We need to consider our attitudes from the point of view of those who suffer by them, and by the practices that follow from them. If we can make this unaccustomed mental switch we may discover a pattern in our attitudes and practices that operates so as consistently to benefit the same group—usually the group to which we ourselves belong—at the expense of another group. So we come to see that there is a case for a new liberation movement.

The aim of this book is to lead you to make this mental switch in your attitudes and practices toward a very large group of beings: members of species other than our own. I believe that our present attitudes to these beings are based on a long history of prejudice and arbitrary discrimination. I argue that there can be no reason—except the selfish desire to preserve the privileges of the exploiting group—for refusing to extend the basic principle of equality of consideration to members of other species. I ask you to recognize that your attitudes to members of other species are a form of prejudice no less objectionable than prejudice about a person's race or sex.

Source: Singer, Peter. *Animal Liberation: A New Ethics for Our Treatment of Animals.* New York: HarperCollins, 1975. Reprinted with permission from Peter Singer.

DOCUMENT 4: Animal Welfare Act (1966, with amendments through 2014)

Originally titled the Laboratory Animal Welfare Act, the Animal Welfare Act (AWA) was passed by the U.S. Congress in 1966 under the Lyndon Johnson administration. The original act arose out of concerns over the use of dogs and cats in research, especially the theft of pets for such purposes. The AWA established a licensing and record-keeping process for dealers and research facilities to combat these practices and also covered other species such as monkeys and rabbits. It is enforced by the U.S. Department of Agriculture (USDA) and is the nation's only federal law providing basic care standards for animals that are used in research, transported, exhibited, bred for sale, or handled by dealers. Although amended multiple times and currently including standards for providing psychological enrichment and regulating the imposition of pain, many still argue that the AWA's requirements are only minimal, it is not well enforced, and inspections are too infrequent. The following excerpt illustrates that, despite the broad range of covered activities, the AWA does not cover vast numbers of animals by excluding from its definition of "animal" those such as livestock and certain key species used extensively in research.

§ 2131—Congressional statement of policy

The Congress finds that animals and activities which are regulated under this chapter are either in interstate or foreign commerce or substantially affect such commerce or the free flow thereof, and that regulation of animals and activities as provided in this chapter is necessary to prevent and eliminate burdens upon such commerce and to effectively regulate such commerce, in order—

(1) to insure that animals intended for use in research facilities or for exhibition purposes or for use as pets are provided humane care and treatment;

(2) to assure the humane treatment of animals during transportation in commerce; and

(3) to protect the owners of animals from the theft of their animals by preventing the sale or use of animals which have been stolen.

The Congress further finds that it is essential to regulate, as provided in this chapter, the transportation, purchase, sale, housing, care, handling, and treatment of animals by carriers or by persons or organizations engaged in using them for research or experimental purposes or for exhibition purposes or holding them for sale as pets or for any such purpose or use.

[. . .]

§ 2132—Definitions

[. . .]

(g)The term "animal" means any live or dead dog, cat, monkey (nonhuman primate mammal), guinea pig, hamster, rabbit, or such other warm-blooded animal, as the Secretary may determine is being used, or is intended for use, for research, testing, experimentation, or exhibition purposes, or as a pet; but such term excludes (1) birds, rats of the genus Rattus, and mice of the genus Mus, bred for use in research, (2) horses not used for research purposes, and (3) other farm animals, such as, but not limited to livestock or poultry, used or intended for use as food or fiber, or livestock or poultry used or intended for use for improving animal nutrition, breeding, management, or production efficiency, or for improving the quality of food or fiber. With respect to a dog, the term means all dogs including those used for hunting, security, or breeding purposes. . . .

Source: Animal Welfare Act (Laboratory Animal Welfare Act of 1966), P.L. 89-544, 80 Stat. 340. 7 U.S.C. § 2131 et seq.

DOCUMENT 5: The Cambridge Declaration on Consciousness (2012)

The Cambridge Declaration on Consciousness was written collaboratively by a team of scientists from the fields of neurology, psychology, and biology, with neuroscientist Philip Low as lead author. The Declaration is a formal statement from a broad group of scientists declaring that many species of nonhuman animals have, like humans, the structural capacity within their brains and nervous systems to be "conscious," that is, subjectively aware of their experiences and the world around them and, potentially, aware of themselves as beings separate from that world, or "self-conscious." Although for many nonscientists the Declaration's conclusion will appear to state the obvious, animal consciousness has been the subject of considerable debate for centuries in Western science. One of the main hindrances to scientific acceptance of animal consciousness has been the inability, based on accepted standards, to provide absolute proof *of animals' mental states. The Cambridge Declaration is significant in stating that, though not absolute proof, overwhelming evidence exists that nonhumans have many of the same neural structures as humans. Broader societal recognition of the Declaration points toward expanded and improved legal protections for many animals. The following excerpt reflects only minimal deletions from the Declaration.*

On this day of July 7, 2012, a prominent international group of cognitive neuroscientists, neuropharmacologists, neurophysiologists, neuroanatomists and computational neuroscientists gathered at The University of Cambridge to reassess the neurobiological substrates of conscious experience and related behaviors in human and non-human animals. While comparative research on this topic is naturally hampered by the inability of non-human animals, and often humans, to clearly and readily communicate about their internal states, the following observations can be stated unequivocally:

The field of Consciousness research is rapidly evolving. Abundant new techniques and strategies for human and non-human animal research have been developed. Consequently, more data is becoming readily available, and this calls for a periodic reevaluation of previously held preconceptions in this field. Studies of non-human animals have shown that homologous brain circuits correlated with conscious experience and perception can be selectively facilitated and disrupted to assess whether they are in fact necessary for those experiences. Moreover, in humans, new non-invasive techniques are readily available to survey the correlates of consciousness.

The neural substrates of emotions do not appear to be confined to cortical structures. In fact, subcortical neural networks aroused during affective states in humans are also critically important for generating emotional behaviors in animals. Artificial arousal of the same brain regions generates corresponding behavior and feeling states in both humans and non-human animals. . . . Systems associated with affect are concentrated in subcortical regions where neural homologies abound. Young human and nonhuman animals without neocortices retain these brain-mind functions. Furthermore, neural circuits supporting behavioral/electrophysiological states of attentiveness, sleep and decision making appear to have arisen in evolution as early as the invertebrate radiation, being evident in insects and cephalopod mollusks (e.g., octopus).

Birds appear to offer, in their behavior, neurophysiology, and neuroanatomy a striking case of parallel evolution of consciousness. Evidence of near human-like levels of consciousness has been most dramatically observed in African grey parrots. Mammalian and avian emotional networks and cognitive microcircuitries appear to be far more homologous than previously thought. Moreover, certain species of birds have been found to exhibit neural sleep patterns similar to those of mammals, including REM sleep and, as was demonstrated in zebra finches, neurophysiological patterns, previously thought to require a mammalian neocortex. Magpies in particular have been shown to exhibit striking similarities to humans, great apes, dolphins, and elephants in studies of mirror self-recognition.

. . . . Evidence that human and nonhuman animal emotional feelings arise from homologous subcortical brain networks provide compelling evidence for evolutionarily shared primal affective qualia.

We declare the following: "The absence of a neocortex does not appear to preclude an organism from experiencing affective states. Convergent evidence indicates that non-human animals have the neuroanatomical, neurochemical, and neurophysiological substrates of conscious states along with the capacity to exhibit intentional behaviors. Consequently, the weight of evidence indicates that humans are not unique in possessing the neurological substrates that generate consciousness. Nonhuman animals, including all mammals and birds, and many other creatures, including octopuses, also possess these neurological substrates."

Source: Low, Philip et al. The Cambridge Declaration on Consciousness, Publicly proclaimed in Cambridge, UK, on July 7, 2012, at the Francis Crick Memorial Conference on Consciousness in Human and non-Human Animals.

DOCUMENT 6: *The Case for Animal Rights* by Tom Regan (1983)

Tom Regan (1938–) is a founding and major figure in animal rights philosophy. His book, The Case for Animal Rights, *is considered a foundational text for the modern animal rights movement. In the book he utilizes a rights-based ethical argument, previously applied only to humans, to argue for human moral and legal obligations to animals. In* The Case, *Regan argues that it is unethical to use sentient animals for human benefit, including for food or medical experimentation. His logic rests on the concept of inherent value, or individuals' having value in and of themselves, regardless of their value to others. Regan argues that animals who are what he calls "subjects-of-a-life" have inherent value because they have the same qualities (although in varying degrees) that we believe give humans inherent value. The excerpts below outline Regan's definition of moral agents (those who can act morally) and moral patients (those who do not have the capacity to act morally but are still subjects-of-a-life), and how both groups have inherent value and, therefore, deserve ethical treatment.*

A helpful place to begin is to distinguish between moral agents and moral patients. . . . Moral agents are individuals who have a variety of sophisticated abilities, including in particular the ability to bring impartial moral principles to bear on the determination of what, all considered, morally ought to be done and, having made this determination, to freely choose or fail to choose to act as morality, as they conceive it, requires. Because moral agents have these abilities, it is fair to hold them morally accountable for what they do, assuming that the circumstances of their acting as they do in a particular case do not dictate otherwise.

[. . .]

In contrast to moral agents, *moral patients* lack the prerequisites that would enable them to control their own behavior in ways that would make them morally accountable for what they do. A moral patient lacks the ability to formulate, let alone bring to bear, moral principles in deliberating about which one among a number of possible acts it would be right or proper to perform. Moral patients, in a word, cannot do what is right, nor can they do what is wrong. . . . Only moral agents can do what is wrong. Human infants, young children, and the mentally deranged or enfeebled of all ages are paradigm cases of human moral patients.

[. . .]

Individuals who are moral patients differ from one another in morally relevant ways. Of particular importance is the distinction between (a) those individuals who are conscious and sentient (i.e., can experience pleasure and pain) but who lack other mental abilities, and (b) those individuals who are conscious, sentient, and possess . . . other cognitive and volitional abilities (e.g., belief and memory). Some animals, for reasons already advanced, belong in category (b); other animals quite probably belong in category (a). . . . Our primary interest . . . concerns the moral status of animals in category (b). . . . [T]the notion of a *moral patient* is . . . understood as applying to *animals in category (b) and to those other moral patients like these animals in the relevant respects*. . . .

[. . .]

Individuals are subjects-of-a-life if they have beliefs and desires; perception, memory, and a sense of the future, including their own future; an emotional life together with feelings of pleasure and pain; preference- and welfare-interests; the ability to initiate action in pursuit of their desires and goals; a psychophysical identity over time; and an individual welfare in the sense that their experiential life fares well or ill for them, logically independently of their utility for others and logically independently of their being the object of anyone else's interests. Those who satisfy the subject-of-a-life criterion themselves have a distinctive kind of value—inherent value—and are not to be viewed or treated as mere receptacles.

[. . .]

All moral agents and *all* those moral patients with whom we are concerned *are* subjects of a life that is better or worse for them, in the sense explained, logically independently of the utility they have for others and logically independently of their being the object of the interests of others.

[. . .]

We are to treat those individuals who have inherent value in ways that respect their inherent value.

Source: The Case for Animal Rights, by Tom Regan, © 2004 by the Regents of the University of California. Published by the University of California Press. Reprinted with permission from University of California Press and Tom Regan.

DOCUMENT 7: *Commission of Patents v. President and Fellows of Harvard College* (2002)

After the U.S. Patent and Trademark Office (USPTO) interpreted federal patent law to allow multicellular organisms to be patentable as novel compositions of matter, they issued the first patent for OncoMouse™ in 1988. OncoMouse was deemed a novel composition of matter because it is a transgenic mouse that, through biotechnological intervention, had its genome altered to include human cancer-promoting genes. Created by researchers at Harvard University with funding from DuPont Corporation, the patent granted the creators exclusive right to own and license (and thereby earn money from) OncoMouse. While several other countries updated their patent laws to allow organisms created through biotechnologies to be novel and made by humans, Canada's Supreme Court made the opposite decision. After the Canadian Patent Office rejected Harvard's application in 1993, on the grounds that multicellular organisms were not inventions, a series of court cases brought the question to their Supreme Court. In a 5-4 decision, the justices determined that animals were not patentable matter. The case is important because it legally constructs animals as lifeforms who cannot be "inventions" of humans. This excerpt provides a description of what the Canadian justices were asked to determine and the main logic of their ruling.

B. The Definition of Invention: Whether a Higher Life Form Is a "Manufacture" or a "Composition of Matter"

The sole question in this appeal is whether the words "manufacture" and "composition of matter", within the context of the *Patent Act*, are sufficiently broad to include higher life forms. . . . Comparisons with the patenting schemes of other countries will therefore be of limited value. The best reading of the words of the Act supports the conclusion that higher life forms are not patentable.

(1) The Words of the Act

For a higher life form to fit within the definition of "invention", it must be considered to be either a "manufacture" or a "composition of matter". While the definition of "invention" in the *Patent Act* is broad, Parliament did not define "invention" as "anything new and useful made by man". The choice of an exhaustive definition signals a clear intention to exclude certain subject matter as being outside the confines of the Act. The word "manufacture" ("*fabrication*"), in the context of the Act, is commonly understood to denote a non-living mechanistic product or process, not a higher life form. The words "composition of matter" ("*composition de matières*") as they are used in the Act do not include a higher life form such as the OncoMouse. The words occur in the phrase "art, process, machine, manufacture or composition of matter". A collective term that completes an enumeration is often restricted to the same genus as the terms which precede it, even though the collective term may ordinarily have a much broader meaning. Just as "machine" and "manufacture" do not imply a living creature, the words "composition of matter" are best read as not including higher life forms. While a fertilized egg injected with an oncogene may be a mixture of various ingredients, the body of a mouse does not consist of ingredients or substances that have been combined or mixed together by a person. Moreover, "matter" captures only one aspect of a higher life form, generally regarded as possessing qualities and characteristics that transcend the particular genetic material of which it is composed. Higher life forms cannot be conceptualized as mere "compositions of matter" within the context of the *Patent Act*. Just because all inventions are unanticipated and unforeseeable, it does not necessarily follow that they are all patentable. It is possible that Parliament did not intend to include higher life forms in the definition of "invention". It is also possible that Parliament did not regard cross-bred plants and animals as patentable because they are better regarded as "discoveries". Because patenting higher life forms would involve a radical departure from the traditional patent regime, and because the patentability of such life forms is a highly contentious matter that raises a number of extremely complex issues, clear and unequivocal legislation is required for higher life forms to be patentable. The current Act does not clearly indicate that higher life forms are patentable.

Source: Harvard College v. Canada (Commissioner of Patents), (2002) 4 S.C.R. 45, 2002 SCC 76.

DOCUMENT 8: Convention on Biological Diversity (1992)

The Convention on Biological Diversity (CBD) is an international treaty that emerged from a series of negotiations that began in 1988 and were organized by the UN Environment Programme (UNEP). The CBD opened for signature at the 1992 "Rio Earth Summit" of the UN Conference on Environment and Development in Brazil and was ratified (met the threshold for countries signing the treaty) in 1993. As of early 2016, 196 parties have ratified the CBD, but the United States has not. The CBD has three main objectives: conservation of biodiversity (plant and animal), sustainable use of natural resources, and equitable sharing of these resources. It is the first international treaty to focus on protecting Earth as a whole and to recognize that maintaining and protecting the biodiversity of nonhuman animal species, in particular, is essential for the survival of both the planet and humans. A 2000 addition to the CBD, called The Cartagena Protocol on Biosafety and addressing specific concerns about biotechnologies, was ratified and entered into force in 2003. This excerpt shows what participating countries have agreed to do to promote the three main objectives inside their borders.

Article 8 In-situ Conservation: Each Contracting Party shall, as far as possible and as appropriate: (a) Establish a system of protected areas or areas where special measures need to be taken to conserve biological diversity; (b) Develop, where necessary, guidelines for the selection, establishment and management of protected areas or areas where special measures need to be taken to conserve biological diversity; (c) Regulate or manage biological resources important for the conservation of biological diversity whether within or outside protected areas, with a view to ensuring their conservation and sustainable use; (d) Promote the protection of ecosystems, natural habitats and the maintenance of viable populations of species in natural surroundings; (e) Promote environmentally sound and sustainable development in areas adjacent to protected areas with a view to furthering protection of these areas; (f) Rehabilitate and restore degraded ecosystems and promote the recovery of threatened species, inter alia, through the development and implementation of plans or other management strategies; (g) Establish or maintain means to regulate, manage or control the risks associated with the use and release of living modified organisms resulting from biotechnology which are likely to have adverse environmental impacts that could affect the conservation and sustainable use of biological diversity, taking also into account the risks to human health; (h) Prevent the introduction of, control or eradicate those alien species which threaten ecosystems, habitats or species; (i) Endeavour to provide the conditions needed for compatibility between present uses and the conservation of biological diversity and the sustainable use of its components; (j) Subject to its national legislation, respect, preserve and maintain knowledge, innovations and practices of indigenous and local communities embodying traditional lifestyles relevant for the conservation and sustainable use of biological diversity and promote their wider application with the approval and involvement of the holders of such knowledge, innovations and practices and encourage the equitable sharing of the benefits arising from the utilization of such knowledge, innovations and practices; (k) Develop or maintain necessary legislation and/or other regulatory provisions for the protection of threatened species and populations; (l) Where a significant adverse effect on biological diversity has been determined pursuant to Article 7, regulate or manage the relevant processes and categories of activities; and (m) Cooperate in providing financial and other support for in-situ conservation outlined in subparagraphs (a) to (l) above, particularly to developing countries.

Source: "Text of the CBD." Convention on Biological Diversity, 1992. Available at: https://www.cbd.int/convention/text/default.shtml

DOCUMENT 9: Convention on International Trade in Endangered Species of Wild Fauna and Flora (CITES) (1975)

CITES is an international treaty administered by the UN Environment Programme (UNEP). It grew out of concerns expressed by members of the World Conservation Union (IUCN) in the 1960s about the growing impact of global trade of wild animals and plants. Recognizing the need for international cooperation, the treaty was drafted in 1973 and ratified (officially agreed upon by individual countries) in 1975, when it went into force. As of 2016, there are 181 countries, including the United States, who are party to the treaty. CITES provides a three-appendix framework for management. Appendix 1 listing means a species is so endangered that trade must essentially stop. Appendix II listing covers trade in species that are possibly becoming threatened. Appendix III covers trade in species that may be impacted within only one country. Species qualify for one of the three appendix listings by meeting specific biological criteria (e.g., population decline, geographic concentration) as set forth in the agreement. Approximately 5,600 species of animals are protected by CITES from overexploitation. This excerpt is the language used to define the highest, or most critical, listing and how trade for these species must be handled by member parties.

Article II: 1. Appendix I shall include all species threatened with extinction which are or may be affected by trade. Trade in specimens of these species must be subject to particularly strict regulation in order not to endanger further their survival and must only be authorized in exceptional circumstances.

Article III: 1. All trade in specimens of species included in Appendix I shall be in accordance with the provisions of this Article.

2. The export of any specimen of a species included in Appendix I shall require the prior grant and presentation of an export permit. An export permit shall only be granted when the following conditions have been met:

(a) a Scientific Authority of the State of export has advised that such export will not be detrimental to the survival of that species;

(b) a Management Authority of the State of export is satisfied that the specimen was not obtained in contravention of the laws of that State for the protection of fauna and flora;

(c) a Management Authority of the State of export is satisfied that any living specimen will be so prepared and shipped as to minimize the risk of injury, damage to health or cruel treatment; and

(d) a Management Authority of the State of export is satisfied that an import permit has been granted for the specimen.

3. The import of any specimen of a species included in Appendix I shall require the prior grant and presentation of an import permit and either an export permit or a re-export certificate. An import permit shall only be granted when the following conditions have been met:

(a) a Scientific Authority of the State of import has advised that the import will be for purposes which are not detrimental to the survival of the species involved; (b) a Scientific Authority of the State of import is satisfied that the proposed recipient of a living specimen is suitably equipped to house and care for it; and (c) a Management Authority of the State of import is satisfied that the specimen is not to be used for primarily commercial purposes.

4. The re-export of any specimen of a species included in Appendix I shall require the prior grant and presentation of a re-export certificate. A re-export certificate shall only be granted when the following conditions have been met:

(a) a Management Authority of the State of re-export is satisfied that the specimen was imported into that State in accordance with the provisions of the present Convention;

(b) a Management Authority of the State of re-export is satisfied that any living specimen will be so prepared and shipped as to minimize the risk of injury, damage to health or cruel treatment; and (c) a Management Authority of the State of re-export is satisfied that an import permit has been granted for any living specimen.

5. The introduction from the sea of any specimen of a species included in Appendix I shall require the prior grant of a certificate from a Management Authority of the State of introduction.

Source: Convention on International Trade in Endangered Species of Wild Fauna and Flora, July 1, 1975. 27 UST 1087; TIAS 8249; 993 UNTS 243.

DOCUMENT 10: U.S. Endangered Species Act (ESA) (1973)

The ESA was passed into law by the U.S. Congress in 1973. It was one of the government's responses to growing public concern about the environment and reflected a desire to protect the nation's biological assets. The purpose is to protect both species and ecosystems. It is administered by the Department of the Interior's U.S. Fish and Wildlife Service, having primary responsibility for terrestrial and freshwater species, and the Commerce Department's National Marine Fisheries Services, having primary responsibility for marine wildlife. The ESA employs a two-layer system whereby a species may be listed as 1) endangered or 2) threatened with extinction in the near future. Nearly 1,500 species are currently listed. The aim is to manage a species to "recovery" so that it may be removed from the list. While environmental advocates support this protection, it has been opposed by those who believe in personal property rights and do not want the government managing their land or development opportunities for it. This excerpt is from Section 4 of the Act, dealing with the Determination of Endangered Species and Threatened Species. It shows the language used to define what is required to be listed and what the government must do for protection and recovery.

SEC. 4. (a) GENERAL.—(1) The Secretary shall by regulation promulgated in accordance with subsection (b) determine whether any species is an endangered species or a threatened species because of any of the following factors: (A) the present or threatened destruction, modification, or curtailment of its habitat or range; (B) overutilization for commercial, recreational, scientific, or educational purposes; (C) disease or predation; (D) the inadequacy of existing regulatory mechanisms; or (E) other natural or manmade factors affecting its continued existence.

[...] (d) PROTECTIVE REGULATIONS.—Whenever any species is listed as a threatened species pursuant to subsection (c) of this section, the Secretary shall issue such regulations as he deems necessary and advisable to provide for the conservation of such species. The Secretary may by regulation prohibit with respect to any threatened species any act prohibited under section 9(a)(1), in the case of fish or wildlife, or section 9(a)(2), in the case of plants, with respect to endangered species; except that with respect to the taking of resident species of fish or wildlife, such regulations shall apply in any State which has entered into a cooperative agreement pursuant to section 6(c) of this Act only to the extent that such regulations have also been adopted by such State.

[...] (f)(1) RECOVERY PLANS.—The Secretary shall develop and implement plans (hereinafter in this subsection referred to as "recovery plans") for the conservation and survival of endangered species and threatened species listed pursuant to this section, unless he finds that such a plan will not promote the conservation of the species. The Secretary, in developing and implementing recovery plans, shall, to the maximum extent practicable—(A) give priority to those endangered species or threatened species, without regard to taxonomic classification, that are most likely to benefit from such plans, particularly those species that are, or may be, in conflict with construction or other development projects or other forms of economic activity; (B) incorporate in each plan—(i) a description of such site-specific management actions as may be necessary to achieve the plan's goal for the conservation and survival of the species; (ii) objective, measurable criteria which, when met, would result in a determination, in accordance with the provisions of this section, that the species be removed from the list; (iii) estimates of the time required and the cost to carry out those measures needed to achieve the plan's goal and to achieve intermediate steps toward that goal.

Source: The Endangered Species Act of 1973 (ESA), 16 U.S.C. § 1531 et seq.

DOCUMENT 11: The Five Freedoms (1979)

The following excerpt (with only administrative information deleted) is from an original document listing five aspects of animal welfare that would evolve into what are now known as the Five Freedoms—a set of basic, broadly stated welfare components. These five aspects put forth by Britain's Farm Animal Welfare Council (FAWC) were the result of the 1965 Report of the Technical Committee to Enquire into the Welfare of Animals kept under Intensive Livestock Husbandry Systems (known as the Brambell Report). The impetus for the Committee's formation is largely credited to the publication of British animal advocate and author Ruth Harrison's book, Animal Machines *(1964), which detailed the inhumane treatment of animals on Britain's increasingly industrialized farms (leading Harrison to coin the term "factory farm"). As stated today, the Freedoms are: 1) from hunger and thirst, 2) from discomfort, 3) from pain, injury or disease, 4) to express normal behavior, and 5) from fear and distress. Although originally applied only to farmed animals, the Freedoms are today frequently applied more broadly to any captive animal and are recognized worldwide by a number of organizations and govern-*

ments as the minimum components of good welfare. However, unless enacted into law, providing animals with these freedoms is not mandatory.

Farm Animal Welfare Council Press Statement

In his statements on the 25th July and 4th December, the Minister of Agriculture, Fisheries and Food announced that the terms of reference of the Farm Animal Welfare Council would be to keep under review the welfare of farm animals on agricultural land, at markets, in transit and at the place of slaughter, and to advise the Agriculture Ministers of any legislative or other changes that may be necessary. The Council is free to publicise its views and will do so whenever, as now, circumstances make it appropriate. Ministers have asked the Council to advise as speedily as possible on revisions to the Welfare Codes for Cattle, Pigs, Domestic Fowls and Turkeys and to undertake this revision in such a way as to reflect advances in scientific knowledge and husbandry practice since the Brambell Committee reported in 1965. In preparing revisions of these Codes, the Council will build on the valuable work already done by the Farm Animal Welfare Advisory Committee, now disbanded.

The Codes are not mandatory but intended to create the best possible standards of welfare for animals in all systems of livestock husbandry, both intensive and extensive. The Council wishes the revised Codes to provide farm animals with the following:

1. freedom from thirst, hunger or malnutrition;
2. appropriate comfort and shelter;
3. prevention, or rapid diagnosis and treatment, of injury and disease;
4. freedom to display most normal patterns of behaviour;
5. freedom from fear.

The Codes are intended to contain more specific recommendations than formerly and should, in particular, place more emphasis on behavioural needs.

The Council may not necessarily be able to endorse every husbandry practice used on livestock farms but, until changes can be achieved, believes that it is in the interest of the animals to continue to give advice on the best possible management within those systems.

In addition to advising on revisions of the Codes, the Council will give careful consideration to the adequacy of farm animal welfare legislation in all the areas to which its remit extends, on the farm, in markets, during domestic transport, during export and at the place of slaughter.

Throughout its work the Council recognises the need for increased knowledge of the physiological and behavioural needs of farm animals, and, where research and development on this work appears to be insufficient, it will ask for work to be undertaken. Moreover, the Council also accepts that animal welfare raises certain points of ethics which are themselves beyond scientific investigation. The Council will therefore especially wish to encourage alternative systems of livestock husbandry which are ethically acceptable to the concerned public, can be shown to improve the welfare of the livestock in question and be economically competitive with existing systems of intensive production.

[. . .]

Source: Farm Animal Welfare Council, The National Archives, UK. Available at: http://webarchive.nationalarchives.gov.uk/20121007104210/http:/www.fawc.org.uk/freedoms.htm

DOCUMENT 12: Humane Methods of Slaughter Act (1978)

The Humane Methods of Slaughter Act of 1978 (updating the same Act of 1958) was passed by the U.S. Congress under the Jimmy Carter administration. The original Act was passed in response to high public demand for a law to address concerns over animal suffering at the time of slaughter. Therefore, the main component of that Act was the requirement to render animals unconscious prior to slaughter through a blow or gunshot to the head, or electrical or chemical stunning. The 1978 Act added the provision for U.S. Department of Agriculture (USDA) inspections, including the ability for inspectors to halt the slaughtering line if the animals were being improperly handled and until correction of the problem. Birds/poultry are not covered by either Act. Animals slaughtered in accordance with any religious ritual that prohibits unconsciousness at the time of slaughter are excluded under both Acts. Some animal advocates have challenged the efficacy of the law and inspections. The USDA Food Safety and Inspection Service (FSIS) currently enforces the law. With the exception of deletions of administrative/technical language, the following is the Act in its entirety.

An Act

To amend the Federal Meat Inspection Act to require that meat inspected and approved under such Act be produced only from livestock slaughtered in accordance with humane methods, and for other purposes

Be it enacted by the Senate and House of Representatives of the

United States of America in Congress assembled, that this Act may be cited as the "Humane Methods of Slaughter Act of 1978".

SEC. 2. Section 3 of the Federal Meat Inspection Act (21U.S.a. 603) is amended by inserting "(a)" immediately before the first sentence and adding at the end thereof a new subsection (b) as follows:

"(b) For the purpose of preventing the inhumane slaughtering of livestock, the Secretary shall cause to be made, by inspectors appointed for that purpose, an examination and inspection of the method by which cattle, sheep, swine, goats, horses, mules, and other equines are slaughtered and handled in connection with slaughter in the slaughtering establishments inspected under this Act. The Secretary may refuse to provide inspection to a new slaughtering establishment or may cause inspection to be temporarily suspended at a slaughtering establishment if the Secretary finds that any cattle, sheep, swine, goats, horses, mules, or other equines have been slaughtered or handled in connection with slaughter at such establishment by any method not in accordance with the Act of August 27, 1958 (72 Stat. 862; 7 U.S.C. 1901–1906) until the establishment furnishes assurances satisfactory to the Secretary that all slaughtering and handling in connection with slaughter of livestock shall be in accordance with such a method.".

[. . .]

SEC. 4. Section 20(a) of the Federal Meat Inspection Act (21U.S.C.

620) is amended by inserting after the first sentence a new sentence as

follows: "No such carcasses, parts of carcasses, meat or meat food products shall be imported into the United States unless the livestock from which they were introduced was slaughtered and handled in connection with slaughter m accordance with the Act of August 27, 1958

('72 Stat.862;7 U.S.a.1901–1906) .".

[. . .]

SEC. 6. Nothing in this Act shall be construed to prohibit, abridge, or in any way hinder the religious freedom of any person or group. Notwithstanding any other provision of this Act, in order to protect freedom of religion, ritual slaughter and the handling or other preparation of livestock for ritual slaughter are exempted from the terms of this Act. For the purposes of this section the term "ritual slaughter" means slaughter in accordance with section 2(b) of the Act of August 27, 1958 (72 Stat. 862; 7 U.S.C. 1902(b)).

[. . .]

Source: The Humane Methods of Slaughter Act of 1978, P.L. 85–765; 7 U.S.C. 1901 et seq.

DOCUMENT 13: The International Convention for the Regulation of Whaling (ICRW) (1946)

The International Whaling Commission (IWC) was formed to provide international cooperation in managing global whale populations. The ICRW, signed in 1946 and fully enacted in 1948, is recognized as one of the earliest pieces of international environmental regulation. Signatories to the convention voluntarily agree to adhere to the management guidelines, which include limits on hunting, species being hunted, and limits on hunting locations and times. International outcry over whaling led to a 1986 ban on commercial whaling by the IWC. This ban continues, although some hunting is allowed for indigenous (native) peoples and "scientific research." There are currently 88 member countries who contribute to the financial support of the Commission. The Secretariat, or management, of the IWC is headquartered in Cambridge, United Kingdom. The IWC has been criticized for being ineffective—countries such as Iceland and Norway continue commercial hunting in their territorial waters, while Japan hunts in many parts of the oceans. The excerpt below, from the 1946 convention, provides the rationale for undertaking such an agreement and outlines the general form of the "Schedule," which is the main document that outlines specific practices and is evaluated on a regular basis.

The Governments whose duly authorised representatives have subscribed hereto,

Recognizing the interest of the nations of the world in safeguarding for future generations the great natural resources represented by the whale stocks;

Considering that the history of whaling has seen over-fishing of one area after another and of one species of whale after another to such a degree that it is essential to protect all species of whales from further over-fishing;

Recognizing that the whale stocks are susceptible of natural increases if whaling is properly regulated, and that increases in the size of whale stocks will permit increases in the number of whales which may be captured without endangering these natural resources;

Recognizing that it is in the common interest to achieve the optimum level of whale stocks as rapidly as possible without causing widespread economic and nutritional distress;

Recognizing that in the course of achieving these objectives, whaling operations should be confined to those species best able to sustain exploitation in order to give an interval for recovery to certain species of whales now depleted in numbers;

Desiring to establish a system of international regulation for the whale fisheries to ensure proper and effective conservation and development of whale stocks on the basis of the principles embodied in the provisions of the International Agreement for the Regulation of Whaling, signed in London on 8th June, 1937, and the protocols to that Agreement signed in London on 24th June, 1938, and 26th November, 1945; and

Having decided to conclude a convention to provide for the proper conservation of whale stocks and thus make possible the orderly development of the whaling industry;

Have agreed as follows:

Article I

1. This Convention includes the Schedule attached thereto which forms an integral part thereof. All references to "Convention" shall be understood as including the said Schedule either in its present terms or as amended in accordance with the provisions of Article V.

Article II

As used in this Convention:

1. "Factory ship" means a ship in which or on which whales are treated either wholly or in part;

2. "Land station" means a factory on the land at which whales are treated whether wholly or in part;

3. "Whale catcher" means a ship used for the purpose of hunting, taking, towing, holding on to, or scouting for whales;

4. "Contracting Government" means any Government which has deposited an instrument of ratification or has given notice of adherence to this Convention.

Article V

1. The Commission may amend from time to time the provisions of the Schedule by adopting regulations with respect to the conservation and utilization of whale resources, fixing

(a) protected and unprotected species;

(b) open and closed seasons;

(c) open and closed waters, including the designation of sanctuary areas;

(d) size limits for each species;

(e) time, methods, and intensity of whaling (including the maximum catch of whales to be taken in any one season);

(f) types and specifications of gear and apparatus and appliances which may be used;

(g) methods of measurement; and

(h) catch returns and other statistical and biological records.

Source: International Convention for the Regulation of Whaling, Washington, 2nd December 1946. Reprinted with permission. Available at: http://iwc.int/history-and-purpose

DOCUMENT 14: International Union for the Conservation of Nature Red List Categories and Criteria (2001)

The International Union for the Conservation of Nature (IUCN), founded in 1948, is the first international environmental organization. Its members include countries, nongovernmental organizations, and individual volunteers who are experts in their fields. The IUCN holds official observer status at the UN General Assembly and is headquartered in Switzerland. Their goal is to develop practical solutions to both environmental conservation issues and human development challenges (e.g., clean water, food access, economic opportunities). The IUCN publishes a Red List Database of Threatened Species (plants, animals, and fungi), the most comprehensive and scientifically managed catalog globally. The Red List ranks species as being of least concern, near threatened, vulnerable, endangered, and critically endangered. The Red List is not legally binding for member countries but is instead used as justification for their conservation efforts and in relation to other international agreements such as the Convention on the Illegal Trade in Flora and Fauna (CITES). As of 2016, the IUCN has evaluated over 76,000 species and found 22,000 at risk for extinction. This excerpt outlines specific methods scientists use to assess when a species meets the criteria to become listed at the highest level of critically endangered—meaning that without drastic action extinction is likely.

V. The criteria for critically endangered, endangered and vulnerable

CRITICALLY ENDANGERED (CR)

A taxon is Critically Endangered when the best available evidence indicates that it meets any of the following criteria (A to E), and it is therefore considered to be facing an extremely high risk of extinction in the wild:

A. Reduction in population size based on any of the following:

1. An observed, estimated, inferred or suspected population size reduction of ≥90% over the last 10 years or three generations, whichever is the longer, where the causes of the reduction are clearly reversible AND understood AND ceased, based on (and specifying) any of the following:

(a) direct observation

(b) an index of abundance appropriate to the taxon

(c) a decline in area of occupancy, extent of occurrence and/or quality of habitat

(d) actual or potential levels of exploitation

(e) the effects of introduced taxa, hybridization, pathogens, pollutants, competitors or parasites.

2. An observed, estimated, inferred or suspected population size reduction of ≥80% over the last 10 years or three generations, whichever is the longer, where the reduction or its causes may not have ceased OR may not be understood OR may not be reversible, based on (and specifying) any of (a) to (e) under A1.

3. A population size reduction of ≥80%, projected or suspected to be met within the next 10 years or three generations, whichever is the longer (up to a maximum of 100 years), based on (and specifying) any of (b) to (e) under A1.

4. An observed, estimated, inferred, projected or suspected population size reduction of ≥80% over any 10 year or three generation period, whichever is longer (up to a maximum of 100 years in the future), where the time period must include both the past and the future, and where the reduction or its causes may not have ceased OR may not be understood OR may not be reversible, based on (and specifying) any of (a) to (e) under A1.

B. Geographic range in the form of either B1 (extent of occurrence) OR B2 (area of occupancy) OR both:

1. Extent of occurrence estimated to be less than 100 km2, and estimates indicating at least two of a-c:

a. Severely fragmented or known to exist at only a single location.

b. Continuing decline, observed, inferred or projected, in any of the following:

(i) extent of occurrence

(ii) area of occupancy

(iii) area, extent and/or quality of habitat

(iv) number of locations or subpopulations

(v) number of mature individuals.

c. Extreme fluctuations in any of the following:

(i) extent of occurrence

(ii) area of occupancy

(iii) number of locations or subpopulations

(iv) number of mature individuals.

2. Area of occupancy estimated to be less than 10 km2, and estimate indicating at least two of a-c:

a. Severely fragmented or known to exist at only a single location.

b. Continuing decline, observed, inferred or projected, in any of the following:

(i) extent of occurrence

(ii) area of occupancy

(iii) area, extent and/or quality of habitat

(iv) number of locations or subpopulations

(v) number of mature individuals.

c. Extreme fluctuations in any of the following:

(i) extent of occurrence

(ii) area of occupancy

(iii) number of locations or subpopulations

(iv) number of mature individuals.

C. Population size estimated to number fewer than 250 mature individuals and either:

1. An estimated continuing decline of at least 25% within three years or one generation, whichever is longer, (up to a maximum of 100 years in the future) OR

2. A continuing decline, observed, projected, or inferred, in numbers of mature individuals AND at least one of the following (a-b):

a. Population structure in the form of one of the following:

(i) no subpopulation estimated to contain more than 50 mature individuals, OR

(ii) at least 90% of mature individuals in one subpopulation.

b. Extreme fluctuations in number of mature individuals.

D. Population size estimated to number fewer than 50 mature individuals.

E. Quantitative analysis showing the probability of extinction in the wild is at least 50% within 10 years or three generations, whichever is the longer (up to a maximum of 100 years).

Source: IUCN. *IUCN Red List Categories and Criteria: Version 3.1.* 2nd ed. Gland, Switzerland and Cambridge, UK: IUCN, 2012, 16–18. Reprinted with permission.

DOCUMENT 15: Kinshasa Declaration on Great Apes (2005)

The Kinshasa Declaration on Great Apes (KDGA) is the only species-specific protection program of the United Nations (UN). It is co-administered by the UN Environment Programme (UNEP) and the UN Educational, Scientific, and Cultural Organization (UNESCO). It was formalized as a Declaration in 2005, after the founding of the Great Apes Survival Partnership (GRASP) in 2001. The cooperation between dozens of governments, conservation groups, researchers, private companies, and the UN has as its goal to ensure the long-term survival of humanity's closest relatives—bonobos, chimpanzees, gorillas, and orangutans—in the 23 countries of Africa and Asia where they live. All four great ape species are classified as endangered due to human activity. GRASP meets every four years to address conservation issues such as habitat loss and restoration, land-rights, illegal trade, advocacy, disease monitoring, and local and global education. The excerpt provides the justification for the Declaration and outlines the key ways in which they hope to accomplish their goals.

We, the representatives of the great ape range States, donor and other States, international and intergovernmental organizations, academic and scientific communities, non-governmental organizations, industry and the private sector, meeting at Kinshasa, Democratic Republic of the Congo, on 9 September 2005, Aware that there is a high risk of extinction in the wild for all great ape species, due largely to the destruction of forests and other habitat; threats from human activities, including increasing encroachments by human populations on their habitat; civil disturbances and wars; poaching for bushmeat and for the live animal trade; and diseases such as ebola which can decimate ape populations, Recognizing that great apes are flagship species for tropical forests and woodland areas and play a key role in maintaining the health and diversity of their ecosystems, and that their decline and potential extinction may precipitate the decline of other culturally, economically or ecologically important species, Also recognizing the intrinsic value of great apes as part of the world's natural heritage, which we have a moral duty to conserve and share with future generations, Recognizing further that great ape populations and their habitats can provide direct and indirect benefits to local communities and other stakeholders, and contribute to poverty alleviation through the development of carefully regulated ecologically sustainable ecotourism and other non-destructive enterprises and through the environmental services that forests provide, Recognizing moreover that all species of great apes are afforded the highest level of legal protection under relevant wildlife law in their respective range States, Recalling the World Charter for Nature, adopted by the United Nations General Assembly by its resolution 37/7 of 28 October 1982, which underscores the importance of not compromising the genetic viability on the earth.

. . . Reaffirm our commitment to work together to ensure that the Great Apes Survival Project Partnership has the capacity to realize its full potential as a key component of the international effort to save great apes by:

(a) Urging all 23 great ape range States to become or remain active partners of the Great Apes Survival Project Partnership; (b) Also encouraging other States which either already support or participate to a significant extent in programmes for the conservation of great apes and their habitat, or could contribute to such an effort in such a way as to become full partners of the Great Apes Survival Project Partnership; (c) Encouraging other international organizations, . . . to become or remain active partners of the Great Apes Survival Project Partnership; (d) Encouraging non-governmental organizations that have historically either played an important role in efforts to conserve the great apes, . . . to redouble their efforts in that regard and to become or remain partners of the Great Apes Survival Project Partnership; (e) Encouraging the academic and business communities, industry and the private sector, . . . to become full partners of the Great Apes Survival Project Partnership; (f) Forming strategic active partnerships with private sector ecotourism organizations to create sustainable economic development that enhances livelihoods for local communities in the range States.

Source: Great Apes Survival Project, United Nations Educational, Scientific, and Cultural Organization. Available at: http://www.unesco.org/mab/doc/grasp/E_KinshasaDeclaration.pdf

DOCUMENT 16: *Livestock's Long Shadow*—Report by the Food and Agriculture Organization of the United Nations (2006)

The Food and Agriculture Organization of the United Nations (FAO) has as its mission to end world hunger, assist with knowledge production about global agriculture, and to develop best animal husbandry practices for all countries. Established in 1945, it is headquartered in Rome, Italy. As the human population has grown, there has been a concurrent rise in agricultural production in every form (subsistence, herding, small-scale market-based, and industrial methods). The result is increasing conflict between human food needs and environmental health. Out of all types of agriculture, meat production presents a major concern because of livestock's land, water, and food requirements. It is estimated the current global livestock population is around 30 billion animals. In 2006, the FAO published the first-ever report on the environmental impact of livestock. The conclusions were alarming—for example, nearly 18 percent of all greenhouse gas contributions to climate change are from livestock. The FAO suggested urgent, cooperative, global action to mitigate negative impacts and develop sustainable solutions to both feed humans and keep the environment healthy. This excerpt presents some conclusions the FAO drew related to environmental impact differences between species, demonstrating that solutions may involve making hard decisions about which animals to eat in the future.

There are huge differences in environmental impact between the different forms of livestock production, and even the species.

. . . Beef is produced in a wide range of intensities and scales. At both ends of the intensity spectrum there is considerable environmental damage. On the extensive side, cattle are instrumental in degradation of vast grassland areas and are a contributing factor to deforestation (pasture conversion), and the resulting carbon emissions, biodiversity losses and negative impacts on water flows and quality. On the intensive side, feedlots are often vastly beyond the capacity of the surrounding land to absorb nutrients. While in the feedlot state the conversion of concentrate feed into beef is far less efficient than into poultry or pork, and therefore beef has significantly higher resource requirements per unit than pork or poultry. However, taking the total life cycle into account, including the grazing phase, concentrate feed per kilogram of growth is lower for beef than for non-ruminant systems.

. . . Extensive pig production, based on use of household waste and agro-industrial by-products, performs a number of useful environmental functions by turning biomass of no commercial value—and that otherwise would be waste—into high-value animal protein. However, extensive systems are incapable of meeting the surging urban demand in many developing countries, not only in terms of volume but also in sanitary and other quality standards. The ensuing shift toward larger-scale grain-based industrial systems has been associated with geographic concentration, to such extents that land/livestock balances have become very unfavourable, leading to nutrient overload of soils and water pollution. China is a prime example of these trends. Furthermore, most industrial pig production in the tropics and sub-tropics uses waste-flushing systems involving large amounts of water. This becomes the main polluting agent, exacerbating negative environmental impact.

Poultry production has been the species most subject to structural change. In OECD countries, production is almost entirely industrial, while in developing countries it is already predominantly industrial. Although industrial poultry production is entirely based on feed grains and other high value feed material, it is the most efficient form of production of food of animal origin (with the exception of some forms of aquaculture), and has the lowest land requirements per unit of output. Poultry manure is of high nutrient content, relatively easy to manage and widely used as fertilizer and sometimes as feed. Other than for feedcrop production, the environmental damage, though perhaps locally important, is of a much lower scale than for the other species.

In conclusion, livestock-environment interactions are often diffuse and indirect; and damage occurs at both the high and low end of the intensity spectrum, but it is probably highest for beef and lowest for poultry.

Source: Food and Agriculture Organization of the United Nations, 2006, Henning Steinfeld, *Livestock's Long Shadow*, 274-275. Reproduced with permission.

DOCUMENT 17: *On the Origin of Species* by Charles Darwin (1859)

Intensely curious about the natural world, British naturalist Charles Darwin (1809–1882) joined Her Majesty's Ship "Beagle" on a natural history collecting trip to South America in 1831. The following excerpt from On the Origin of Species *illustrates in part the impact that voyage had on his life's work. The publication of* The Origin *set in motion a revolution in the biological sciences and beyond. A number of professional scientists and amateur naturalists had been pondering and debating the topic of evolution for at least a century before Darwin's work, and therefore the concept of evolution itself was not new. Darwin's contribution to this ongoing field of inquiry was the mechanism—natural selection—by which evolution could take place (although the lesser-known fellow Englishman Alfred Russell Wallace arrived at the same conclusions). Darwin was unable to complete all the details related to his groundbreaking theory because genetic processes were unknown during his time. However, drawing on his own observations, knowledge of domesticated animal breeding, and the fossil record, Darwin theorized that, given appropriately long timespans, living organisms could change sufficiently over innumerable generations to form new species. His work, therefore, challenged long-held Christian ideas about the age of Earth and divine creation.*

Introduction

WHEN on board H.M.S. 'Beagle,' as naturalist, I was much struck with certain facts in the distribution of the inhabitants of South America, and in the geological relations of the present to the past inhabitants of that continent. These facts seemed to me to throw some light on the origin of species—that mystery of mysteries, as it has been called by one of our greatest philosophers. On my return home, it occurred to me, in 1837, that something might perhaps be made out on this question by patiently accumulating and reflecting on all sorts of facts which could possibly have any bearing on it (p. 1).

[. . .]

In considering the Origin of Species, it is quite conceivable that a naturalist, reflecting on the mutual affinities of organic beings, on their embryological relations, their geographical distribution, geological succession, and other such facts, might come to the conclusion that each species had not been independently created, but had descended, like varieties, from other species. Nevertheless, such a conclusion, even if well founded, would be unsatisfactory, until it could be shown how the innumerable species inhabiting this world have been modified, so as to acquire that perfection of structure and coadaptation which most justly excites our admiration. Naturalists continually refer to external conditions, such as climate, food, &c., as the only possible cause of variation. In one very limited sense, as we shall hereafter see, this may be true; but it is preposterous to attribute to mere external conditions, the structure, for instance, of the woodpecker, with its feet, tail, beak, and tongue, so admirably adapted to catch insects under the bark of trees (p. 3).

[. . .]

I am fully convinced that species are not immutable; but that those belonging to what are called the same genera are lineal descendants of some other and generally extinct species, in the same manner as the acknowledged varieties of any one species are the descendants of that species. Furthermore, I am convinced that Natural Selection has been the main but not exclusive means of modification (p. 6).

[. . .]

Conclusion

The belief that species were immutable productions was almost unavoidable as long as the history of the world was thought to be of short duration . . . (p. 481).

[. . .]

The mind cannot possibly grasp the full meaning of the term of a hundred million years; it cannot add up and perceive the full effects of many slight variations, accumulated during an almost infinite number of generations (p. 481).

[. . .]

There is grandeur in this view of life, with its several powers, having been originally breathed into a few forms or into one; and that, whilst this planet has gone cycling on according to the fixed law of gravity, from so simple a beginning endless forms most beautiful and most wonderful have been, and are being, evolved (p. 490).

Source: Darwin, Charles. *On the Origin of Species.* London: John Murray, 1859.

DOCUMENT 18: U.S. Pets Evacuation and Transportation Standards Act (2006)

The Pets Evacuation and Transportation Standards Act (PETS) was the direct result of the disastrous impact of Hurricane Katrina's category five (the highest) direct hit on New Orleans, Louisiana, in 2005. The breach of a key levee resulted in major flood damage and stranded residents. Inadequate government planning and response contributed an additional level of crisis as authorities demanded that people leave their homes but would not allow them to take their pets. An estimated 600,000 animals were involuntarily left behind, with roughly 250,000 animals dying. Media images of people being forced to leave animals on the sides of highways or hiding with their pets from authorities in their homes helped contribute to a public backlash. Congress responded to the outrage and passed PETS to require that emergency preparedness plans include animals. PETS is managed by the Federal Emergency Management Agency (FEMA). Since PETS passage, 30 states have passed their own versions—with some requiring planning for all animals (pets, livestock, zoos) and others just pets. Plans include forming animal response teams, sheltering and care of animals, identifying rescued animals, and reconnecting them to their guardians. This excerpt is the entirety of H.R. Bill 3858 as it was entered into law.

An Act To amend the Robert T. Stafford Disaster Relief and Emergency Assistance Act to ensure that State and local emergency preparedness operational plans address the needs of individuals with household pets and service animals following a major disaster or emergency.

Be it enacted by the Senate and House of Representatives of the United States of America in Congress assembled,

SECTION 1. SHORT TITLE.

This Act may be cited as the 'Pets Evacuation and Transportation Standards Act of 2006'.

SEC. 2. STANDARDS FOR STATE AND LOCAL EMERGENCY PREPAREDNESS OPERATIONAL PLANS.

Section 613 of the Robert T. Stafford Disaster Relief and Emergency Assistance Act (42 U.S.C. 5196b) is amended—

(1) by redesignating subsection (g) as subsection (h); and

(2) by inserting after subsection (f) the following:

'(g) Standards for State and Local Emergency Preparedness Operational Plans- In approving standards for State and local emergency preparedness operational plans pursuant to subsection (b)(3), the Director shall ensure that such plans take into account the needs of individuals with household pets and service animals prior to, during, and following a major disaster or emergency.'.

SEC. 3. EMERGENCY PREPAREDNESS MEASURES OF THE DIRECTOR.

Section 611 of the Robert T. Stafford Disaster Relief and Emergency Assistance Act (42 U.S.C. 5196) is amended—

(1) in subsection (e)—

(A) in paragraph (2), by striking 'and' at the end;

(B) in paragraph (3), by striking the period and inserting '; and'; and

(C) by adding at the end the following:

'(4) plans that take into account the needs of individuals with pets and service animals prior to, during, and following a major disaster or emergency.'; and

(2) in subsection (j)—

(A) by redesignating paragraphs (2) through (8) as paragraphs (3) through (9), respectively; and

(B) by inserting after paragraph (1) the following:

'(2) The Director may make financial contributions, on the basis of programs or projects approved by the Director, to the States and local authorities for animal emergency preparedness purposes, including the procurement, construction, leasing, or renovating of emergency shelter facilities and materials that will accommodate people with pets and service animals.'.

SEC. 4. PROVIDING ESSENTIAL ASSISTANCE TO INDIVIDUALS WITH HOUSEHOLD PETS AND SERVICE ANIMALS FOLLOWING A DISASTER.

Section 403(a)(3) of the Robert T. Stafford Disaster Relief and Emergency Assistance Act (42 U.S.C. 5170b(a)(3)) is amended—

(1) in subparagraph (H), by striking 'and' at the end;

(2) in subparagraph (I), by striking the period and inserting '; and'; and

(3) by adding at the end the following:

'(J) provision of rescue, care, shelter, and essential needs—

'(i) to individuals with household pets and service animals; and

'(ii) to such pets and animals.'.

Source: Pets Evacuation and Transportation Standards Act of 2006, 42 U.S.C.A. § 5196a-d.

DOCUMENT 19: *The Sexual Politics of Meat* by Carol Adams (1990)

Carol Adams (1951–) is a feminist who writes about and advocates for animal rights. Her book, The Sexual Politics of Meat, *has contributed an essential perspective to ethical, political, and lifestyle discussions around the treatment of animals and humans. Adams was the first writer to make detailed and convincing arguments about the ways in which systems of oppression, especially those of women and animals, are linked together. In this book, she demonstrates the parallels of a patriarchal (male-dominated) social structure that devalues both women and animals: in other words, a system that sees women and animals as objects rather than individual subjects. She documents the ways in which we experience this "objectification," or turning beings into passive objects rather than active subjects, in culture through advertisements, language, and meat-eating practices. She argues that feminists must also be animal advocates and refuse to participate in meat eating and other animal exploitation to avoid reinforcing a system of power that is oppressive rather than respectful. In the excerpt below, she outlines the concept of "absent referent"—a term she coined to express the methods by which women and animals become, either literally or figuratively, dismembered, and thereby erased as whole beings under patriarchy.*

Through butchering, animals become absent referents. Animals in name and body are made absent as animals for meat to exist. Animals' lives precede and enable the existence of meat. If animals are alive they cannot be meat. Thus a dead body replaces the live animal. Without animals there would be no meat eating, yet they are absent from the act of eating meat because they have been transformed into food.

. . . There are actually three ways by which animals become absent referents. One is literally: as I have just argued, through meat eating they are literally absent because they are dead. Another is definitional: when we eat animals we change the way we talk about them, for instance, we no longer talk about baby animals but about veal or meat. As we will see even more clearly in the next chapter, which examines language about eating animals, the word meat has an absent referent, the dead animals. The third way is metaphorical. Animals become metaphors for describing people's experiences. In this metaphorical sense, the meaning of the absent referent derives from its application or reference to something else.

. . . This chapter posits that a structure of overlapping but absent referents links violence against women and animals. Through the structure of the absent referent, patriarchal values become institutionalized. Just as dead bodies are absent from our language about meat, in descriptions of cultural violence women are also often the absent referent. Rape, in particular, carries such potent imagery that the term is transferred from the literal experience of women and applied metaphorically to other instances of violent devastation, such as the "rape" of the earth in ecological writings of the early 1970s. The experience of women thus becomes a vehicle for describing other oppressions. Women, upon whose bodies actual rape is most often committed, become the absent referent when the language of sexual violence is used metaphorically. These terms recall women's experiences but not women.

. . . Sexual violence and meat eating, which appear to be discrete forms of violence, find a point of intersection in the absent referent. Cultural images of sexual violence, and actual sexual violence, often rely on our knowledge of how animals are butchered and eaten. For example, Kathy Barry tells us of "maisons d'abattage (literal translation: houses of slaughter)" where six or seven girls each serve 80 to 120 customers a night. In addition, the bondage equipment of pornography—chains, cattle prods, nooses, dog collars, and ropes—suggests the control of animals. Thus, when women are victims of violence, the treatment of animals is recalled.

Similarly, in images of animal slaughter, erotic overtones suggest that women are the absent referent. If animals are the absent referent in the phrase "the butchering of women," women are the absent referent in the phrase "the rape of animals." The impact of a seductive pig relies on an absent but imaginable, seductive, fleshy woman. Ursula Hamdress is both metaphor and joke; her jarring (or jocular) effect is based on the fact that we are all accustomed to seeing women depicted in such a way. Ursula's image refers to something that is absent: the human female body. The structure of the absent referent in patriarchal culture strengthens individual oppressions by always recalling other oppressed groups.

Source: Copyright © Carol Adams, 1990, *The Sexual Politics of Meat*, 40–43. Bloomsbury Academic US, an imprint of Bloomsbury Publishing Inc.

DOCUMENT 20: Treaty of Amsterdam (1997)

The Treaty of Amsterdam followed the Maastricht Treaty of 1992, which established the European Union (EU). Maastricht included a Declaration that called on EU member countries to pay attention to animal welfare when drafting legislation. It is significant that EU organizing documents contain provisions concerning animal welfare, but Amsterdam is especially significant because it specifically refers to animals as "sentient beings." Additionally, in this Treaty animal protection and welfare is at the level of a Protocol, which holds a higher level of importance than a Declaration in EU treaties. Although the language changed very little, animal protection and welfare were also included in the Treaty of Lisbon, signed in 2007, but the level of importance is increased again, from a Protocol to an Article. Despite increasing regard given to animals in these overarching EU Treaties, neither a Protocol nor an Article requires specific laws to be enacted by member countries.

However, the broader significance is that animal protection and welfare is included in these documents and, in particular, that the EU recognizes animal sentience. The following excerpt from the Treaty is the relevant Protocol.

TREATY OF AMSTERDAM AMENDING THE TREATY ON EUROPEAN UNION, THE TREATIES ESTABLISHING THE EUROPEAN COMMUNITIES AND CERTAIN RELATED ACTS

[. . .]

Protocol on protection and welfare of animals

THE HIGH CONTRACTING PARTIES,

DESIRING to ensure improved protection and respect for the welfare of animals as sentient beings,

HAVE AGREED UPON the following provision which shall be annexed to the Treaty establishing the European Community,

In formulating and implementing the Community's agriculture, transport, internal market and research policies, the Community and the Member States shall pay full regard to the welfare requirements of animals, while respecting the legislative or administrative provisions and customs of the Member States relating in particular to religious rites, cultural traditions and regional heritage.

Source: The Treaty of Amsterdam, European Union, October 2, 1997. 1997 O.J. (C340) 1, 37 I.L.M. 253.

GLOSSARY

Agency: an individual's or group's ability to exert power and/or choice in the world.

Animal Geography: the study of where, when, why, and how nonhuman animals intersect with human societies.

Anthropocene: Earth's current time period that is considered to be dominated by humans.

Anthropocentrism: a worldview that is human-centered and places primary importance on human beings.

Anthropomorphism: attributing what are considered to be human characteristics to nonhumans.

Aquaculture: the farming of aquatic organisms for food under controlled conditions.

Biodiversity: the total of genetic, species, and ecosystem variety on Earth.

Biogeography: the study of how organisms are distributed across the surface and over the history of Earth.

Biotechnology: manipulation of the cellular structures of living organisms.

Black Market Animal Trade: the illegal trade of animals, or their parts, for human purposes.

Breed Specific Legislation: laws that ban, or regulate, specific dog breeds.

CAFO: concentrated animal feeding operation.

Canned Hunting: a type of hunting where hunters pay a fee to shoot animals that are fenced in and unable to escape.

Climate Change: shifts in the statistical distribution of weather patterns worldwide over an extended period, irrespective of cause.

Conservation: the protection and management of natural areas/resources and/or species.

Cruelty: causing pain and/or suffering, either intentionally or unnecessarily.

Culling: the killing of animals that are deemed to cause problems or to be unnecessary.

Designer Breeds: a cross of two different domestic purebred breeds or two different wild species.

Domestication: human manipulation of other species through breeding and behavioral modification to produce desired physical and behavioral traits.

Ecosystem: an environment and all organisms in it that function as a whole.

Emotion: an intense but short-lived mental response to an event, associated with specific body changes.

Empathy: an ability to understand another's situation through experiencing emotions as if one were the other.

Endangered Species: wild animals (and plants) classified as being at risk of extinction in the near future.

Ethics: a concern for what is good, right, or just in our individual and collective lives.

Ethology: the study of animal behavior.

Evolution: change in heritable traits passed on from generation to generation.

Exploit: to utilize for one's own needs.

Factory Farming: the rearing of animals for meat, milk, or eggs using practices geared toward maximum output per animal.

Feedlots: large fenced areas without vegetation in which beef cattle are placed to gain weight prior to slaughter.

Feral: descended from domesticated animals but now primarily living outside of human control and/or without human care.

Flagship Species: animals who evoke emotional responses in humans and thereby arouse public support for their conservation.

Habitat: an area in which a particular species lives and that provides the specific resources needed for survival.

Hoarding: humans' acquiring and attempting to care for more domesticated animals than they are capable of, such that the animals are severely neglected.

Human-Animal Bond: a process that integrates psychological, social, and physical impulses between humans and other animals.

Human-Animal Studies: the study of the spectrum of relations between humans and other species from social science (history, geography, anthropology) and humanities (literature, art, philosophy) perspectives.

Hybrid: a mixture of two or more components into one.

Indicator Species: animals or plants that are sensitive to disturbance and whose health forecasts or reflects changing conditions in their ecosystems.

Indigenous: original inhabitants (animal, human, plant) of an area; native.

Intelligence: ability to learn and understand, including adapting to new or challenging situations.

Invasive Species: a species that is introduced to a new location and becomes dominant over, and detrimental to, pre-existing local species.

Keystone Species: those that maintain the diversity and functions of an ecosystem and whose impacts are typically disproportionate to their relative population size.

Marine Mammal Parks: commercial theme parks that house marine mammals in tanks for the public to view both in and out of shows.

Moral Agent: an individual with the ability to choose between right and wrong actions.

Natural Selection: the means by which evolution takes place; the passing on of favorable genetic traits that have allowed an organism to successfully survive and reproduce.

Personhood: the status of being recognized as a unique individual with moral and legal rights.

Poaching: the illegal taking of wildlife.

Rights: Moral and/or legal claims to respectful treatment, including not being harmed or held captive against one's will.

Selective Breeding: controlling reproduction in order to produce offspring that have the most useful and/or appealing qualities for humans.

Self-conscious: recognizing oneself as an individual that is separate from other individuals and the surrounding world.

Sentience: the ability for conscious awareness of one's experiences.

Service Animals: animals trained to assist people with a wide variety of disabilities.

Social Construction: an idea or practice that exists because a group of people agree that it does.

Species: the smallest unit of the major categories of biological classification (taxonomy).

Speciesism: the giving of unfair preference to humans over all other species and humans' viewing of some species more favorably than others.

Stereotypic Behavior: repetitive behaviors that do not have obvious function or purpose and are not common in natural living environments.

Subsistence: relating to basic survival needs.

Taxidermy: the craft of preparing and mounting animal skins to appear "lifelike."

Taxonomy: the system of biological classification of living organisms.

Traditional Chinese Medicine: the diagnosis, treatment, and prevention of illness based on ideas derived from the Chinese medical system.

Transgenic: made up of the genes of more than one species.

Trophy Hunting: when the main goal of the hunt is not considered essential, such as food, but to take all/part of the animal as a symbol of the hunter's success.

Veganism: a lifestyle choice in which one abstains from eating or using animal-derived food or products.

Vegetarianism: a dietary choice that involves not eating meat but typically includes eating eggs and dairy products.

Vivisection: cutting open of a living being.

Welfare: relating to physical and/or emotional states of being, including actions that have an impact on such states.

Wild: not under human control or not substantially or directly influenced by human processes.

Xenotransplantation: the use of nonhuman animal parts for human medical treatments.

Zoogeomorphology: the study of how animals physically alter their landscape.

SELECTED BIBLIOGRAPHY

Akhtar, A. 2012. *Animals and Public Health: Why Treating Animals Better Is Critical to Human Welfare.* New York: Palgrave Macmillan.

Alberti, S., ed. 2011. *The Afterlives of Animals: A Museum Menagerie.* Charlottesville, VA: University of Virginia Press.

Allen, D. 2013. *The Nature Magpie*. London: Icon Books.

Appleby, M., Mench, J., Olsson, I. A., and Hughes, B. 2011. *Animal Welfare*, 2nd ed. Cambridge, MA: CABI.

Armstrong, S. J., and Botzler, R. G., eds. 2008. *The Animal Ethics Reader,* 2nd ed. New York: Routledge.

Baker, S. 2001. *Picturing the Beast: Animals, Identity, and Representation*. Chicago: University of Illinois Press.

Barnes, S. 2014. *Ten Million Aliens: A Journey Through the Entire Animal Kingdom*. New York: Marble Arch Press.

Beetz, A. M., and Podberscek, A. L. 2005. *Bestiality and Zoophilia: Sexual Relations with Animals*. West Lafayette, IN: Purdue University Press.

Bekoff, M. 2007. *The Emotional Lives of Animals: A Leading Scientist Explores Animal Joy, Sorrow, and Empathy—and Why They Matter.* Novato, CA: New World Library.

Best, S., and Nocella, A. J. 2004. *Terrorists or Freedom Fighters? Reflections on the Liberation of Animals.* New York: Lantern Books.

Bodart-Bailey, B. M. 2006. *The Dog Shogun: The Personality and Policies of Tokugawa Tsunayoshi*. Honolulu: University of Hawaii Press.

Bradbury, J., and Vehrencamp, S. 2011. *Principles of Animal Communication,* 2nd ed. Sunderland, MA: Sinauer Associates, Inc.

Bradshaw, G. A. 2009. *Elephants on the Edge: What Animals Teach Us about Humanity*. New Haven, CT: Yale University Press.

Bradshaw, J. 2011. *In Defence of Dogs: Why Dogs Need Our Understanding*. London: Allen Lane.

Broom, D. M. 2014. *Sentience and Animal Welfare*. Wallingford, UK: CABI.

Brown, J. H., and Lomolino, M. V. 1998. *Biogeography*, 2nd ed. Sunderland, MA: Sinauer Associates.

Brown, T. 1986. *Tom Brown's Field Guide to Nature Observation and Tracking*. New York: Berkley Books.

Butler, D. R. 1995. *Zoogeomorphology: Animals as Geomorphic Agents*. Cambridge, UK and New York: Cambridge University Press.

Caro, T. 2010. *Conservation by Proxy: Indicator, Umbrella, Keystone, Flagship, and Other Surrogate Species*. Washington, DC: Island Press.

Carr, N. 2014. *Dogs in the Leisure Experience.* Wallingford, UK: CABI.

Cavalieri, P., and Singer, P., eds. 1994. *The Great Ape Project: Equality beyond Humanity*. New York: St. Martin's Press.

Chaline, E. 2011. *Fifty Animals That Changed the Course of History*. Hove, UK: David & Charles.

Clutton-Brock, J. 2012. *Animals as Domesticates: A World View through History.* East Lansing, MI: Michigan State University Press.

Coleman, J. T. 2004. *Vicious: Wolves and Men in America.* New Haven, CT and London: Yale University Press.

Copeland, M. W. 2003. *Cockroach.* London: Reaktion Books.

Cox, C. B., and Moore, P. D. 2010. *Biogeography: An Ecological and Evolutionary Approach.* Hoboken, NJ: Wiley.

Crane, E. 1999. *The World History of Beekeeping and Honey Hunting.* London: Gerald Duckworth & Co.

Cronon, W. 1992. *Nature's Metropolis: Chicago and the Great West.* New York: W.W. Norton.

Darwin, C. 1871. *The Descent of Man, and Selection in Relation to Sex.* New York: A.L. Burt Company.

Darwin, C. 1964 [1859]. *On the Origin of Species.* Cambridge, MA: Harvard University Press.

Darwin, C. 2009 [1872]. *The Expression of the Emotions in Man and Animals.* London and New York: Penguin Classics.

Davies, B. 2005. *Black Market: Inside the Endangered Species Trade in Asia.* San Rafael, CA: Earth Aware Editions.

Davies, G., and Brown, D. 2007. *Bushmeat and Livelihoods: Wildlife Management and Poverty Reduction.* Oxford: Blackwell Publishing Ltd.

De Mello, M. 2012. *Animals and Society: An Introduction to Human-Animal Studies.* New York: Columbia University Press.

De Waal, F. 2009. *The Age of Empathy: Nature's Lessons for a Kinder Society.* New York: Harmony Books.

Dolin, E. J. 2010. *Fur, Fortune, and Empire: The Epic History of the Fur Trade in America.* New York: W.W. Norton & Company.

Donaldson, S., and Kymlicka, W. 2011. *Zoopolis: A Political Theory of Animal Rights.* Oxford: Oxford University Press.

Duffy, R. 2010. *Nature Crime: How We're Getting Conservation Wrong.* New Haven, CT: Yale University Press.

Dunlop, R. H., and Williams, D. J. 1996. *Veterinary Medicine: An Illustrated History.* St. Louis: Mosby.

Ellis, R. 2005. *Tiger Bone and Rhino Horn: The Destruction of Wildlife for Traditional Chinese Medicine.* Washington, DC: Island Press.

Emel, J., and Neo, H., eds. 2015. *The Political Ecologies of Meat.* London: Earthscan.

Estes, J. A., DeMaster, D. P., Doak, D. F., Williams, T. M., and Brownell, Jr., R. L. 2006. *Whales, Whaling, and Ocean Ecosystems.* Berkeley: University of California Press.

Fleig, D. 1996. *The History of Fighting Dogs.* Neptune, NJ: TFH Publications.

Franzen, J. L. 2010. *The Rise of Horses: 55 Million Years of Evolution.* Translated by Kirsten M. Brown. Baltimore: The Johns Hopkins University Press.

Fudge, E. 2008. *Pets.* Stocksfield, UK: Acumen.

Galdikas, B. 1995. *Reflections of Eden: My Years with the Orangutans of Borneo.* Boston, New York, Toronto and London: Little, Brown.

Gaston, K. J., and Spicer, J. I. 2004. *Biodiversity: An Introduction,* 2nd ed. Malden, MA: Blackwell.

Glen, S., and Moore, M. T. 2001. *Best Friends: The True Story of the World's Most Beloved Animal Sanctuary.* New York: Kensington Books.

Goodall, J. 1971. *In the Shadow of Man.* New York: Houghton Mifflin Company.

Goodall, J. 1990. *Through a Window: My Thirty Years with the Chimpanzees of Gombe*. Boston: Houghton Mifflin.

Griffin, D. R. 1981. *The Question of Animal Awareness: Evolutionary Continuity of Mental Experience*. New York: The Rockefeller University Press.

Griffin, E. 2007. *Blood Sport: Hunting in Britain since 1066*. New Haven, CT and London: Yale University Press.

Hancocks, D. 2001. *A Different Nature: The Paradoxical World of Zoos and Their Uncertain Future.* Berkeley: University of California Press.

Harari, Y. 2015. *Sapiens: A Brief History of Humankind.* New York: Harper Collins Publishers.

Harbolt, T. L. 2003. *Bridging the Bond: The Cultural Construction of the Shelter Pet*. West Lafayette, IN: Purdue University Press.

Harrison, R. 1964. *Animal Machines: The New Factory Farming Industry.* London: Vincent Stuart Publishers.

Hickman, C., Keen, S., Larson, A., and Roberts, L. 2013. *Integrated Principles of Zoology.* 16th ed. New York: McGraw-Hill Education.

Holt-Jensen, A. 2009. *Geography: History and Concepts, A Student's Guide.* London and Thousand Oaks, CA: Sage Publications.

Hornsby, A. 2000. *Helping Hounds*. Lydney, UK: Ringpress Books.

Ingold, T. 2011. *Being Alive: Essays on Movement, Knowledge and Description*. New York: Routledge.

Jensen, P. 2002. *The Ethology of Domestic Animals: An Introductory Text.* Oxford: CABI.

Kalof, L. 2007. *Looking at Animals in Human History*. Chicago: University of Chicago Press.

Kemmerer, L. 2012. *Animals and World Religions.* New York: Oxford University Press.

Kirby, D. 2013. *Death at SeaWorld: Shamu and the Dark Side of Killer Whales in Captivity.* New York: St. Martin's Press.

Kisling, V., ed. 2000. *Zoo and Aquarium History: Ancient Animal Collections to Zoological Gardens.* Boca Raton, FL: CRC Press.

Kolbert, E. 2014. *The Sixth Extinction: An Unnatural History*. New York: Henry Holt and Co.

Kurlansky, M. 1997. *Cod: A Biography of the Fish That Changed the World.* London: Walker and Company.

LaDuke, W. 1999. *All Our Relations: Native Struggles for Land and Life*. Cambridge, MA: South End Press.

Lavigne, D. 2013. *Elephants and Ivory*. Yarmouth Port, MA: IFAW.

Le Chêne, E. 2009. *Silent Heroes: The Bravery and Devotion of Animals in War*. London: Souvenir Press.

Leigh, D., and Geyer, M. 2003. *One at a Time: A Week in an American Animal Shelter*. Santa Cruz, CA: No Voice Unheard.

Lockwood, J. L., Hoopes, M. F., and Marchetti, M. P. 2013. *Invasion Ecology.* 2nd ed. Oxford, UK: Wiley-Blackwell.

Maehr, D. S., Noss, R. F., and Larkin, J. L., eds. 2001. *Large Mammal Restoration: Ecological and Sociological Challenges in the 21st Century*. Washington, DC: Island Press.

McManus, P., Albrecht, G. and Graham, R. 2013. *The Global Horseracing Industry: Social, Economic, Environmental and Ethical Perspectives.* New York: Routledge.

Mech, D., and Boitani, L., eds. 2003. *Wolves: Behavior, Ecology, and Conservation*. Chicago and London: The University of Chicago Press.

Midgley, M. 1998. *Animals and Why They Matter*. Athens, GA: University of Georgia Press.

Montgomery, S. 1991. *Walking with the Great Apes: Jane Goodall, Dian Fossey, Birute Gildikas*. Boston, New York, and London: Houghton Mifflin.

Nagy, K., and Johnson II, P. D., eds. 2013. *Trash Animals: How We Live with Nature's Filthy, Feral, Invasive, and Unwanted Species*. Minneapolis: University of Minnesota Press.

Neme, L. A. 2009. *Animal Investigators: How the World's First Wildlife Forensics Lab is Solving Crimes and Saving Endangered Species*. New York: Simon and Schuster.

Newkirk, I. 2000. *Free the Animals: The Story of the Animal Liberation Front.* New York: Lantern Books.

Nierenberg, D. 2005. *Happier Meals: Rethinking the Global Meat Industry.* Danvers, MA: Worldwatch Institute.

Niman, N. H. 2014. *Defending Beef: The Case for Sustainable Meat Production*. White River Junction, VT: Chelsea Green Publishing.

Olmert, S. 2009. *Made for Each Other: The Biology of the Human-Animal Bond.* Cambridge, MA: Da Capo Press.

Perlo, K. 2009. *Kinship and Killing: The Animal in World Religions*. New York: Columbia University Press.

Phillips, C. J. C. 2015. *The Animal Trade*. Wallingford, UK: CABI.

Philo, C., and Wilbert, C. 2004. *Animal Spaces, Beastly Places: New Geographies of Human-animal Relations*. New York and London: Routledge.

Posewitz, J. 1994. *Beyond Fair Chase: The Ethic and Tradition of Hunting*. Guilford, CT and Helena, MT: Morris Book Publishing, LLC.

Rachels, J. 1990. *Created from Animals: The Moral Implications of Darwinism*. New York: Oxford University Press.

Rader, K. 2004. *Making Mice: Standardizing Animals for American Biomedical Research. 1900–1955.* Princeton, NJ: Princeton University Press.

Raffles, H. 2011. *Insectopedia*. New York: Pantheon Books.

Rauch, A. 2013. *Dolphins.* London: Reaktion Books.

Regan, T. 1983. *The Case for Animal Rights.* Berkeley: The University of California Press.

Rupke, N., ed. 1987. *Vivisection in Historical Perspective.* London: Croom Helm.

Russell, E. 2011. *Evolutionary History: Uniting History and Biology to Understand Life on Earth.* New York: Cambridge University Press.

Sanders, C. 1999. *Understanding Dogs: Living and Working with Canine Companions*. Philadelphia: Temple University Press.

Schiebinger, L. 2008. *Nature's Body: Gender in the Making of Modern Science.* New Brunswick, NJ: Rutgers University Press.

Schweid, R. 1999. *The Cockroach Papers: A Compendium of History and Lore*. New York and London: University of Chicago Press.

Scully, M. 2003. *Dominion: The Power of Man, the Suffering of Animals, and the Call to Mercy*. London: St. Martin's Griffin.

Simon, L. 2014. *The Greatest Show on Earth: A History of the Circus*. London: Reaktion Books.

Singer, P. 1975. *Animal Liberation: A New Ethics for Our Treatment of Animals.* New York: Random House.

Smil, V. 2013. *Should We Eat Meat? Evolution and Consequences of Modern Carnivory*. West Sussex, UK: John Wiley & Sons.

Sorenson, J. 2009. *Ape*. London: Reaktion Books.

Steinfeld, H., Gerber, P., Wassenaar, T., Castel, V., Rosales, M., and De Haan, C. 2006. *Livestock's Long Shadow*. Rome: FAO.

Stirling, I. 2011. *Polar Bears: A Natural History of a Threatened Species*. Markham, Ontario: Fitzhenry & Whiteside.

Taylor, N. 2013. *Humans, Animals, and Society: An Introduction to Human-Animal Studies*. New York: Lantern Books.

Thayer, G. A. 2013. *Going to the Dogs: Greyhound Racing, Animal Activism, and American Popular Culture*. Lawrence, KA: University of Kansas Press.

Tinbergen, N. 1951. *The Study of Instinct*. Oxford: Clarendon Press.

Urbanik, J. 2012. *Placing Animals: An Introduction to the Geography of Human-Animal Relations*. Lanham, MD: Rowman & Littlefield Publishers.

Wagner, A. 2013. *The Gaddi Beyond Pastoralism: Making Place in the Indian Himalayas*. London: Berghahn.

Waldau, P., and Patton, K., eds. 2006. *A Communion of Subjects: Animals in Religion, Science, and Ethics*. New York: Columbia University Press.

Watson, J. B. 1919. *Psychology from the Standpoint of a Behaviorist*. Philadelphia and London: J.P Lippincott Company.

Wenzel, G. 1991. *Animal Rights, Human Rights: Ecology, Economy, and Ideology in the Canadian Arctic*. Toronto: University of Toronto Press.

Wilkie, R., and Inglis, D., eds. 2007. *Animals and Society: Critical Concepts in the Social Sciences*. New York: Routledge.

Wilkie, R. M. 2010. *Livestock/Deadstock: Working with Farm Animals from Birth to Slaughter.* Philadelphia: Temple University Press.

Williams, E. E., and DeMello, M. 2007. *Why Animals Matter: The Case for Animal Protection*. Amherst, NY: Prometheus Books.

Wilson, E. O. 1984. *Biophilia.* Cambridge, MA: Harvard University Press.

Wilson, E. O. 1992. *The Diversity of Life.* Cambridge, MA: Harvard University Press.

Wilson, R. M. 2010. *Seeking Refuge: Birds and Landscapes of the Pacific Flyway*. Seattle: University of Washington Press.

Winograd, N. J. 2007. *Redemption: The Myth of Pet Overpopulation and the No Kill Revolution in America*. Los Angeles: Almaden Books.

Wise, S. 2000. *Rattling the Cage: Toward Legal Rights for Animals*. Cambridge, MA: Perseus Books.

Wolch, J., and Emel, J., eds. 1998. *Animal Geographies: Place, Politics, and Identity in the Nature-Culture Borderlands*. New York: Verso.

ABOUT THE CONTRIBUTORS

Stefanie Georgakis Abbott received her PhD in Public and International Affairs from Virginia Tech in 2014. Her research focuses on border studies, post-structuralism, and critical animal studies. Currently, she is working on a paper examining the political economy of conservation, focusing specifically on the market for permits to hunt endangered species.

Daniel Allen, PhD, is a teaching fellow in geography at Keele University, UK, a fellow of the Royal Geographical Society (RGS), and member of the All-Party Parliamentary Group for Animal Welfare (APGAW). He is author of two books, *Otter* (Reaktion Books 2010) and *The Nature Magpie* (Icon Books 2013).

Maan Barua holds a PhD in geography from the University of Oxford and is currently a postdoctoral fellow at the School of Geography and the Environment, University of Oxford. Maan's past and ongoing research projects include work on animals' geographies, nonhuman labor and the economy, and postcolonial urban ecologies.

Pratyusha Basu is an associate professor of geography in the Department of Sociology and Anthropology at the University of Texas-El Paso. Her research focuses on rural economies and cultures in the global South, especially on gender and small-scale dairying in India and Kenya.

Barbara Hardy Beierl teaches in the RISE program at Rivier University. She holds a master's degree from San Francisco State University and a doctoral degree from Wayne State University. Her fields of study include English and American literature, Shakespeare and English Renaissance studies, and Human-Animal Studies.

Marc Bekoff is professor emeritus of Ecology and Evolutionary Biology at the University of Colorado, Boulder. He is a fellow of the Animal Behavior Society and a former Guggenheim fellow. Marc has published more than 1,000 scientific and popular essays and 30 books, and his homepage is marcbekoff.com.

Valerie Benka received graduate degrees in conservation biology and public policy from the University of Michigan, where she studied human-wildlife conflict in Kenya. She attended the Animals and Public Policy Program at Tufts University, focusing on management of free-roaming dogs in Nepal. She works for an animal welfare nonprofit.

Jordan Fox Besek is a graduate teaching fellow in the sociology department at the University of Oregon. His work engages classical and contemporary theory in order to interrogate the social, material, and historical bases through which ecological knowledge is built, and how this knowledge filters through political, economic, and cultural systems.

Sarah M. Bexell is a research associate professor in the Institute for Human-Animal Connection at the University of Denver. She is also the director of Conservation Education at the Chengdu Research Base of Giant Panda Breeding, China. Her research investigates the intersections of human overpopulation, overconsumption, and biodiversity decline.

Dawn Biehler is associate professor of geography and environmental studies at the University of Maryland, Baltimore County. Her research addresses historical geography, environmental justice, urban public health, and urban nature. She is the author of *Pests in the City: Flies, Bedbugs, Cockroaches, and Rats* (University of Washington Press 2013).

Courtney Brown received a master's in social work (MSW), focusing on sustainable development and global practice, and a certificate in animal assisted social work from the University of Denver in 2015. Courtney's professional interests include the human-animal bond, One Health, and conservation social work, specifically related to sustainable development.

David R. Butler (PhD, University of Kansas) is the Texas State University System Regents' Professor, and University Distinguished Professor of Geography at Texas State University, located in San Marcos, Texas. His research interests are in the fields of geomorphology and biogeography, with special emphases on zoogeomorphology and mountain environments.

Christopher J. Byrd is a graduate student in the Department of Animal Sciences at Purdue University, located in West Lafayette, Indiana. His research, focused on the well-being of agricultural species, uses animal behavior to develop quantitative methodologies for assessing and improving animal welfare on farms.

Neil Carr is an associate professor in the Department of Tourism, University of Otago and the editor of *Annals of Leisure Research*. His research focuses on tourism and leisure, with a particular emphasis on animals, children, and families. His recent publications include *Dogs in the Leisure Experience* (CABI 2014).

Nichole Chapel is a PhD student at Purdue University in the animal sciences department, working in farm animal welfare. She received her bachelor's (2012) and master's (2014) degrees from North Dakota State University. Her previous work has included lameness in dairy cows and she is currently working with swine.

Nathan Clay is a PhD candidate in geography at Penn State University, writing a dissertation on agri-environmental governance and livelihoods in Rwanda. His research interests center on the interactions among rural livelihoods, environmental conservation, and agricultural development, particularly in tropical forests and mountains in the global South.

John Clayton is an ESRC-funded PhD Researcher at the School of Planning and Geography, Cardiff University. His research utilizes innovative methodologies to investigate novel relationships between fish, anglers, and piscivorous predators in rural England. He also holds an MA (with distinction) in critical global politics from Exeter University.

Rosemary-Claire Collard is an assistant professor in geography, planning and environment at Concordia University in Montreal. She is interested in the relationship between capitalism and animal life. Her recent research tracked the exotic pet trade across six countries and is forthcoming as a book, *Animal Traffic*, with Duke University Press.

Marion W. Copeland, an independent scholar of literary animal studies, is affiliated with the Center for Animals and Public Policy at Tufts University and the Humane Society University. She serves on the boards of the Dakin Humane Society, in Springfield and Leverett, Massachusetts, and NILAS (Nature in Legend and Story).

J. Keri Cronin (kericronin.com) is an associate professor in the Department of Visual Arts at Brock University (Ontario, Canada), where she teaches classes in the history of art and visual culture. Her research focuses on the ways in which imagery has been used in animal advocacy.

Margo DeMello teaches anthrozoology at Canisius College, and is the program director for Human-Animal Studies at the Animals and Society Institute. She has written a number of books on Human-Animal Studies, including *Teaching the Animal: Human Animal Studies Across the Disciplines* and *Animals and Society: An Introduction to Human-Animal Studies*.

Peter Derbyshire received his bachelor of science from the University of Western Australia before completing research projects focused on fetal pig kidney development and cardiac function in ornate dragons. Peter is currently a PhD candidate at the University of Western Australia, examining the cardiac function and innervation of reptiles.

Karin Dirke, PhD, is an assistant professor in history of ideas at Stockholm University. Her research interests concern the historic relationship between humans and other animals, both wild and domestic.

Kalli F. Doubleday conducts research on human-carnivore conflict as a doctoral candidate in the Department of Geography and the Environment—University of Texas, Austin. She has taught courses at Texas Christian University and works with Sariska Tiger Conservation Organization as a wildlife rescuer and educator during summer months.

Debra Durham has a PhD in animal behavior and has studied chimpanzees in their natural habitats and in captivity. She collaborates on a range of chimpanzee conservation and welfare projects.

Ian Edwards, PhD, teaches for the Department of Anthropology at the University of Oregon. He also serves as a social scientist for the World Conservation Union's Commission on Ecosystem Management for North America and the Caribbean, Resilience Task Force, Dryland Ecosystems Thematic Group, and Red List of Ecosystems Thematic Group.

Erica Elvove, MSW, assistant director of the Institute for Human-Animal Connection and an adjunct faculty at University of Denver's Graduate School of Social Work, focuses on the promotion of social justice through a human-animal-environmental connection lens and providing innovative educational opportunities in the human-animal interaction field for practitioners.

Jean Estebanez is a lecturer at Université Paris-Est Créteil (UPEC). His research relates to zoos and human-animal relationships in the context of work.

Anita Hagy Ferguson is a PhD candidate in environmental social science and project manager of the Center for Biodiversity Outcomes at Arizona State University. As a human-animal mediator, philosopher, and geographer, she studies how humans think about and share space with large predators. She works collaboratively to enable human-predator coexistence.

Russell Fielding is an assistant professor of environmental studies at the University of the South in Sewanee, Tennessee. He is trained as a geographer and has research interests in questions of subsistence, cultural tradition, and resource conservation.

Katherine Fogelberg, DVM, PhD, is assistant professor and director of the Environmental & Occupational Health Sciences MPH program at the UNT Health Science Center in Fort Worth, Texas. She teaches public health and also instructs medical students about zoonotic disease, food safety and security, and sociological aspects of animals in public health.

Kathryn Gillespie, PhD, is a geographer dedicated to understanding the hierarchies of power and privilege operating in multispecies relationships with a particular emphasis on farmed animals. She is currently working on a book, *The Cow with Ear Tag #1389*, about the role of animals in the U.S. dairy industry.

Brooke Harland is a dual master's degree student at the Graduate School of Social Work and the Josef Korbel School of International Studies. Her focus is sustainable development and global practice. Her passion is to share and develop sustainable practices within the framework of One Health.

Wendi A. Haugh is associate professor of anthropology and African studies at St. Lawrence University in Canton, New York.

Philip Howell is an historical and cultural geographer at the University of Cambridge. He has written about prostitution, social regulation, and imperialism. He also specializes in historical animal geography and is the author of *At Home and Astray: The Domestic Dog in Victorian Britain* (University of Virginia Press 2015).

Jenny R. Isaacs is a PhD candidate at Rutgers University, New Brunswick, NJ. She won the Rutgers Geography Graduate Teaching Award for 2014. She has worked with the National Park Service, UNESCO, Conserve Wildlife Foundation of New Jersey, the New Jersey Audubon, and the Wildlife Conservation Society's Central Park Zoo.

Stephanie Johnson received her masters of social work from the University of Denver, Colorado, in 2015. Her area of focus included sustainable development and global practice with a certificate in animal assisted social work. Her professional interests include conservation social work and the human-animal bond.

Connie L. Johnston, PhD, is an adjunct professor with the DePaul University Department of Geography. Prior to this, she spent a year as a fellow with the Science, Technology, and Society Program at Harvard University. She has published a number of journal articles, book chapters, and book reviews on the topic of human-animal relations and was awarded a National Science Foundation grant for her research on the geography of farmed animal welfare in the United States and Europe. She earned her master's degree from Duke University and her doctorate from Clark University.

Erik Jönsson is a postdoctoral fellow at the Department of Human Geography, Lund University. His current research explores future visions and research environments built up around in vitro meat production. His previous research has primarily centered on the political ecology of upscale golf.

Jeffrey C. Kaufmann is professor of anthropology at the University of Southern Mississippi. He has published widely on Madagascar, especially on Mahafale pastoralism, on human-plant-animal relationships, and on the conservation of nature.

Dawna Komorosky is an associate professor of criminal justice at California State University, East Bay. Areas of specialization include the link between domestic violence and animal cruelty, humane education, and women in the criminal justice system. She has created and taught courses that examine both family violence and animal cruelty.

Stasja Koot, PhD, works at the Wageningen University as an ecological anthropologist. His research is focused on indigenous people, nature conservation, tourism, land conflicts, and natural resource management. He is interested in extraction, wildlife crime, and the effects of online developments on nature conservation, with a focus on southern Africa.

Helen Kopnina (PhD, Cambridge University, 2002) is currently employed at both the Leiden University, Institute of Cultural Anthropology and Development Sociology, and at The Hague University of Applied Science. Kopnina is the (co)author of over 50 articles and 10 books, including *Culture and Conservation: Beyond Anthropocentrism* (2015).

Yu-ling Kung is a doctoral student in Cultural Studies at the University of Canterbury, New Zealand. Her research explores animals in comics. Her current work focuses on animals in the kung-fu genre.

Anthony M. Levenda is a PhD candidate in the Toulan School of Urban Studies and Planning at Portland State University. His research and publications lie at the intersection of urban geography, political economy, and science and technology studies. His dissertation examines the political economy and ecology of networked urban infrastructures.

John Lupinacci teaches pre-service teachers and graduate students in the Cultural Studies and Social Thought in Education (CSSTE) program using an eco-anarchist approach that advocates for the development of scholar-activist educators.

Erin Luther is a PhD candidate in environmental studies at York University, specializing in communication about human-animal relationships. She has also worked as an education specialist for over a decade at Toronto Wildlife Centre and is the author of a manual for NGO educators on resolving human-wildlife conflicts.

William S. Lynn is a research scientist in the George Perkins Marsh Institute at Clark University, senior fellow for Ethics and Public Policy in the Center for Urban Resilience at Loyola Marymount University, and former director of the Masters in Animals and Public Policy (MAPP) program at Tufts University.

Linda Madden is a PhD candidate at the University of Auckland. She studies the animal geographies of Auckland City, focusing on encounters between species and the spaces in which these occur, as well as looking at the way distinctions between species can be perpetuated through methodologies used to study them.

John T. Maher is an animal lawyer and adjunct professor of animal law at Touro Law Center in Central Islip, New York. He is attempting to change the law to embody the interests of all nonhuman life and objects.

Troy A. Martin has previously written about visual pedagogy and factory farm images in the *Journal of Critical Animal Studies.* He holds a PhD in educational studies with a concentration in cultural studies from the University of North Carolina at Greensboro.

William L. McKeithen is a doctoral student in geography at the University of Washington. His research explores the intersections of sexuality, gender, and nature-society relations.

Phil McManus is professor of urban and environmental geography, and head of the School of Geosciences, The University of Sydney, Australia. He is a coauthor of *The Global Horseracing Industry* (Routledge, 2013).

Alicia McNorth holds a bachelor of science in natural resources from the University of Massachusetts and a juris doctor degree from the University of Oregon School of Law. While earning her law degree she focused on captive animal issues and clerked at the Animal Legal Defense Fund.

Debra Merskin, PhD, is associate professor of media studies at the University of Oregon. Her research focuses on the intersections of racism, sexism, and speciesism in media and popular culture. She is writing a book, *Seeing Species* (Peter Lang) and advocates for inclusive representations of animals at animalsandmedia .org.

Eric Daryl Meyer is a postdoctoral faculty fellow at Loyola Marymount University in Los Angeles. His research explores the ways that Christians have used animals (or ideas about animals) in order to define and shape human nature.

Mara Miele is professor in human geography in the School of Geography and Planning at Cardiff University, UK. Her work addresses the geographies of ethical food consumption and the role of animal welfare science and technology in challenging the role of farmed animals in current agricultural practices and policies.

Randy Moore is H. T. Morse-Alumni Distinguished Teaching Professor of Biology at the University of Minnesota.

Skye Naslund is a doctoral candidate in the Department of Geography at the University of Washington. Bringing together health geography and critical animal studies, she examines parasite-host relationships and the neglect of parasites and other noncharismatic organisms in biomedical research and practice.

Harvey Neo is an assistant professor of geography at the National University of Singapore. His main research interests are the political economy of the livestock industry, nature/society issues, and the discourse and development of eco-cities in Asian contexts. He is also an editor of *Geoforum*, a leading geography journal.

Anthony J. Nocella II is an assistant professor of sociology and criminology at Fort Lewis College, executive director of the Institute for Critical Animal Studies, national co-coordinator of Save the Kids, co-editor of the Critical Animal Studies and Theory Book Series with Lexington Books, and editor of the *Peace Studies Journal*.

Angela Dawn Parker has an MSc (geography) from Concordia University, Montreal, where she studied human-animal relationships within safe spaces. She runs an animal rescue in Ontario, Canada, and has been teaching about horses for over 30 years. She also maps feral cat populations and studies the effects of "Trap-Neuter-Return" initiatives.

Merle Patchett is a lecturer in human geography, and her research focuses on theories, histories, and geographies of practice. She engages empirically with a range specialized skills (e.g., taxidermy), practitioners (e.g., artists and architects), and places of practice (e.g., museums, galleries and archives) to develop practice-based methodologies.

Heather Pospisil is a doctoral student in environment and sustainability at the University of Saskatchewan and an emergency responder for International Bird Rescue. She holds a master of arts in philosophy from the California Institute of Integral Studies.

Dr. **Núria Querol i Viñas** is a cell and genetics biologist and a medical doctor. She is a professor of criminal investigation (UB) and anthrozoology (UAB), fellow at IHAC (DU), advisory board member at the National Law Enforcement Center on Animal Abuse (NSA) and on the Board of Directors of NCOVAA.

Jessica Bell Rizzolo is a PhD student in sociology at Michigan State University, specializing in animal studies, environmental science and policy, and conservation criminology. She has published on both captive elephant issues and the ivory trade and is currently conducting a project on elephant welfare in Thailand's tourism industry.

Emma Roe teaches and carries out research within the area of food and animal welfare. She is based in the Department of Geography and Environment, in the Faculty of Social and Human Sciences, at the University of Southampton, UK.

Rosibel Roman is a doctoral student in geography, with a focus on political ecology, environmental history, and Russian studies at Florida International University in Miami, Florida.

Camilla Royle is a postgraduate student in geography at King's College London.

Chie Sakakibara is an assistant professor of environmental studies at Oberlin College. Her academic backgrounds are in cultural geography and Native American studies, and she has been working with Iñupiaq whalers of northern Alaska in order to explore their cultural response to climate change as a form of social resilience.

Ivan Sandoval-Cervantes is a PhD candidate in the department of anthropology at the University of Oregon. In September 2014, Ivan published "For the Love of Dogs: Approaching Animal-Human Interactions in Mexico," which appeared in *Anthropology News*.

Avi Sapkota was born and raised in Nepal. He obtained his undergraduate degree in veterinary medicine in Nepal and PhD in animal science (focusing in animal behavior, physiology and welfare) from Texas Tech University, Lubbock. Currently, he works as a postdoctoral research associate at Purdue University.

Kenneth Shapiro is cofounder of the Animals and Society Institute; founder and editor of *Society and Animals: Journal of Human-Animal Studies*; cofounder and coeditor of *Journal of Applied Animal Welfare Science*; and editor of the Brill Human-Animal Studies book series. He has also published a number of books.

Michelle L. Shuey, PhD, is an instructor in the geography and environmental sciences department at University of Hawaii at Hilo. Shuey conducts research on how habitat conversion to human land uses contributes to conflicts between humans and wildlife.

Anna C. Sloan is a doctoral student in anthropology at the University of Oregon and the cofounder of its Human-Animal Research Interest Group, which fosters interdisciplinary communication between animal studies scholars on campus. She studies intersections between gender and human-animal relations in archaeological and contemporary Yup'ik communities.

Jan-Erik Steinkrüger is working as a scientific assistant at the geographical department of the University of Bonn. After his graduate degree in philosophy, political science, and geography, he wrote a PhD thesis on the historical and cultural geographies of zoological gardens and amusement parks.

Professor **Philip Tedeschi** is executive director of the Institute for Human-Animal Connection at the University of Denver within the Graduate School of Social Work.

Philip's work focuses on the connection between people and animals. He teaches graduate level courses in human-animal interaction, animal welfare, human ecology, and international social work.

Jennifer E. Telesca is assistant professor of environmental justice at Pratt Institute. She holds dual MAs in anthropology (UConn-Storrs) and in law (NYU), and a PhD in media studies (NYU). Her work appears in the journals *Cambridge Anthropology* and *Humanity*. She is finalizing her book manuscript, tentatively titled *Red Gold*.

Breanna Ten Eyck graduated from the University of Minnesota with a BS in genetics, cell biology, and development. She has special interest in pediatric onset genetic diseases and genetic counseling and has spent time researching and writing about the evolution versus creationism controversy, specifically with regard to the Grand Canyon.

Gwyneth Anne Thayer earned her BA in art history from Brown University, her MA in art history from the University of Texas at Austin, and her PhD in public history from Middle Tennessee State University. She is currently the associate head and curator of Special Collections at North Carolina State University Libraries.

Mary Trachsel is an associate professor of rhetoric at the University of Iowa. Her PhD, from the University of Texas, is in English composition and rhetoric. She now studies human-animal communication and human-animal relationships in the context of environmental studies and ecological ethics.

Bernard Unti is senior policy adviser and special assistant to the president and CEO of the Humane Society of the United States. He works on strategic, policy, program, and communications priorities for the HSUS and its affiliated entities. Unti received a PhD in U.S. history from American University.

Julie Urbanik, PhD, is an independent scholar, creative geographer, and cofounder of the animal geography specialty group of the Association of American Geographers. In addition to publishing articles, book chapters, and book reviews related to animals and geography, she is the author of *Placing Animals: An Introduction to the Geography of Human-Animal Relations*, which was deemed an essential geography title for 2013 by *Choice Library Journal*. Urbanik also produced the first animal geography–based documentary *Kansas City: An American Zoöpolis*. She holds a master's degree in gender studies from the University of Arizona and a doctorate in geography from Clark University.

Andy VanderLinde holds a master of arts degree in international development studies and aims to promote community participation, decrease populations' vulnerability, and improve empowerment. With over 20 years of work with horses, she develops and coordinates animal assisted activities and animal assisted therapy services for low-income youth and vulnerable adults.

Stephen Vrla is a doctoral student at Michigan State University, earning a dual PhD in sociology and curriculum, instruction, and teacher education, with graduate specializations in animal studies and environmental science and policy. His research involves developing and implementing curricula that prepare secondary students to become critically ecoliterate citizens.

Gabe Wigtil is a wildlife conservation policy researcher, analyst, and communicator with a master of public policy degree from Oregon State University.

Sharon Wilcox holds a PhD in geography from the University of Texas at Austin, where she studied the human dimensions of jaguar (*Panthera onca*) conservation. She has taught at the University of Texas, San Antonio and the University of Texas, Austin, and has worked with the organizations Defenders of Wildlife and Ocean Conservancy.

Robert M. Wilson is an associate professor in the Department of Geography at Syracuse University. He has written extensively on the historical geography and environmental history of wildlife and wildlife management. He is also the author of *Seeking Refuge: Birds and Landscapes of the Pacific Flyway*.

Drew Robert Winter is an anthropologist, activist, and author. Formerly the director of publications and a board member at the Institute for Critical Animal Studies, he is currently a PhD student at Rice University, where his work is supported by the Culture & Animals Foundation.

INDEX

Note: Page numbers listed in **bold** indicate main encyclopedia entry for term. Page numbers in *italics* indicate illustrations.